# Residential Construction Academy

# Masonry

# Residential Construction Academy

# Masonry

## Brick and Block Construction

## Robert B. Ham

DELMAR
CENGAGE Learning

Australia • Brazil • Japan • Korea • Mexico • Singapore • Spain • United Kingdom • United States

**Residential Construction Academy: Masonry: Brick and Block Construction**
Robert B. Ham

Vice President, Technology and Trades ABU:
  David Garza

Director of Learning Solutions: Sandy Clark

Managing Editor: Larry Main

Senior Acquisitions Editor: James DeVoe

Development: Ohlinger Publishing Services

Marketing Director: Deborah S. Yarnell

Marketing Manager: Kevin Rivenburg

Marketing Specialist: Mark Pierro

Director of Production: Patty Stephan

Production Manager: Stacy Masucci

Content Project Manager: Jennifer Hanley

Senior Art Director: B. Casey

Editorial Assistant: Tom Best

Cover Image: ©Ivan Barta/Alamy

For product information and technology assistance, contact us at
**Cengage Learning Customer & Sales Support, 1-800-354-9706**
For permission to use material from this text or product,
submit all requests online at **www.cengage.com/permissions**
Further permissions questions can be emailed to
**permissionrequest@cengage.com**

Library of Congress Control Number: 2007025085

ISBN-13: 978-1-4180-5284-3

ISBN-10: 1-4180-5284-1

**Delmar**
Executive Woods
5 Maxwell Drive
Clifton Park, NY 12065
USA

Cengage Learning is a leading provider of customized learning solutions with office locations around the globe, including Singapore, the United Kingdom, Australia, Mexico, Brazil, and Japan. Locate your local office at **www.cengage.com/global**

Cengage Learning products are represented in Canada by Nelson Education, Ltd.

To learn more about Delmar, visit **www.cengage.com/delmar**

Purchase any of our products at your local bookstore or at our preferred online store **www.cengagebrain.com**

**Notice to the Reader**

Publisher does not warrant or guarantee any of the products described herein or perform any independent analysis in connection with any of the product information contained herein. Publisher does not assume, and expressly disclaims, any obligation to obtain and include information other than that provided to it by the manufacturer. The reader is expressly warned to consider and adopt all safety precautions that might be indicated by the activities described herein and to avoid all potential hazards. By following the instructions contained herein, the reader willingly assumes all risks in connection with such instructions. The publisher makes no representations or warranties of any kind, including but not limited to, the warranties of fitness for particular purpose or merchantability, nor are any such representations implied with respect to the material set forth herein, and the publisher takes no responsibility with respect to such material. The publisher shall not be liable for any special, consequential, or exemplary damages resulting, in whole or part, from the readers' use of, or reliance upon, this material.

Printed in the United States of America
Print Number: 09     Print Year: 2022

# Table of Contents

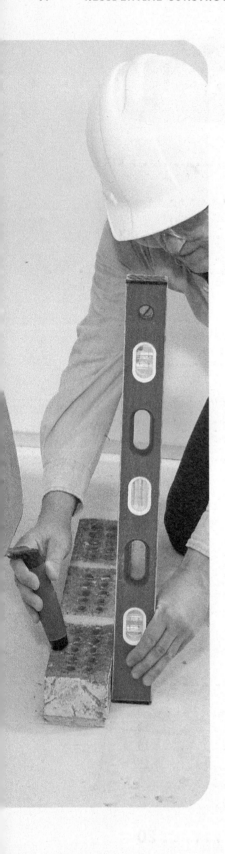

## Chapter 4   Masonry Construction Equipment . . 49

## Chapter 23    Cleaning Brick and Concrete Masonry . . . . . . . . . . . . . . . . . . . 333

# Preface

## Home Builders Institute Residential Construction Academy: Masonry

### About the Residential Construction Academy

One of the most pressing problems confronting the building industry today is the shortage of skilled labor. It is estimated that the construction industry must recruit 200,000 to 250,000 new craft workers each year to meet future needs. This shortage is expected to continue well into the next decade because of projected job growth and a decline in the number of available workers. At the same time, the training of available labor is becoming an increasing concern throughout the country. The lack of training opportunities has resulted in a shortage of 65,000 to 80,000 skilled workers per year. This crisis is affecting all construction trades and is threatening the ability of builders to build quality homes.

These are the reasons for the creation of the innovative *Residential Construction Academy Series*. The *Residential Construction Academy Series* is the perfect way to introduce people of all ages to the building trades while guiding them in the development of essential workplace skills, including carpentry, electrical, HVAC, masonry, plumbing, and facilities maintenance. The products and services offered through the *Residential Construction Academy* are the result of cooperative planning and rigorous joint efforts between industry and education. The program was originally conceived by the National Association of Home Builders—the premier association of more than 235,000 members in the residential construction industry—and its workforce development arm, the Home Builders Institute.

For the first time, construction professionals and educators created National Standards for the construction trades. Since the summer of 2001, the National Association of Home Builders (NAHB), through the Home Builders Institute (HBI), has been developing residential craft standards in these six trades: carpentry, electrical wiring, HVAC, plumbing, masonry, and facilities maintenance. Groups of masonry employers from across the country met with an independent research and measurement organization to begin the development of new craft training standards. The guidelines from the National Skills Standards Board were followed in developing the new standards. In addition, the process met or exceeded the American Psychological Association standards for occupational credentialing.

Then, through a partnership between HBI and Delmar Cengage Learning, learning materials—textbooks, videos, and instructor's curriculum and teaching tools—were created to effectively teach these standards. A foundational tenet of this series is that students *learn by doing*. A constant focus of the *Residential Construction Academy* is teaching the skills needed to be successful in the construction industry and constantly applying the learning to real-world applications.

Perhaps most exciting to learners and industry is the creation of a National Registry of students who have successfully completed courses in the *Residential Construction Academy Series*. This registry or transcript service provides an easy way to verify skills and competencies achieved. The Registry links construction industry employers and qualified potential employees together in an online database that facilitates student job searches and the employment of skilled workers.

# About This Book

A home is an essential part of life. It provides protection, security, and privacy to its occupants. It is often viewed as the single most important thing a family can own. This book is written for students who want to learn how to become brickmasons or bricklayers, targeting the residential construction market where typically concrete masonry units are used for foundation walls and anchored brick veneer is used for exterior wall facades.

*Masonry: Brick and Block Construction* covers different types of brick and brick pattern bonds, different types of concrete masonry units and architectural concrete masonry units; building brick and block leads; laying brick and block to the line; residential foundations; anchored brick veneer facades; fireplaces and chimneys; brick paving, porches, and steps; brick arches, brick piers, columns, pilasters, and chases; composite and cavity wall construction; estimating materials and labor; cleaning new masonry; safety for masons; working drawings and specifications; and career opportunities. Masonry practices that are commonly used in today's residential construction market are discussed in detail and presented in a way that not only *tells* you what needs to be done but also *shows* you how to do it. General and task-specific safety issues are addressed throughout the textbook, as well as in a special chapter devoted exclusively to safety and OSHA requirements.

This textbook provides a valuable resource for the knowledge and skills required of an entry-level mason, although those actively involved in the industry will also benefit from the material covered. The basic "hands-on" skills as well as the procedures outlined in this book will help individuals gain proficiency in this trade.

## Organization

This textbook is organized so that those new to the industry as well as those already working in the field can gain maximum benefit from its content. The four main sections of the book cover the major aspects of the masonry industry as they affect residential construction:

- **Materials, Products, Tools, and Equipment**
- **Basic Brick and Block Laying Procedures**
- **Procedures for Residential Masonry Construction**
- **Safety Training and Career Opportunities**

# Features

This innovative series was designed with input from educators and industry and informed by the curriculum and training objectives established by the Standards Committee. The following features aid learning:

**Learning features** such as the **Introduction, Objectives,** and **Glossary** set the stage for the coming body of knowledge and help the learner identify key concepts and information. These learning features serve as a road map for the chapter. The learner also may use them as a reference later.

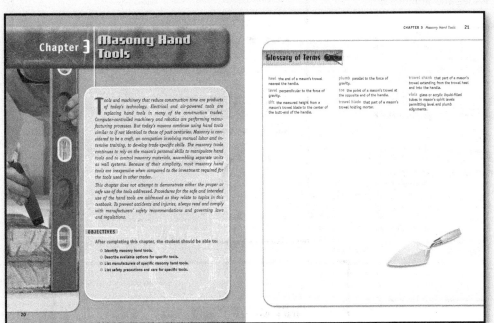

**Active Learning** is a core concept of the *Residential Construction Academy Series*. Information is heavily illustrated to provide a visual of new tools and tasks encountered by the learner. In the *Procedures* sections, various tasks used in brick and block construction are grouped in a step-by-step approach. The overall effect is a clear view of the task, making learning easier.

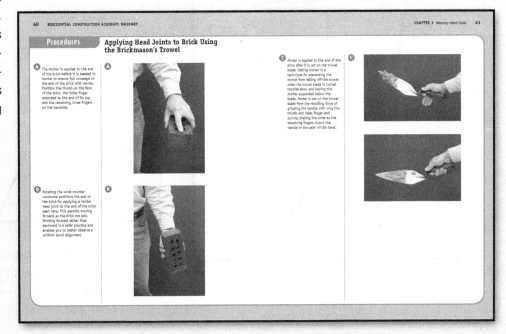

**Safety** is featured throughout the text to instill safety as an "attitude" among learners. Safe jobsite practices by all workers is essential; if one person acts in an unsafe manner, all workers on the job are at risk of being injured too. Learners will come to appreciate that safety is a blend of ability, skill, and knowledge that should be continuously applied to all they do in the construction industry.

**Caution** features highlight safety issues and urgent safety reminders for the trade.

**From Experience** provides tricks of the trade and mentoring wisdom that make a particular task a little easier for the novice to accomplish.

**Examples** of applied mathematical formulas and calculations are provided to give the student more exposure to the calculations and their use.

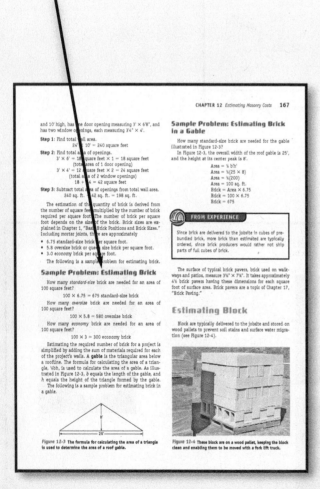

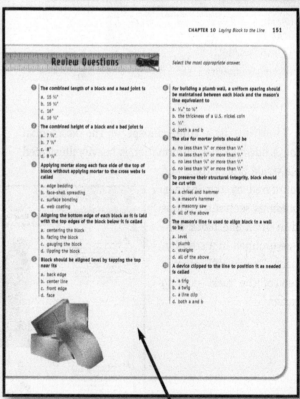

**Review Questions** complete each chapter. These are designed to reinforce the information learned in the chapter as well as give students the opportunity to think about what they have learned and accomplished.

# Turnkey Curriculum and Teaching Material Package

We understand that a text is only one part of a complete, turnkey educational system. We also understand that instructors want to spend their time on teaching, not preparing to teach. The *Residential Construction Academy Series* is committed to providing thorough curriculum and preparatory materials to aid instructors and alleviate some of their heavy preparation commitments. An integrated teaching solution is ensured with the text, including the Instructor's e.resource™, print Instructor's Resource Guide, Student Workbook, and DVDs.

## e.resource™

Delmar Cengage Learning's **e.resource™** is a complete guide to classroom management. The CD-ROM contains lecture outlines, notes to instructors with teaching hints, cautions, answers to review questions, and other aids for the instructor using this series. Designed as a complete and integrated package, the instructor is also provided with suggestions for when and how to use the accompanying **PowerPoint®** presentations, **Computerized Testbank, Workbook,** and **DVD** package components. There is also a print **Instructor's Resource Guide** available.

## PowerPoint®

The series includes a complete set of PowerPoint presentations providing lecture outlines that can be used to teach the course. Instructors may teach from this outline or can make changes to suit individual classroom needs.

## Computerized Testbank

The Computerized Test Bank contains hundreds of questions that can be used for in-class assignments, homework, quizzes, or tests. Instructors can edit the questions in the testbank or create and save new questions.

## Online Testing

An online testing and grading tool has been developed to enhance the *Residential Construction Academy Series*. This robust system provides pretesting and posttesting options that can help teachers track their students' grades, completion time, and level of success. The self-grading tests are driven by the National Skill Standards developed by HBI.

## DVDs

The *Masonry DVD Series* is an integrated part of the *Residential Construction Academy Masonry* package. The series contains a set of four, 20-minute lessons that provide step-by-step masonry instruction. All the essential information is covered in this series, including safety for masons, building brick leads, laying brick to the line, and building block leads.

# About the Author

**R**obert Benjamin Ham is a licensed and insured masonry contractor who has taught industrial arts, agricultural education, and masonry to high school students and adult learners for more than 33 years. Mr. Ham holds a B.S. degree in Agricultural Education from Virginia Tech and has completed the requirements of the Virginia Apprenticeship Council as a journeyman bricklayer. He is issued a Collegiate Professional License by The Commonwealth of Virginia and is endorsed to teach both Agricultural Education and Masonry-Bricklaying.

Mr. Ham served as the lead masonry teacher for the Virginia Department of Education's Trade and Industrial Education from 1994 until his retirement in 2007. In 1995, under the direction of Mr. Ham, Valley Tech's Masonry Program was recognized by the Virginia Chapter, Valley District of The Associated General Contractors of America as meeting the construction industry training standards established by the AGC. In 2000, Mr. Ham chaired the writing team that developed the instructional framework for the Virginia Department of Education's Masonry Programs.

Mr. Ham is the 2002 recipient of The Virginia Association of Trade and Industrial Education's Outstanding Teacher of the Year. The National Association of State Directors of Career and Technical Education Consortium recognizes him for significant contributions to the progress of career technical education in the United States.

In 2006, Mr. Ham completed the West Virginia University's course requirements as an OSHA Construction Outreach Trainer. He is authorized to conduct Construction Outreach Training in accordance with guidelines provided by the OSHA Office of Training and Education.

As a volunteer, Mr. Ham has been joined by his students, alumni, and advisory committee members in building 19 house foundations for the Staunton-Augusta Chapter of Habitat for Humanity. In 2005, the Staunton-Augusta-Waynesboro Chapter of Habitat for Humanity presented him with their Distinguished Service Award.

# Acknowledgments

## Masonry National Skill Standards

The NAHB and the HBI would like to thank the many individual members and companies that participated in the creation of the Masonry National Skills Standards. Special thanks are extended to the following individuals and companies:

Steve Daly
Dawn Faull Evaluation & Certification Services

Dennis Graber
Director of Technical Publications,
National Concrete Masonry Association

Beatrix Kerkoff
Portland Cement Association

Bob McKay, ESSROC Cement Corporation

William Panarese
Ronald Rodgers
Wasdyke Associates

Christopher Sieto
ESSROC Cement Corporation

David Smith
Interlocking Concrete Pavement Institute

Pieter VanderWerf
Building Works

Ray Wasdyke
Wasdyke Associates
Mike Weber, PCA

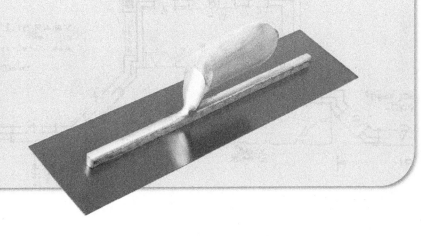

## Manuscript Review

In addition to the standards committee, Thomson Delmar Learning and the author extend our sincere gratitude to the following manuscript reviewers:

Chijioke E. Akamiro
Norfolk State University, Norfolk, VA

Joe Dusek
Triton College, Chicago, IL

Gary Edwards
State University of New York at Delhi, Delhi, NY

Merritt Johnson
Northeast Technology Center, Afton, OK

Todd A. Larson
Wisconsin Indianhead Technical College, Rice Lake, WI

Robert McKay
Essroc Cement Corporation, Nazareth, PA

Charles McRaven
Piedmont Virginia Community College and Technical School, Charlottesville, VA

Brian Moser
Upper Bucks County Area Vocational Technical School, Perkasie, PA

Al Rusch
Waukesha County Technical College, Pewaukee, WI

Jason Talbot
Home Builders Institute, Ogden, UT

## From the Author

I am thankful for each of the hundreds of students that I have had the privilege and opportunity to teach. Each of you has been a blessing to me, giving me opportunities to develop teaching strategies for you and others to become motivated, challenged, and successful. Your influences on my life impacted this book. I give thanks to my parents, the late Raleigh G. Ham, Sr. and Margaret Garber Ham. It was my father, a journeyman bricklayer and masonry contractor, who gave me a passion for masonry. I credit both of my parents for exemplifying patience, kindness, and a concern for the welfare of others, qualities that have enabled me to become an effective teacher. I give thanks to my loving wife, Donna, and son, Ben, for their encouragement. And above all, I give thanks to the One who is always with me. As the reader, you are the ultimate evaluator of the usefulness of this book. I welcome your comments and suggestions, for as with my purpose for living, I wish to improve.

# Chapter 1

# Basic Brick Positions and Brick Sizes

Due to their rectangular shapes, brick can be oriented with the face of a wall in a variety of positions. Varying the orientation of brick with the face of a wall is a way to both create unique designs that enhance a wall's appearance and satisfy the structural requirements for masonry walls. The various brick sizes permit building walls that have distinctive characters. Together, the variety of brick positions and sizes allow walls to be designed with unlimited architectural effects.

## OBJECTIVES

After completing this chapter, the student should be able to:

- Identify the six brick positions.
- Explain how extruded brick and wood-mold brick are formed.
- Identify the different sizes of brick available as standard production units.

# Glossary of Terms

**cored brick** a brick whose cross-section is partially hollow.

**double-wythe wall** two adjacent masonry walls, joined together with brick or joint reinforcement.

**extruded brick** brick that are shaped by forcing wet clay through a die, a form used to impress or shape an object.

**face brick** brick intended to be used on the exposed surface of walls.

**frog** a recessed area formed on the bottom side of wood-mold brick, created at the bottom of a brick mold when the mold is filled with wet clay.

**header** the position for a brick having its end oriented with the face of a wall and its bottom side bedded in mortar.

**hollow brick** brick in which the combined surface area of the holes is more than 25% of the total bed surface area of the brick.

**paving brick** abrasion-resistant brick intended for pedestrian and/or vehicular traffic.

**shiner** the position for a brick having its face side or back side bedded in mortar with its top side or bottom side oriented with the face of the wall.

**single-wythe brick wall** a brick wall having a bed depth equivalent to the width of one brick.

**soldier** the position for a brick having its end bedded in mortar and its face side oriented with the face of the wall.

**solid masonry unit** a masonry unit whose net cross-sectional area parallel to its bedding area is 75% or more of its gross cross-sectional area measured in the same plane.

**stretcher** the position for a brick observed in a wall having its bottom side bedded in mortar and its face side oriented with the face of the wall

**wood-mold brick** brick whose shapes are created by placing wet clay in wooden forms or molds.

The orientation or positioning of kiln-fired clay and shale brick enables walls to be functionally strong and attractively designed. By imagining brick assimilated in different positions or orientations, masons can create unique and attractive wall patterns (see Figure 1-1).

# The Manufacturing of Brick

Most of today's mass-produced brick are made using one of two manufacturing processes: extrusion or molding. Each process produces brick that have similar characteristics but also notable differences.

## Extruded Brick

Brick are formed from clay or shale that is mixed with water and possibly other wetting agents until it has a consistency similar to modeling clay. **Extruded brick** are shaped by forcing wet clay through a die, which is a form used to impress or shape an object. For brick, a rectangular-shaped die gives the rectangular-shaped form to brick as exerted pressure forces a continuous column of clay from the mixing chamber that is called the pug mill (see Figure 1-2).

Optional texturing and coloring may be applied to the top side of the extruded column, becoming the face side of the brick, and ends of the column. Afterwards, the column is divided, forming individual brick (see Figure 1-3). The part of the extruded column of clay resting on the moving conveyor belt becomes the back side of the brick because it receives no additional texturing or coloring.

Because they have optional texturing and coloring, both ends are suitable for finished wall exposure. There is no distinction as to the top side or bottom side of an extruded brick.

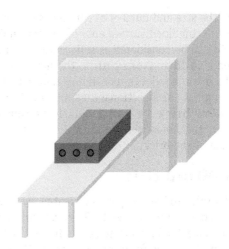

**Figure 1-2 Similar to squeezing toothpaste from a tube, a column of clay or shale is extruded from the pug mill.**

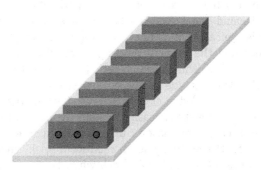

**Figure 1-3 The column of clay is parted to create individual brick of equal height.**

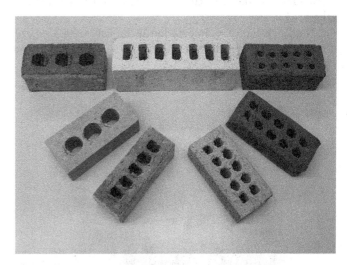

**Figure 1-4 The shapes and sizes of the openings or cores vary.**

Extruded brick can be solid, having no hollow spaces, or cored. A **cored brick** has holes entirely through the brick called core holes (see Figure 1-4). In addition to conserving natural resources and reducing production and delivery costs, the core holes reduce brick weight, improve the uniformity of drying and firing, and allow brick and mortar to

**Figure 1-1 The variety of brick positions, or brick orientation of different brick, creates a unique wall design.**

interlock. The size and number of core holes varies. Depending upon the combined surface area of the core holes, many cored brick can be considered as solid masonry units. To be considered a solid masonry unit, the net cross-sectional area of the brick parallel to its bedding area must be 75% or more of its gross cross-sectional area measured in the same plane. Hollow brick are brick in which the combined surface area of the holes is more than 25% of the total bed surface area of the brick.

## Wood Mold Brick

Wood-mold brick shapes are created by placing wet clay in wooden forms or molds. They do not have cores. Unlike extruded brick, there is no distinction between the face side and back side for wood-mold brick. However, there is a distinction between the top side and bottom side of a wood-mold brick. The top of a wood-mold brick is typically a relatively flat surface, some with a noticeable protruding edge beyond the faces. The bottom side of a wood-mold brick is typically not as flat as the top side, a condition resulting from the forming process, which gives wood-mold brick a distinguishing characteristic (see Figure 1-5).

Either the top side or the bottom side of a wood-mold brick can be bedded in mortar. However, bedding the bottom side in mortar makes it easier to align the flatter, top edge of the brick to the mason's line. A frog, a recessed area formed on the bottom side of wood-mold brick, created in the bottom of a wood-mold brick when the mold is filled with wet clay, is included in the forming process of many wood-mold brick (see Figure 1-6). The frog increases surface area and can improve the bond between the mortar and the brick.

The Brick Industry Association Technical Notes 3A states that brick are fired in kilns at temperatures ranging from 1800°F to 2100°F.

**Figure 1-5** The top side of a wood-mold brick is flatter when compared with its bottom side. A slightly protruding edge is noticeable between the four sides and the top.

**Figure 1-6** The frog in each of these bricks is a different size and shape.

Face brick, brick intended to be used on the exposed surface of walls, are typically classified by one of three "grades" and one of three "types." The three grades of face brick are Grade SW, Grade MW, and Grade NW. Grade SW brick, also called severe weathering brick, are intended to withstand freeze-thaw weather cycles and are required for walls exposed to temperatures at or below freezing. Grade MW brick, also called moderate weathering, are used where freezing is not an issue. Brick Industry Association Technical Notes 9B recommends Grade NW brick for "interior applications, or where they are protected from water absorption and freezing." Paving brick, abrasion-resistant brick intended for pedestrian and/or vehicular traffic, are classified similarly as being Grade SX, Grade MX, and Grade NX.

The type of brick refers to the appearance of the brick. Allowable chippage, cracks, and size variations are factors determining the type of brick. The three type classifications of brick are FBX, FBS, and FBA. FBX brick permit smaller dimensional tolerances and less noticeable chips and cracks than types FBS and FBA. FBX brick are often specified for commercial construction projects. FBS brick are typically used for the exterior wall facades of brick homes. FBA brick are considerably less uniform than either type FBX or FBS brick. Refer to Table 4, Dimensional Tolerances, in the appendix for more information.

## The Six Brick Positions

Bedding a different side of a brick in mortar or reorienting a brick's position with the face of a wall results in a defined brick position. Optional brick positions permit using brick not only to create decorative wall patterns but also for structural patterns that strengthen walls. The six brick positions are stretcher, header, rowlock, shiner, soldier, and sailor.

Figure 1-7 **The face side of a brick is oriented with the face of a wall when laid in the stretcher position.**

## Stretcher Position

A brick positioned as a stretcher is observed in a wall having its bottom side bedded in mortar and its face side oriented with the face of the wall (see Figure 1-7). Most brick are laid in the stretcher position. Most extruded brick have only one face side, which is optionally textured and colored, as are the two ends of the brick. Avoid mistaking the back side of an extruded brick for its face. In most cases, the back side appears noticeably different from the face side, appearing smooth, slick, and "shiny" when compared to brick correctly faced in a wall. For this reason, a brick that has been mistakenly laid in a wall with its back side exposed to the face of a wall is sometimes called a shiner.

**FROM EXPERIENCE**

It helps to remember the stretcher position by associating it with "arms stretched wide apart."

## Header Position

A header is the position for a brick that has its end oriented with the face of a wall and its bottom side bedded in mortar (see Figure 1-8).

Brick headers can be used to join two adjacent single-wythe brick walls, which are brick walls having a bed depth equivalent to the width of one brick. This procedure creates a double-wythe wall, two adjacent masonry walls joined together with brick or joint reinforcement (see Figure 1-9). Either end of the brick can be oriented to the face of the wall.

**FROM EXPERIENCE**

It helps to remember the header position by associating it with two vehicles meeting "head on."

Figure 1-8 **When laid in a header position, a brick's end is oriented with the wall's face, and its bottom side is bedded in mortar.**

Figure 1-9 **Brick laid in the header position join two single-wythe walls to create a stronger double-wythe wall.**

## Rowlock Position

A *rowlock* is the position for a brick having its end oriented with the face of a wall and its back side bedded in mortar (see Figure 1-10).

Used to cap a double-wythe wall, the face side of the brick in the rowlock position becomes the finished top of the wall (see Figure 1-11). This is called a brick rowlock cap.

**FROM EXPERIENCE**

It helps to remember the rowlock position by associating it with a "row of brick locking together two adjacent walls."

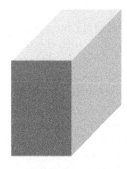

Figure 1-10 **In the rowlock position, a brick's end is oriented with the wall's face while the backside of the brick is bedded in mortar.**

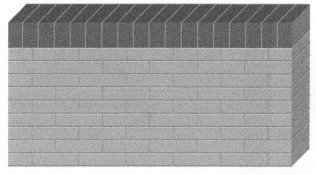

Figure 1-11 **Brick laid in the rowlock position create a cap called a brick rowlock.**

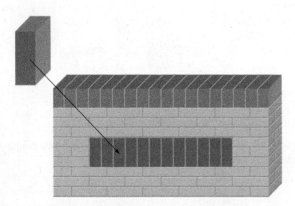

Figure 1-12 A brick laid in the soldier position is standing on end with its face side oriented with the face of the wall.

## Soldier Position

A soldier is the position for a brick having its end bedded in mortar and its face side oriented with the face of the wall (see Figure 1-12). The soldier position is used primarily for creating brick wall patterns rather than as an element for structural bonding.

**FROM EXPERIENCE**

It helps to remember the soldier position by associating it with a soldier standing at attention.

## Sailor Position

A sailor is the position for a brick having its end bedded in mortar and its top or bottom side oriented with the face of the wall (see Figure 1-13). The sailor position is used primarily to create decorative patterns. Not being solid and permitting water migration, extruded cored brick cannot be used in the sailor position for exterior walls. Cored brick may occasionally be laid in the sailor position for decorative patterns on interior walls or as masonry fill where the brick are not seen. The bottom side of wood-mold brick having a frog depression is typically not exposed to the wall's face either.

Figure 1-13 In the sailor position, a brick's top is oriented with the wall's face, and the brick's end is bedded in mortar.

**FROM EXPERIENCE**

It helps to remember the sailor position by associating it with a sailor, a member of a navy, standing at attention on a ship and appearing wider than a soldier as the sea winds blow the clothing.

Figure 1-14 In the shiner position, a brick's top side or bottom side is oriented with the face of a wall, and its backside is bedded in mortar.

## Shiner Position

A shiner is the position for a brick having its face side or back side bedded in mortar with its top side or bottom side oriented with the face of the wall (see Figure 1-14). Do not confuse the shiner position with a brick laid in the stretcher position having its back side mistakenly oriented with the face of a wall, also called a shiner by many.

Not being solid and permitting water migration, extruded cored brick cannot be used in the shiner position for exterior walls. Cored brick may occasionally be laid in the shiner position for decorative patterns on interior walls or as masonry fill where the brick are not seen. The bottom side of wood-mold brick having a frog depression should not be exposed to the wall's face either. The special firebrick sidewalls of a fireplace firebox are laid in the shiner position (see Figure 1-15).

Figure 1-15 Fireplace brick for the sidewalls of this fireplace are laid in the shiner position.

**CAUTION**

**Refer to manufacturer's printed instructions as to the recommended position for bedding firebrick in mortar.**

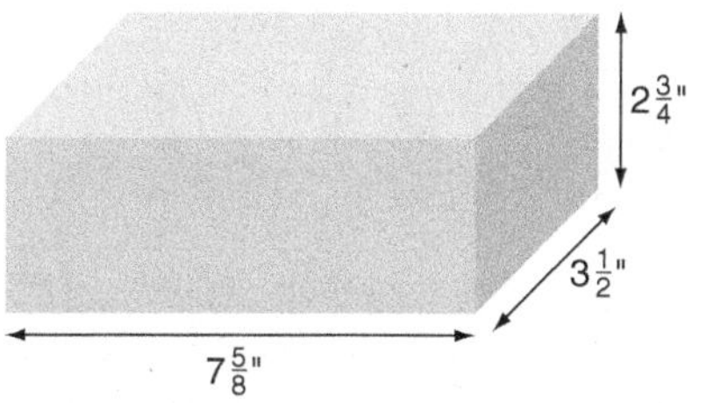


**Figure 1-17 The dimensions of an engineered or oversize brick are 7⅝" long × 2¾" high × 3½" wide.**

# Brick Sizes

Standard productions of brick are available in many sizes, varying in length, width, and height. Common sizes of face brick, brick used on the exposed surface of a wall, include 1) standard-modular size brick, 2) engineered or oversize brick, 3) economy or jumbo utility brick, 4) hollow masonry clay units, and 5) queen-size brick.

## Standard Modular Size Brick

*Standard modular size brick* have a length equivalent to 7⅝", a face height of 2¼", and a width or bed depth equivalent to 3½" (see Figure 1-16). These dimensions may vary slightly between different brick produced at different times and in different heat firing kilns. Standard modular size brick are a popular choice for residential single-family and multifamily homes, commercial and industrial buildings, schools, and churches.

## Engineered or Oversize Brick

*Engineered or oversize brick* have the same length, 7⅝", and width, 3½" as do standard modular size brick. The face height of engineered or oversize brick is 2¾", ½" more than the height of standard size modular brick (see Figure 1-17). As with most brick, these dimensions may vary slightly between different brick produced at different times and in different heat firing kilns. They too are a popular choice for residential single-family and multifamily homes, commercial and industrial buildings, schools, and churches. Their height enhances the architectural styling for those desiring colonial looking brickwork.

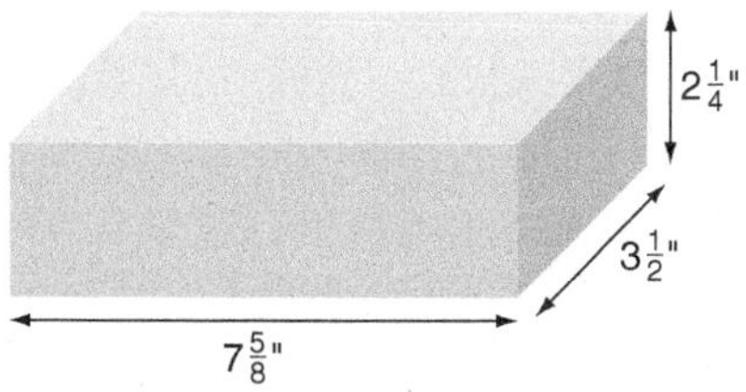


**Figure 1-16 The dimensions of a standard modular size brick are 7⅝" long × 2¼" high × 3½" wide.**

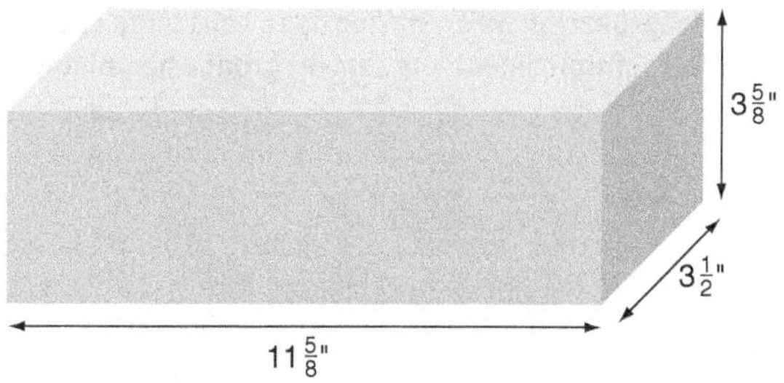


**Figure 1-18 Often called utility brick, economy or jumbo brick are 11⅝" long × 3⅝" high, × 3½" wide.**

## Economy or Jumbo Utility Brick

*Economy or jumbo utility brick* have a length equivalent to 11⅝", a face height of 3⅝", and a width or bed depth equivalent to 3½" (see Figure 1-18). Utility brick offer savings in construction costs because walls can be constructed in less time than if standard modular size or oversize brick are used. They are typically used on larger buildings, such as those seen in shopping malls and hotels.

## Hollow Masonry Clay Units

Hollow masonry clay units are typically available in lengths of 11⅝" and 15⅝"; face heights of 3⅝", 5⅝", and 7⅝"; and widths of 3⅝", 5⅝", and 7⅝". Unlike other brick sizes, hollow masonry clay units are considered to be load-bearing units, capable of supporting their own weight and the weight of other building materials, like those used for floor and roof systems. Their open cells are designed to accommodate strengthening a wall with cement grout and steel reinforcement (see Figure 1-19).

## Queen Size Brick

*Queen size brick* have a length equivalent to 7⅝", a face height of 2¾", and a width or bed depth equivalent to 2¾" (see Figure 1-20). Requiring less material to make, queen size brick offer brick production savings. Being approximately ¾" narrower than standard size modular brick, oversize brick, or utility brick, requires special considerations for establishing a half-lap course pattern of brickwork.

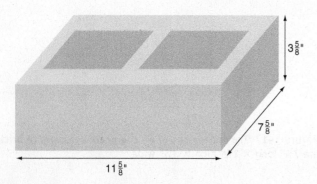

**Figure 1-19** Many hollow masonry clay units are used to construct structural, load-bearing walls, having open cells large enough for optional steel reinforcement and cement grout when required.

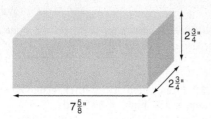

**Figure 1-20** The dimensions of a queen size brick are 7⅝" long × 2¾" high × 2¾" wide.

## Summary

- The opportunities for varying a brick's position as it is laid in a wall provide benefits for both improving a project's structural strength and enhancing its appearance.
- Changing the orientation of a brick or groups of brick in a wall can create decorative patterns without using more costly, special-shaped brick.
- The positions for laying cored brick may be limited to only those places where the cores are not exposed.

# Review Questions

*Select the most appropriate answer.*

1. Brick that are shaped by forcing wet clay through a die, a form used to impress or shape an object are called

   a. extruded brick
   b. pressed brick
   c. repressed brick
   d. wood-mold brick

2. A brick whose cross-section is partially hollow is called a

   a. cored brick
   b. hollow brick
   c. open-cell brick
   d. webbed brick

3. For a brick to be considered a solid unit rather than a hollow unit, the minimum percentage of its net cross-sectional area considered as material rather than openings must be at least

   a. 40%
   b. 50%
   c. 75%
   d. 90%

4. Wood-mold brick

   a. are totally solid brick
   b. have no distinguishing characteristics or identifying differences between a front side and a back side.
   c. are no longer produced
   d. both a and b

5. A recessed area formed on the bottom side of wood-mold brick, created at the bottom of a brick mold when the mold is filled with wet clay, is called a

   a. frog
   b. groove
   c. keyway
   d. relief

6. A brick positioned in a wall with its bottom side bedded in mortar and its face side oriented with the face of the wall is called a

   a. header
   b. sailor
   c. shiner
   d. stretcher

7. A brick positioned in a wall with its bottom side bedded in mortar and its end oriented with the face of the wall is called a

   a. header
   b. sailor
   c. shiner
   d. stretcher

8. A brick positioned in a wall with its back side bedded in mortar and its top- or bottom side oriented with the face of the wall is called a

   a. header
   b. sailor
   c. shiner
   d. stretcher

9. A brick positioned in a wall with one end bedded in mortar and its face side oriented with the face of the wall is called a

   a. rowlock
   b. soldier
   c. sailor
   d. shiner

10  A brick positioned in a wall with its back side bedded in mortar and its end oriented with the face of the wall is called a

a. rowlock

b. soldier

c. sailor

d. shiner

11  Brick with a length of 7⅝" are

a. oversize brick

b. queen size brick

c. standard size brick

d. all of the above

12  Brick with a width of 3½" are

a. oversize brick

b. standard size brick

c. utility brick

d. all of the above

13  Brick with a face height of 2¾" are

a. oversize brick

b. queen size brick

c. utility brick

d. both a and b

14  Brick with a length of 11⅝" are

a. hollow brick masonry units

b. utility brick

c. oversize brick

d. both a and b

15  Another name for a utility brick is

a. economy brick

b. jumbo brick

c. oversize brick

d. both a and b

# Chapter 2 | Brick Pattern Bonds

Two walls with identical brick and mortar can be made to look much different simply by altering the bricks' position in the wall to create different patterns of brickwork. Different patterns of brickwork are used to replicate different times in history, enhancing the recreation of the intended architectural era. In addition, brick patterns emphasize the art of masonry, permitting masons to create wall patterns limited only by the imagination. More importantly, brick patterns are frequently intended to strengthen walls and integrate single-building units into colossal walls capable of supporting massive weights and resisting substantial forces.

## OBJECTIVES

After completing this chapter, the student should be able to:

- Define the term *pattern bond*.
- Identify the five brick pattern bonds.
- Dry bond each of the five pattern bonds.
- Lay out and build a brick wall for each of the pattern bonds.

# Glossary of Terms

**American bond** a pattern consisting of a course of brick laid in the header position typically between every five or six courses of brick laid in the stretcher position.

**bat** another term for half of a full brick length.

**Dutch corner** a styling for corners or ends of walls constructed in either the English bond pattern or Flemish bond pattern characterized by an approximately 6"-long brick at the end.

**English bond** a pattern consisting of alternating courses of brick laid as all stretchers and all headers.

**English corner** also called a queen closure, a styling for corners or ends of walls constructed in either the English bond pattern or Flemish bond pattern characterized by an approximately 2"-long brick piece adjacent to the end of the first brick of a course.

**Flemish bond** a brick pattern consisting of alternating stretchers and headers for each course.

**Flemish garden wall bond** a pattern of brickwork typically having three headers between each stretcher, creating the appearances of diagonal lines and diamond patterns.

**garden walls** brick walls or fencing typically built to add elegance to outdoor living areas.

**pattern bond** the pattern created by the arrangement of masonry units on a masonry wall.

**queen closure** also called a plug, a brick piece aproximately 2" long sometimes used at the ends of walls having Flemish bond and English bond brick patterns.

**running bond** a brick pattern where every brick is laid in the stretcher position, and the brick of alternate courses forms a uniform overlap with the brick below.

**screen wall** a brick wall constructed in the Flemish bond pattern with the headers omitted to create openings through the wall.

**single-wythe brick wall** a brick wall with a bed depth equivalent to the width of one brick.

**snap header** half of a brick length laid in the header position for 4"-wide Flemish bond walls.

**stack bond** also called the stacked bond, a pattern consisting of brick aligned vertically in each consecutive course, in straight columns with no overlap.

**wythe** a term used to express the bed depth of a masonry wall or thickness of a masonry wall in masonry units; each unit of bed depth is considered a wythe.

# Brick Masonry Pattern Bonds

A **pattern bond** is the pattern created by the arrangement of masonry units on a masonry wall. Pattern bonds are created as each brick is oriented in a specific brick position when building masonry walls. Most pattern bonds are a result of individual masonry units overlapping the brick of the previous course. Once assembled, the individual units form a wall capable of supporting weights and forces imposed upon it. This is why pattern bonds are also called structural bonds. Most structural bonds are intended to have a plumb alignment, having the masonry units and their adjoining mortar joints aligned vertically on every other course. The five pattern bonds typically used to build brick walls include the running bond, the Flemish bond, the English bond, the American bond, and the stack bond.

## Running Bond

A **running bond** pattern is created by laying a full row or course of brick in the stretcher position, while the brick of alternate courses form a uniform overlap with the brick below (see Figure 2-1).

Although brick laid in the running bond pattern usually are half-lapped with the brick below it, walls with consecutive courses with one-third lapped brick are also built. Such a one-third lap pattern is typical when jumbo utility brick are specified. Because all brick are aligned in the stretcher position, the running bond is sometimes called the stretcher bond. When using either standard size or engineered brick, the dimensions of the brick permit successive courses to form half-lap patterns at corners, a pattern continuing for the length of each course. When laying utility brick, queen

**Figure 2-2** When laid in the running bond, the only brick requiring cutting are those adjacent to door and window openings or at the ends of courses in the gable.

size brick, or hollow masonry units, the brick of every other course at the corner or end of the lead requires cutting to maintain a half-lap, running bond pattern.

The running bond appears more often than any other pattern bond. The running bond is the pattern typically found on twentieth and twenty-first century, **single-wythe brick walls,** that is, brick walls that have a bed depth equivalent to the width of one brick. It permits the weights or loads carried by 4", single-width brick walls to be distributed throughout the wall more uniformly than other pattern bonds. Of all the pattern bonds, it provides the opportunity for greatest productivity and therefore lowest labor costs, because the longest dimension of the brick is always aligned with the face of the wall. The only brick requiring cutting are those adjacent to openings, windows and doors, or at the ends of a wall that is racked or stepped back, for example a roof gable (see Figure 2-2).

> ⚠️ **CAUTION**
>
> When queen size brick or economy size brick are used, a brick at each end of a wall must be cut shorter if the stretchers between courses are to be aligned in a half-lap pattern.

## Flemish Bond

Originally intended to structurally bond together two adjacent brick walls, each of which is called a **wythe;** the **Flemish bond** pattern consists of alternating stretchers and headers for each course that form a double-wythe wall (see Figure 2-3). It is a pattern of brickwork typically used by nineteenth-century and earlier masons. For successive courses, brick laid in the header position are to align

**Figure 2-1** Brick in the running bond are laid in the stretcher position. The brick and joints between them of every other course align vertically.

**Figure 2-3** Alternating between brick laid in the stretcher position and the header position and having brick of consecutive courses centered on those below creates a Flemish bond pattern.

**Figure 2-4** Full 8"-length header brick join the two adjacent wythes, creating the double-wythe brick wall.

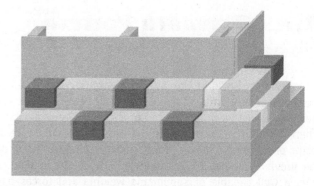

**Figure 2-5** Half-brick permit constructing 4"-wide, Flemish bond pattern walls that become the exterior faces of wood-framed or steel-framed structural walls.

**Figure 2-6** A Flemish bond pattern wall with an English style corner requires a queen closure or plug, a brick approximately 2" long.

**Figure 2-7** A Flemish bond pattern wall with a Dutch style corner requires a brick approximately 6" long at the end of the wall.

centered above brick laid in the stretcher position. Likewise, brick laid in the stretcher position are to align centered above brick laid in the header position.

Using full-length brick for the headers permits cross-lapping them above two adjacent wythes, bonding the wythes to create an 8"-wide bearing wall capable of supporting imposed loads (see Figure 2-4). Before the advent of twentieth-century wood-frame homes, brick walls supported the weight loads of entire structures, including the floor and roof framing. To do this, the brick walls were two, three, or more wythes thick. The Flemish bond allowed any two adjacent wythes to be joined.

Today's typical homes have either wood or steel structural framing. Brick walls are no more than a facade, that is, a protective, attractive, and maintenance-free exterior wall. This facade is only a single-wythe, 3"- or 4"-wide brick wall that is anchored to the structural framing. Flemish bond patterns for 4"-wide walls only replicate the bonded multi-wythe walls of earlier time periods. Each header is a 4"-long half-brick, also called a **bat.** Bats or half-brick seen in 4"-wide Flemish bond walls are called **snap headers** (see Figure 2-5).

To establish the Flemish bond pattern, a shorter brick is required at the ends of walls. Two procedures are recognized for beginning the ends of walls, each creating a different design. The **English corner** requires a 2"-long brick piece sometimes called either a **queen closure** or plug adjacent to the end of the first brick of each course (see Figure 2-6).

The **Dutch corner** requires a 6"-long brick piece, or three-quarter brick as it is sometimes called, at the end of each course. Each is illustrated in Figure 2-7.

**CAUTION**

**Protective eyewear meeting the requirements of OSHA standards should be worn when cutting brick.**

**Garden walls,** brick walls or fencing typically built to add elegance to outdoor living areas, can be Flemish bond wall designs. Omitting the headers for courses creates openings through the wall, resulting in a screen wall design (see Figure 2-8).

Examples of screen walls are found enclosing patios and swimming pools. They are practical for enclosing industrial machinery, heating and air conditioning units, and other types of industrial and commercial equipment. Screen walls enhance the appearance of areas requiring such equipment, masking its appearance, and providing protection to equipment without blocking necessary air flow or ventilation (see Figure 2-9).

**FROM EXPERIENCE**

Use brick without cores to build screen walls to enhance their appearance and to prevent debris from collecting in open brick cores.

**Figure 2-8** This single-wythe brick screen wall is laid in the Flemish bond pattern. Headers are omitted to create the open screen design.

**Figure 2-9** This screen wall, a modification of the Flemish bond pattern omitting the headers, encloses a building's cooling and ventilation equipment. Screen walls can be built either as a single-wythe or a double-wythe wall.

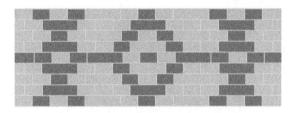

**Figure 2-10** These Flemish garden wall bond patterns all have the same brick position arrangements. Different color patterns give each a unique appearance.

Variations of the Flemish bond pattern, called the Flemish garden wall bond pattern, typically have three headers between each stretcher to create the appearance of diagonal lines and diamond patterns. Using two or more brick colors permits endless pattern designs (see Figure 2-10).

## English Bond

Whereas the Flemish bond pattern is viewed as alternating stretchers and headers in each course, the English bond pattern consists of alternating courses of brick laid as all stretchers and all headers (see Figure 2-11). Like the

**Figure 2-11** Alternating courses of brick that are laid in the stretcher position and the header position creates an English bond pattern.

**Figure 2-12** An English bond pattern wall with an English style corner requires a 2" piece called a queen closure or plug.

**Figure 2-13** An English bond pattern wall with a Dutch style corner requires a 6"-piece at the end of the wall.

Flemish bond, it is a pattern of brickwork typically used by nineteenth-century and earlier masons.

Every other header is centered above a stretcher. Like the Flemish bond, the English bond was first used in walls of two or more brick wythes with full 8"-length brick joining two adjacent wythes. For residential construction, where single-wythe brick walls are anchored to structural framing, the headers are half-brick, also called snap headers.

A shorter brick is required at the ends of walls. Both English corner (see Figure 2-12) and Dutch corner (see Figure 2-13) are design choices for the English bond walls.

## American Bond

The American bond consists of a course of brick laid in the header position typically between every five or six courses of brick that are laid in the stretcher position (see Figure 2-14).

Like the Flemish and English bond patterns, the American bond was first used for joining brick walls of two or more wythes. Becoming more apparent in mid- and late nineteenth-century American architecture, it is an alternative for the Flemish bond or English bond. The American bond can be used on composite walls, or walls with wythes of different materials, as brick and block. The American bond is sometimes referred to as the common bond. However, since one may mistake common to mean "ordinary" or "commonly used" and therefore imply the running bond, it is advisable to refer to this bond only as the American bond.

A 6" piece, or three-quarter brick, is laid in each direction at the corner to begin the header course. The three-quarter brick are needed so that the headers will center on the brick below.

A variation of the American bond, called the American bond with Flemish headers, has alternating headers and

**Figure 2-14** At first glance, an American bond pattern appears similar to a running bond pattern. The header course of the American bond distinguishes the two.

**Figure 2-15** The pattern bond for this wall is called an American bond with Flemish headers.

stretchers instead of all headers between five or six courses of stretchers (see Figure 2-15). This is a more contemporary method of using full 8"-length brick to join two masonry wythes.

As with the Flemish bond and English bond, snap headers are used for 4"-wide walls replicating multi-wythe solid masonry walls.

**FROM EXPERIENCE**

Many mid- and late nineteenth-century homes exhibit the Flemish bond pattern on their front sides and the American bond on the remaining walls. Study period-appropriate regional brickwork when desiring to accurately replicate a home of a specific time period.

## Stack Bond

The **stack bond** pattern (also called the *stacked* bond pattern) consists of brick aligned vertically in each consecutive course in straight columns with no overlap (see Figure 2-16). Evidenced frequently in 1950s American architecture, it lends a contemporary or modern fashion to brickwork when compared to the other bond patterns.

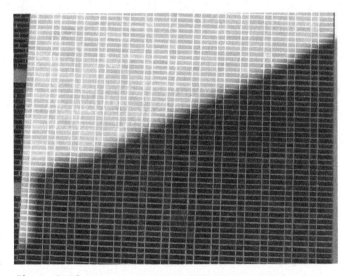

**Figure 2-16 The corner of this building has a stack bond pattern, created with both stretchers and headers, while the rest of the walls are laid in the running bond pattern.**

The stack or stacked bond pattern is mostly used for decorative patterns, enhancing parts of walls otherwise laid in the running bond pattern. Because there is no overlapping of units between brick courses, walls constructed in the stack bond pattern must be reinforced with wire joint reinforcement bedded in the mortar of horizontal bed joints.

## Summary

- Brick pattern bonds are sometimes referred to as brick structural bonds, permitting multi-wythe wall designs that are time-proven to be capable of supporting loads and resisting forces.
- The running bond pattern is the most frequently used pattern in the United States, permitting single-wythe, 4" walls anchored to backup walls.
- To replicate historically correct, time period architectural styling and design, one must be familiar with all brick pattern bonds.
- Some brick pattern bonds require more time than others when building walls because of the attention required for the arrangement of brick.

# Review Questions

*Select the most appropriate answer.*

**1** A brick pattern where every brick is laid in the stretcher position and the brick of alternate courses forms a uniform overlap with the brick below is called the

a. American bond
b. English bond
c. Flemish Bond
d. running bond

**2** The pattern bond permitting the weights or loads carried by 4", single-width brick walls to be distributed throughout the wall more uniformly is the

a. American bond
b. English bond
c. Flemish Bond
d. running bond

**3** A term used to express the bed depth or thickness of a masonry wall in terms of masonry units is

a. cladding
b. column
c. course
d. wythe

**4** The pattern bond that consists of alternating stretchers and headers along each course is called the

a. American bond
b. English bond
c. Flemish Bond
d. running bond

**5** Another term for half of a full brick length is a

a. bat
b. cap
c. closure
d. header

**6** A styling for corners or ends of walls constructed in either the English bond pattern or Flemish bond pattern characterized by an approximately 6"-long brick at the end is called

a. American corner
b. English corner
c. Dutch corner
d. queen corner

**7** Brick walls or fencing typically built to add elegance to outdoor living areas are called

a. cavity walls
b. garden walls
c. parapet walls
d. retaining walls

**8** A brick wall constructed in the Flemish bond pattern with the headers omitted to create openings through the wall is called a

a. hedge wall
b. pudlock wall
c. screen wall
d. serpentine wall

**9** A pattern consisting of alternating courses of brick laid as all stretchers and all headers is called the

a. American bond
b. English bond
c. Flemish Bond
d. running bond

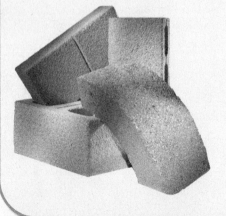

10. A pattern consisting of a course of brick laid in the header position typically between every five or six courses of brick laid in the stretcher position is called the

    a. American bond
    b. English bond
    c. Flemish Bond
    d. running bond

11. A pattern consisting of brick aligned vertically in each consecutive course in straight columns with no overlap is called the

    a. stack bond
    b. stacked bond
    c. running bond
    d. both a and b

12. A pattern of brickwork typically used by nineteenth-century and earlier masons is the

    a. English bond
    b. Flemish Bond
    c. stack bond
    d. both a and b

13. The English style of corner used at the ends of English bond and Flemish bond pattern brick walls requires a 2"-long brick piece sometimes called a

    a. plug
    b. queen closure
    c. snap header
    d. both a and b

14. The pattern bond not likely seen on eighteenth- and nineteenth century multi-wythe exterior masonry walls is the

    a. running bond
    b. stack bond
    c. Flemish bond
    d. both a and b

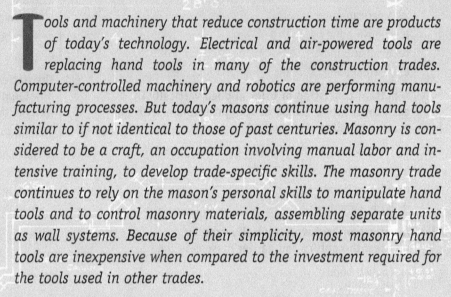

# Chapter 3 | Masonry Hand Tools

Tools and machinery that reduce construction time are products of today's technology. Electrical and air-powered tools are replacing hand tools in many of the construction trades. Computer-controlled machinery and robotics are performing manufacturing processes. But today's masons continue using hand tools similar to if not identical to those of past centuries. Masonry is considered to be a craft, an occupation involving manual labor and intensive training, to develop trade-specific skills. The masonry trade continues to rely on the mason's personal skills to manipulate hand tools and to control masonry materials, assembling separate units as wall systems. Because of their simplicity, most masonry hand tools are inexpensive when compared to the investment required for the tools used in other trades.

This chapter does not attempt to demonstrate either the proper or safe use of the tools addressed. Procedures for the safe and intended use of the hand tools are addressed as they relate to topics in this textbook. To prevent accidents and injuries, always read and comply with manufacturers' safety recommendations and governing laws and regulations.

## OBJECTIVES

After completing this chapter, the student should be able to:

- ⊗ Identify masonry hand tools.
- ⊗ Describe available options for specific tools.
- ⊗ List manufacturers of specific masonry hand tools.
- ⊗ List safety precautions and care for specific tools.

# Glossary of Terms

**heel** the end of a mason's trowel nearest the handle.

**level** perpendicular to the force of gravity.

**lift** the measured height from a mason's trowel blade to the center of the butt-end of the handle.

**plumb** parallel to the force of gravity.

**toe** the point of a mason's trowel at the opposite end of the handle.

**trowel blade** that part of a mason's trowel holding mortar.

**trowel sha** ... trowel extend... part of a mason's ... and into the ha... the trowel heel

**vials** glass or acry... ...uid-filled tubes in mason's spir... ...vels permitting level and pl...b alignments.

## Brickmaso... owel

The *brickmason's tr...* (Figure 3-1) is used for laying both brick and block...h referred to as a brick trowel so as not to be confu...The parts of a plastering, tile-setting, or cement-finishing t...d handle. a trowel include the blade, shank, fer...

The trowel bl...d from a single piece of high-grade tool grade trowel are...el. The shank length (A) and angle be-steel or stainles...ndle determine lift (B) (see Figures 3-1 tween it and th... lift may reduce wrist fatigue. and 3-2). A hig...

Thinner ste... permitting the blade to flex without bend-ing, is desira... The tip of the blade is called the **toe**, and the blade's ...el is below the shank.

The th... patterns of **trowel blades** are the wide London, narro... London, and Philadelphia patterns (see Figure 3-3).

The *narrow London* pattern is a popular choice for laying brick. Its narrow shape sufficiently spreads a ribbon of mortar for a bed joint, and a backup wall does not interfere with furrowing of the bed joint.

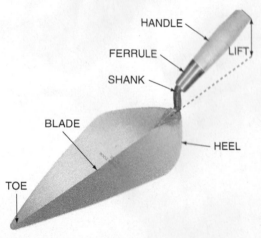

Figure 3-1 **Parts of the brick mason's trowel.** *(Courtesy of Bon Tool Company)*

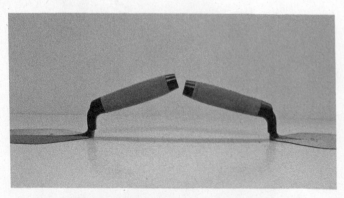

Figure 3-2 **The trowel on the left has a taller shank and a higher lift than the trowel on the right.**

Figure 3-3 **The three trowel patterns: wide London (left), narrow London (center), and Philadelphia (right).** *(Courtesy of Bon Tool Company)*

*For step-by-step procedures for applying bed joints to brick, see the Procedures section at the end of this chapter.*

*For step-by-step procedures for applying head joints to brick, see the Procedures section at the end of this chapter.*

The *wide London* pattern holds more mortar for face shell spreading when laying concrete masonry units or block, and it holds more mortar overall.

The *Philadelphia* pattern is similar to the wide London pattern. Both the Philadelphia and the wide London pattern are approximately 1" wider near the heel than the narrow London pattern. The Philadelphia pattern trowel has a distinct sharp angle between the heel and the sides of the blade, whereas the wide London pattern appears to be more curved at this point.

The length of a trowel is measured from the blade's heel to its toe. Depending on manufacturer and pattern, blades measure in ½" increments from 9" to 13". The author prefers an 11" trowel.

Several choices of trowel handles are available. Wooden handle trowels have a metal band, or ferrule, securing the shank and handle (see Figure 3-4).

A leather handle trowel is actually a combination of the wood handle and leather rings (see Figure 3-5). It too has the metal ferrule.

Figure 3-4 **Select hardwoods are used for wooden handles.** *(Courtesy of Bon Tool Company)*

**Figure 3-5 A leather handle trowel is a combination of wood and leather rings.** *(Courtesy of Marshalltown Company)*

Requiring no ferrule, plastic handles are very durable (see Figure 3-6).

A rubber or nylon handle, possibly having a finger guard behind the shank along with a hard polymer bumper to absorb the impact of tapping masonry units, is becoming more popular (see Figure 3-7).

Some masons prefer a plastic or rubber bumper on the end of the handle (see Figure 3-8). The bumper protects the handle and weights the handle end, making the trowel feel lighter.

**Figure 3-6 Trowels having plastic handles require no ferrules.** *(Courtesy of Marshalltown Company)*

**Figure 3-7 Softer handles claim to reduce user fatigue.** *(Courtesy of Marshalltown Company)*

**Figure 3-8 Pictured is a screw-on and slip-on trowel handle bumper.** *(Courtesy of Bon Tool Company)*

There are several manufacturers of trowels. Marshalltown Trowel Company is more than 100 years old, and W. Rose Inc. has been making trowels for more than 200 years. Some other manufacturers include Bon Tool Company, Goldblatt Tools, and Kraft Tool Company.

## FROM EXPERIENCE

A narrow London trowel blade is recommended for laying brick, and a wide London blade is recommended for laying block. The narrow London and wide London trowels can effectively be used to spread the same amount of mortar for bedding brick when using the procedures for spreading bed joints described in this chapter. The backup wall for brick veneer is not as likely to interfere with the narrow London blade while furrowing mortar as is experienced with the wide London blade. For blocklaying, the wide London trowel holds more mortar for face-shell spreading the bedding of mortar along the edges of the block.

## CAUTION

**Trowel Safety and Care**

- **Avoid contact of the pointed trowel toe with the body or body of others.**
- **Always wear eye protection that meets OSHA standards for protection from mortar, masonry materials, and flying debris.**
- **Do not use trowels with loose or defective handles.**
- **Do not use trowel blades that are bent or chipped.**
- **Do not weld, grind, or reheat trowel blades.**
- **Keep high carbon steel blades oiled or waxed to prevent rust when storing tools.**

## Mason's Hammers

The blade on the mason's hammer is used for breaking and trimming masonry units. There are a variety of blade lengths and widths. The blade should remain dull rather than exhibit a sharp cutting edge. Metal on a sharpened blade edge chips easily and is more likely to cause injuries. The hammer does not cut the masonry unit. Rather, it fractures or breaks the unit at the point of impact. The square-head design of the hammer is intended for breaking, trimming, and cleaning masonry units. Before such action, it is necessary to ensure that the manufacturer's recommendations do not restrict using a specific hammer for striking chisels or other metal. Depending on the manufacturer, the weight of most hammers ranges between 16 and 24 ounces.

Handles are available in wood, fiberglass, or steel (see Figure 3-9). Hickory wood handles are shock absorbent but can split or shrink and separate from the head. Fiberglass handles are less likely to break. Steel handles are the most durable. With some, the head and shaft are forged from one solid piece of steel. Others have a tubular steel handle attached to the head. Both fiberglass and steel handles are available with a vinyl nylon, neoprene fiber, or similar cushion grip.

Some of the brands of mason's hammers are Bon, Estwing, Marshalltown, Stanley, Plumb, Rocket, and Vaughan. Weights, blade designs, and handles vary between manufacturers.

**CAUTION**

**Mason's Hammers Safety and Care**

- **Always wear eye protection that meets OSHA standards.**
- **Never strike against another hammer or hardened metal.**
- **Never use hammers with loose or broken handles.**
- **Drive common nails only, never hardened nails unless recommended by the manufacturer.**
- **Do not weld, machine grind, or treat by reheating.**
- **Reshape the blade using a file intended for metal.**

Figure 3-9 From left to right, masons' hammers with steel, fiberglass, and wooden handles. *(Courtesy of Bon Tool Company)*

# Jointers

A *jointer*, sometimes referred to as a striking iron or striking tool, creates a mortar joint's appearance. Creating the desired appearance with a jointer is known as striking or tooling the joints. Most jointers are made of tempered, hardened, and polished carbon steel. Some are shaped from a hardened acrylic or Plexiglas so as not to discolor custom-colored mortars. All but the rake jointer compact the mortar, making the mortar joint more resistant to water penetration. There are several joint finishes from which to choose. For additions to existing structures, it may be obvious if not specified as to which joint finish is desired. However, because of the differences in their sizes and manufacturers' specific designs, it is just as important that every mason on a specific project use similar jointers that create matching joint finishes. Bon, Marshalltown, Goldblatt, and Rose are familiar brand names for jointers.

## Convex Jointers

The convex jointers (see Figures 3-10 and 3-11) form concave joints. Almost always used for concrete masonry units and frequently with brickwork, the sunken-appearing concave joints are highly recommended for exterior joints because they create a smooth joint resistant to moisture penetration. Convex jointer diameter sizes are typically available in ⅛" increment from ⅜" to 1". Each end of the curved convex jointer is a different size. The ½" by ⅝" jointer

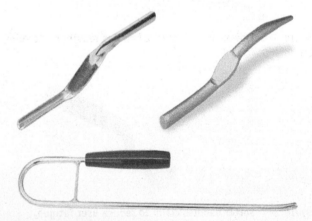

Figure 3-10 Pictured are a curved convex jointer (left), convex bullhorn (right), and convex sled runner (bottom). *(Courtesy of Bon Tool Company)*

Figure 3-11 (A) Two more styles of convex jointers are the barrel jointer with replaceable steel barrels *(Courtesy of Bon Tool Company)*, and (B) a tempered acrylic jointer that eliminates discoloring colored mortars. *(Courtesy of Marshalltown Company)*

is a common size. Depending on its size, the convex jointer size forms the depth of the concave joint. Wider jointers form shallower joints, and narrower jointers create deeper concave joints. Another style of convex jointer is the sled runner. Typical sled runners range from 14" to 20" long and are used for the longer bed joints. By bridging irregular and chipped edges of masonry units, sled runners produce straighter, more uniform appearing joints.

 **FROM EXPERIENCE**

A ½" × ⅝" curved jointer works well for striking most mortar joints. Slide the jointer back and forth in the joints until the jointer compresses the mortar to a depth where the jointer and edges of the brick meet. Keep the tip end of the jointer from digging into the joints to avoid gouging mortar from joints and leaving rough joint finishes.

## Grapevine Jointers

The grapevine jointers have a raised bit, either square or round, in the center in order to create a recessed groove in the middle of mortar joints (see Figure 3-12). As with convex jointers, curved grapevine jointers permit two different sizes. Wooden- or rubber-handled grapevine jointers provide a comfortable grip but limit the tool to a single size. Grapevine sled runners are also available.

The finished joints resemble an American colonial period practice of scribing a line through a joint's center. For this reason, the grapevine jointer is sometimes called the colonial jointer, and the resulting joint is called a colonial joint. However, colonial joint finishes were created with the mason's trowel long before development of the grapevine jointer.

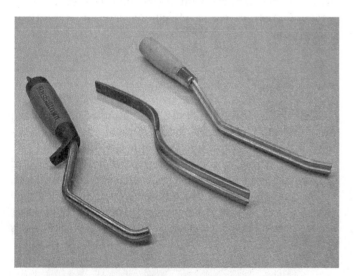

**Figure 3-12** With their center bead, a variety of grapevine jointers permit differently shaped grooves along the joint's center.

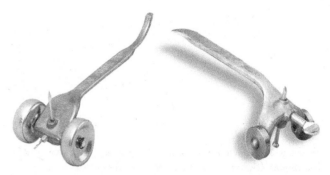

**Figure 3-13** Two styles of joint rakers: the skate wheel (left) and cart wheel (right). *(Courtesy of Bon Tool Company)*

 **FROM EXPERIENCE**

Applying steady pressure, pull or push the jointer through the joint. Be certain that there is no dry mortar buildup at the shoulders of the bead. Excessive buildup of mortar along the center bead of the jointer deforms the joint's intended appearance.

## Joint Rakers

The joint raker or rake jointer (see Figure 3-13) removes rather than compresses the mortar. Adjusting the depth for removing mortar with a nail accomplishes a raked joint. The rake jointer's metal frame is equipped with wheels that travel across the face of the brick as the nail head removes the mortar. The resulting joints are called raked joints. This is the only tooled joint finish produced by removing the mortar. All other joint tooling compresses the mortar within the joints. Raked joints are not recommended for exterior masonry in most regions of the country due to excessive water penetration.

 **FROM EXPERIENCE**

Two heads are better than one. A double-headed 8-penny form nail removes mortar from the joints better than a conventional single-headed nail. Be sure the nail head scrapes all edges of the brick for proper, fully raked joints.

## "V" Jointers

The "V" jointer is made from a square piece of tempered steel (see Figure 3-14). Two sides of the steel form a "V" pattern in the joints.

**Figure 3-14 Shown are a curved "V" jointer (left) and a "V" sled runner (right).** *(Courtesy of Bon Tool Company)*

**Figure 3-16 Shown are a ⅜" × ½" half-round concave beader (left), a ¼" × ¾" square bead jointer (center), and a ¾" square bead jointer (right).** *(Courtesy of Bon Tool Company)*

"V" jointers are available as curved jointers and sled runners. The ½" × ⅝" curved "V" jointer is ideal for striking most head joints. The sled runner creates a straighter and more uniform depth "V" joint. Becoming deeper at the middle, the "V" joint resembles the concave joint. However, the "V" joint forms approximately a 90-degree angle at the joint's center, whereas the concave joint appears as a radial or curving line. "V" joints are struck using the same procedures as used for concave joints.

## Struck/Weathered Jointers

The struck/weathered jointer is used to replicate two of the bed joint finishes found on American colonial brickwork: the trowel-struck bed joint and the weather joints bed joint. Trowel-struck bed joints taper from the bottom front edge of one course inward behind the top front edge of the course below it. This creates a shelf along the top front edge of the brick that allows water penetration. The weathered bed joint tapers inward from the top front edge of one course and behind the bottom edge of the course above it, allowing water to drain down the wall's face rather than to accumulate on the tops of brick as trowel-struck joints allow. Although desirable for duplicating accurate reproductions of eighteenth- and nineteenth-century brickwork, in most regions of the country neither trowel-struck nor weathered joints are recommended for exterior walls. Wherever trowel-struck or weathered bed joints are used, the head joints appear to have a compressed, flat surface. The struck/weathered jointer (see Figure 3-15), the mason's trowel, or the smaller pointing trowel is used for tooling trowel-struck and weathered joints.

**Figure 3-15 This struck/ weathered jointer is made of an abrasive resistant polyethylene.** *(Courtesy of Bon Tool Company)*

**FROM EXPERIENCE**

A pointing trowel makes struck or weathered joints easier to form than the full size mason's trowel.

## Concave Beaders

The concave beader (see Figure 3-16) produces a raised bead along the joint's center. Depending on preference, the bead can appear either square or half-round. Both are used for stonemasonry, but smaller concave beaders are sometimes used for mortar joints of brick facades. As with other curved jointers, the curved beader is a different size at each end. Available in different sizes, the beaded jointer must be small enough to fit in the mortar joint between the brick. Beaded joints are not recommended in severe weathering regions.

Flush cut joints and the extruded joints do not require tooling with jointers. Simply cutting the mortar off the face of the brick produces flush joints. This joint is common on structures built before the twentieth century, especially on the interiors of solid masonry walls. Leaving mortar protrusions as the finished joint appearance creates extruded joints. Neither is recommended for exterior masonry in severe or moderate weathering regions.

## Mason's Levels

The mason's level is used to align masonry units level and plumb. A **level** surface is a flat, horizontal plane perpendicular to the force of gravity (see Figure 3-17).

A **plumb** line is a vertical line parallel to the force of gravity and perpendicular to a level surface (see Figure 3-18).

The mason's level is sometimes referred to as a plumb rule, a plumb level, or simply a level. The stability of select straight grain mahogany makes it a popular choice of woods

**Figure 3-17 These brick are aligned level.**

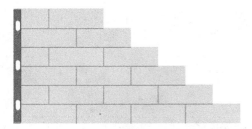

**Figure 3-18 These blocks are aligned vertically plumb.**

for mason's levels. Recent environmental issues have caused tool manufacturers to use other woods, laminated wood products, plastics, or metal box beams for level production. The two types of levels are spirit levels and electronic levels.

## Spirit Levels

*Spirit level* have glass or acrylic sealed tubes (**vials**) partially filled with alcohol, oil, or other liquids not susceptible to freezing. Partially filling the tubes or vials leaves an air bubble. Level and plumb alignments are confirmed when the air bubble is centered in the vials. The vials are positioned in openings referred to as windows. Glass or acrylic window lenses protect the vials. Some vials are slightly curved rather than straight. Vials with a slight curvature can only be referenced when the level is positioned so that the middle of the vial is higher than either end. Curved vials better control the stability of the bubble. Levels with curved vials include two vials in each window because reference is possible only to the vial whose curvature is higher in the center. This is the bottom vial of the pair (see Figure 3-19).

 **FROM EXPERIENCE**

Glass cover lenses are easily broken. Local glass dealers can cut cover lenses to fit most levels. Use an adhesive caulk to seal the glass in the level's frame. Using gloves that are resistant to caulk penetration and prevent skin contact with the caulk, apply a small bead around the perimeter of the glass and smooth with protected finger.

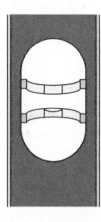

**Figure 3-19 Slight curvature of the vials permits referring to the bottom vial only.**

Levels with straight vials require only one vial per window because reference is made to the bubble regardless of the level's edge positioned against the work. Levels with only one straight vial in each window are referred to as *monovial levels*. There are different lengths of levels, typically ranging from 18" to 72".

 **FROM EXPERIENCE**

Having both a 24" level and a 48" level is desirable. For aligning leads plumb in the beginning and for smaller projects, a 24" level is easier to handle than a 48" level. Although the accuracy of the shorter 24" level is the same as the longer 48" level, a longer level eliminates alignment errors committed by the mason commonly associated with the shorter level when aligning work beyond the reach of a 24" level.

Edge protection for wooden levels is one of two types, brass bound or stainless steel edge rails. Bound levels have brass inserts affixed in grooves on each of the four edges, and screws attach stainless steel strip rails to the level's edges (see Figure 3-20).

A 48", mahogany, brass bound level is shown in Figure 3-21.

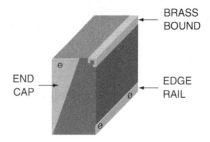

**Figure 3-20 Securely affixed metal bindings or edge rails attached with screws are precisely fit to the level's edges.**

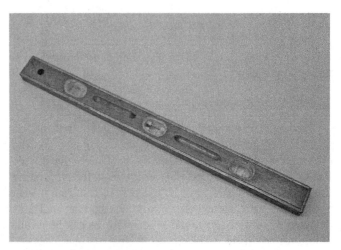

**Figure 3-21 Brass edge rails and rubber end caps protect this wooden level.**

**Figure 3-22 Wood or urethane is inlaid on both sides of the metal "I"-beam.** *(Courtesy of Bon Tool Company)*

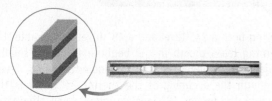

**Figure 3-23 Laminated hardwoods and stainless steel edge rails create this professional-quality mason's level.** *(Courtesy of Bon Tool Company)*

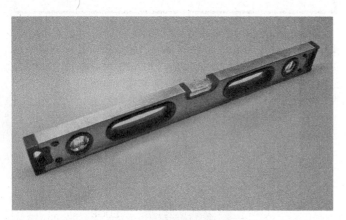

**Figure 3-24 This metal level, designed not to warp or twist, offers extreme accuracy.**

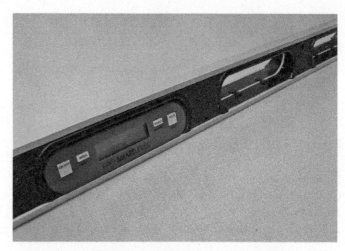

**Figure 3-25 This level features an electronic sensor module along with level and plumb vials.**

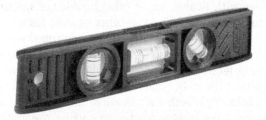

**Figure 3-26 This 7½ torpedo level simplifies level and plumb alignment of a single brick.** *(Courtesy of Bon Tool Company)*

**Figure 3-27 Hooks at each end of this line level permit it to be attached to a line.**

The "I" beam level has either an aluminum or magnesium "I" beam with inlaid wood or urethane on both sides (see Figure 3-22).

Laminated hardwoods, that is several pieces of glued-up wood, are also used in making levels (see Figure 3-23).

The metal box-beam level is a professional-quality level with mono vials (see Figure 3-24). The top-mounted level vial is very visible.

## Electronic Levels

Today's technology permits electronic levels that rely on a battery operated, digital electronic sensor module. The electronic module is attached to rails of various lengths (see Figure 3-25). Some rails include spirit vials for performing level and plumb operations. In addition to confirming level and plumb conditions, the electronic module indicates degrees, percent of slope, and pitch. Modules can be recalibrated for accuracy.

A short torpedo level can be used to align a single brick level and plumb (see Figure 3-26).

A line level (see Figure 3-27) can be hooked to the mason's line to confirm a level tensioned line.

Some of the level brands are American, Bon, Crick, Empire, Marshalltown, Mays, National, Sand's, Smith, Stabila, and Starrett.

*For step by step instructions on checking the accuracy of a mason's spirit level, see the Procedures section at the end of this chapter.*

Figure 3-28 **Suspended from a string, the plumb bob creates a plumb line.** *(Courtesy of Bon Tool Company)*

# Plumb Bobs

Before the spirit level, or levels with vials first filled with alcohol, masons relied on plumb bobs. Still used today, a plumb bob is a pointed weight hung from a string (see Figure 3-28). A plumb line exits from the held end of the string and the tip of the plumb bob.

 **FROM EXPERIENCE**

Avoid attaching a twisted line with the plumb bob because it untwists when suspended from the plumb bob, which causes the plumb bob to uncontrollably spin rather than remain still. Use braided string lines instead of twisted lines.

Before spirit levels, suspending a plumb bob from a line attached to a board's center was used to plumb a wall line (see Figure 3-29).

 **CAUTION**

**Mason's Levels Safety and Care**

- Avoid putting fingers in the round hang hole because fingers might be broken when the level is moved.
- Remove or replace broken vials.
- Replace broken window glasses.
- Periodically apply a nontoxic wood preservative to wood levels and wax to metal levels.
- Never pound levels with other tools.
- Wear eye protection and head protection that meet OSHA standards when working below a plumb bob.

Figure 3-29 **An early example of a plumb rule is shown in this photo.**

# Mason's Rules and Tapes

Folding rules are used for measuring and course spacing masonry walls (see Figure 3-30). Masonry course spacing scales are covered in Chapter 7. The 6' folding rule is divided into $\frac{1}{16}$" increments. The opposite side of the folding rule includes masonry scales. Masonry course scales are equal increments used for uniform course spacing. The three sets of scales are 1) the standard size brick spacing scales, 2) the oversize brick spacing scales, and 3) the modular

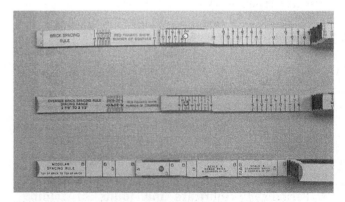

Figure 3-30 **Shown are the brick spacing rule for standard size brick (top), the oversize brick spacing rule for oversize or engineered brick (center), and the modular spacing rule for modular masonry construction (bottom).**

**Figure 3-31** Modular spacing scales are on this 10' retractable steel tape measurer. *(Courtesy of Bon Tool Company)*

**Figure 3-32** This 100' steel tape measurer is marked in ⅛" increments.

spacing scales. Ten standard size brick scales, numbered 1-9 and 0, are identified on the brickmason's rule, permitting different bed joint widths for standard size brick. The oversize brickmason's rule includes 11 scales lettered A through K, permitting different bed joint widths for oversize or engineered brick. The modular spacing rule includes 6 scales. Each scale represents a different masonry unit and permits wall height increments of 16" for all masonry units.

The masonry scales are included on some steel tape measurers (see Figure 3-31).

Steel or cloth, 50' and 100' tape measurers permit dimensional layout of larger projects (see Figure 3-32).

**CAUTION**

**Rule and Tape Safety and Care**

- Avoid pinching fingers between spacing rule joint hinges.
- Avoid contact between fingers and the edges of steel tapes.
- Periodically lubricate the joints of folding rules, and wax steel tapes.
- Discard broken rules and tapes.

**Figure 3-33** Shown are bonded-braided line (top left), twisted line (top right), and braided line and reel (bottom). *(Courtesy of Bon Tool Company)*

## Mason's Lines

The mason's string line is used to align straight each course of masonry units when building walls (see Figure 3-33). Available in a variety of colors and lengths, the mason's line is made of either nylon or polypropylene. Pulled tightly and held under tension by line fasteners to prevent sagging, these lines resist breaking under tension, resist cutting by the trowel blade, and do not rot or mildew. Three separate strands of line are either twisted or braided to form the mason's line. Braided line is more durable and more expensive than twisted line. A bonded braided line is superior to twisted or braided line. The test strength of a line refers to the pounds of force required to break the line. Regardless of type, the safe working load for any line is 20% of the rated breaking strength.

**CAUTION**

**Line Safety and Care**

- Always wear eye protection in the presence of tensioned lines.
- Remove twigs before tensioning lines.
- Use only those fasteners designed to secure lines, and never use nails to secure tensioned lines.
- Discard frayed or damaged lines.

## Line Fasteners

Several items are used to attach or position the mason's line. These items include line blocks, line pins, adjustable line stretchers, and line twigs.

**Figure 3-34 Line blocks secure the tensioned mason's line to the end of walls or corner poles.** *(Courtesy of Bon Tool Company)*

**Figure 3-35 Placed securely in mortar head joints, steel line pins secure the mason's line.** *(Courtesy of Bon Tool Company)*

**Figure 3-36 Shown are two styles of adjustable line dogs.** *(Courtesy of Bon Tool Company)*

## Line Blocks

A line block (see Figure 3-34), also called a corner block, attaches to an outside corner of a masonry wall or a mason's corner pole. Line is taken through the slot, then wrapped around the line block, and secured together. Line tension holds the line block at the corner. Line blocks are made of wood, plastic, or metal.

## Line Pins

A line pin (see Figure 3-35) is a flat, tapered piece of metal approximately 4" long to which a line is attached. Inserted into a head joint, it secures a tensioned line. It is especially useful for attaching a line to inside corners where corner blocks cannot be used.

**Figure 3-37 Besides standard size concrete masonry units, the sliding adjustable line dog can be adjusted to fit a variety of architectural concrete masonry units.** *(Courtesy of Bon Tool Company)*

## Line Stretchers

The line stretcher, or line dog (see Figure 3-36), is adjustable for 4", 6", 8", 10", and 12" concrete masonry units. It is attached across the top of the block. Line tension secures the line dog. Line dogs are made of cast aluminum or plastic.

The sliding adjustable line dog accommodates both standard size and specialty concrete masonry units (see Figure 3-37).

## Line Twigs

A line twig, or trig (see Figure 3-38), is used to accurately position the mason's line. Made from spring steel, its formed slot grips the line. When weighted on top of a masonry unit, the twig positions a line as desired. On longer walls, twigging the line eliminates line sagging near the middle. In addition, twigging the line prevents excessive line movement caused by other masons working on the same wall.

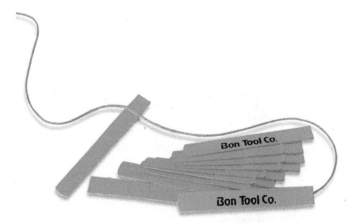

**Figure 3-38 The line twig clips to the mason's line.** *(Courtesy of Bon Tool Company)*

**CAUTION**

**Line Fasteners Safety and Care**

- Always wear eye protection meeting OSHA standards in the presence of line fasteners to which lines are tensioned.
- Discard wood line blocks that have wood splits, bent line pins and twigs, and damaged line stretchers.
- Never use a nail to secure a tensioned line.
- Remove twigs before tensioning or removing lines.

# Chisels

Chisels are used for breaking and trimming masonry units. Chisels are produced by metal-forging processes using hardened, tempered tool steel. Although masons often refer to the process as cutting the brick or block, the mason's chisel scores and breaks masonry units, reducing their size for the intended application. Because the chisel is used to score rather than cut, it has a dull edge. A good selection of chisels includes the brick set, the mason's chisel, the flat chisel, the plugging chisel, assorted sizes of star chisels, and the concrete chisel.

## Brick Set

Used for cutting brick, the brick set has a blade about 4" wide (see Figure 3-39). When struck with a hammer, the resulting break is precise and straight.

**FROM EXPERIENCE**

A sand-filled bucket permits the sand to both cushion the brick and steady it as hammer blows are delivered to the chisel. Cushioning the brick in sand allows for more accurate breaks.

Figure 3-39 **Rather than a sharp cutting edge, striking the brick set's dull blade with a hammer precisely breaks masonry units along line of impact.** *(Courtesy of Bon Tool Company)*

## Plugging Chisel

The plugging chisel, sometimes called a tuckpointer's chisel, is used for removing mortar joints (see Figure 3-40).

## Mason's Chisel

The mason's chisel is designed for cutting brick, block, and stone. Its blade is narrower and more tapered than the brick set (see Figure 3-41).

## Flat Chisel

The flat chisel is a general-purpose or utility chisel with a thin blade about 1" wide (see Figure 3-42). The flat chisel is good for trimming hardened mortar from brick and block.

## Star Chisels

The star chisel, also known as the star drill, is used to cut a round hole in masonry materials (see Figure 3-43).

Figure 3-40 **When impacted with a hammer, the narrow, tapered point of the plugging chisel is designed to remove mortar joints.** *(Courtesy of Bon Tool Company)*

Figure 3-41 **Similar to the brick set, a mason's chisel cuts masonry units.** *(Courtesy of Bon Tool Company)*

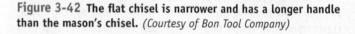

Figure 3-42 **The flat chisel is narrower and has a longer handle than the mason's chisel.** *(Courtesy of Bon Tool Company)*

Figure 3-43 **Impacted with a hammer, a star chisel creates a hole in masonry units.** *(Courtesy of Bon Tool Company)*

**Figure 3-44** **The concrete chisel has a pointed tip.** *(Courtesy of Bon Tool Company)*

**Figure 3-45** **The small size of the pointing trowel permits precise placement of mortar when filling joints and creating mortar wash caps.** *(Courtesy of Bon Tool Company)*

Rotating the four-pointed tip between hammer strikes improves performance. Star chisels are available in ⅛" increment sizes typically ranging from ¼" to ¾" diameter.

**FROM EXPERIENCE**

Clamping the star chisel in vice-grip pliers makes rotating it easier as the chisel penetrates beyond the surface.

## Concrete Chisel

The pointed concrete chisel breaks concrete and other masonry materials (see Figure 3-44).

**CAUTION**

**Chisel Safety and Care**

- Wear eye protection meeting the intent of OSHA standards when using chisels.
- Wear abrasive-resistant gloves to protect the hands.
- Remove metal burs from the heads of chisels with a metal-cutting file.
- Strike the heads of chisels only with those hammers recommended by the hammers' manufacturers.

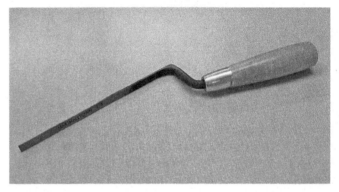

**Figure 3-46** Mortar is compacted into joints with the narrow blade of the tuckpointer.

gives the thin, narrow tool the proper flexing for filling mortar joints. A curved design with two blade widths at opposite ends is available.

## Margin Trowel

The margin trowel (see Figure 3-47) has a rectangular-shaped blade and is used to place or remove mortar in small spaces where the mason's trowel is too big. Blade widths are typically either 1½" or 2" with blade lengths of 5", 6", and 8".

## Pointing Trowels

The pointing trowel (see Figure 3-45) is used for filling mortar joints. Trowel-struck or weathered joints can be tooled using a pointing trowel. It is also used to form a mortar wash cap below a course of brick set back or "racked" from the face of the brick course below it. It is similar to the mason's trowel, only smaller. Pointing trowels are available in ½" increment lengths from 4½" to 7".

## Tuckpointer

The tuckpointer, also called the slicker, is used to compact mortar into joints (see Figure 3-46). They are available in 1/16" increment widths from 3/16" to 1". High carbon steel

**CAUTION**

**Pointing and Margin Trowels, Tuckpointer Safety and Care**

- Avoid contact of the trowel tips with the body or body of others.
- Keep high carbon steel blades oiled or waxed to prevent rusting when storing tools.
- Discard defective tools.

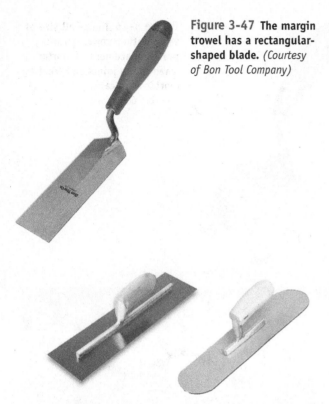

Figure 3-47 **The margin trowel has a rectangular-shaped blade.** *(Courtesy of Bon Tool Company)*

Figure 3-48 **Above the blade of a cement finishing trowel is a center-mounted handle. Shown here are a square-end trowel (left) and a round-end trowel (right).** *(Courtesy of Bon Tool Company)*

# Finishing Trowels

Cement finishing trowels are used to smooth concrete surfaces and to apply cement or mortar plaster to the faces of masonry walls. Masons use finishing trowels when parging, that is, coating the exterior faces of foundation walls with mortars or cements. Having the handle centered above the blade makes applying mortar to walls easier than using a brickmason's trowel. Both square-end and round-end trowels are available (see Figure 3-48). For parging, the author prefers a 14" × 4", round-end blade.

 **FROM EXPERIENCE**

A round-end finishing trowel leaves less noticeable lap lines on the surface of parged walls.

# Chalk Line Reels

The chalk line reel (see Figure 3-49) is used when laying out wall lines. The self-chalking cotton line is pulled from the reel box containing colored, powdered chalk.

Figure 3-49 **Using colored chalk, this metal chalk line reel and line permits chalking lines for wall layouts.** *(Courtesy of Bon Tool Company)*

A tensioned line transfers chalk from the line to a surface when the middle of the line is raised above the surface and released.

 **FROM EXPERIENCE**

With the line pulled from the chalk box, snapping a heavily chalk coated line in mid-air before snapping it on a surface creates a sharper image, a better defined line.

 **CAUTION**

**Chalk Reel Safety and Care**

- **Wear eye protection that meets OSHA standards when exposed to chalk dust.**
- **Avoid breathing chalk dust.**
- **Replace frayed or broken lines.**

# Framing Square

The framing square (see Figure 3-50) is used to confirm square or 90-degree angles for masonry corners, columns, and piers. The lengths of its two sides or blades are 24" and 16". Square corners of longer walls are confirmed using the builder's level or mathematical calculations involving triangulation.

# Combination Square

The combination square (see Figure 3-51) has a head that slides along the 12" blade and can be locked at any point. This is helpful when marking multiple units at identical lengths or when gauging racking and corbelling widths for brickwork.

**Figure 3-50 A mason often uses a framing square to confirm square corners.**

## Sliding T-bevel

The sliding T-bevel (see Figure 3-52) has an adjustable steel blade that can be set at a desired angle. It is helpful for marking brick that needs to be cut at angles other than 90 degrees, such as those needing cutting for fireplace fireboxes and brick arches.

## Mason's Brushes

Some masons use the trowel blade's edge for removing fragments of mortar remaining along the edges of brick and block after tooling the mortar joints, whereas others prefer a mason's brush (see Figure 3-53). Soft bristles minimize staining the wall and scuffing the mortar joints.

**Figure 3-52 The steel blade of a sliding T-bevel can be adjusted at any angle, permitting transferring and marking a desired angle onto a masonry unit for cutting.**

**Figure 3-51 Adjusting the head on the combination square permits gauging multiple operations.**

**Figure 3-53 Two types of brushes are the tampico bristle brushes (a) and the horsehair brushes (b).** *(Courtesy of Bon Tool Company)*

## Mason's Tool Bags

The cotton canvas tool bag (see Figure 3-54) is the masons' favorite for hand tool storage. A hinged, steel frame with metal studs, leather handles, a water-proof fiber bottom board, and optional leather reinforced bottom ensure a durable tool bag. Interior pockets organize small tools and accessories. Leather straps with buckles keep the bag closed.

Some masons prefer a canvas tote bag (see Figure 3-55).

Also available are nylon bags, including those with outside pockets and zippered bags for quicker access to the tools (see Figure 3-56).

## Tool Care

Left unprotected and in the presence of moisture, high carbon tool steel corrodes, and wood warps and twists. Exposed to mortar and moisture, masonry tools should be

**Figure 3-54** Many masons use either the leather bottom or heavyweight canvas tool bag. *(Courtesy of Bon Tool Company)*

**Figure 3-55** Another preference of masons is the canvas tote bag. *(Courtesy of Bon Tool Company)*

**Figure 3-56** This nylon tool bag has several outer compartments and a shoulder strap. *(Courtesy of Bon Tool Company)*

cleaned and protected after use. A light coating of oil prevents metal corrosion.

Most importantly, always read and follow all safety precautions and instructions included with every tool.

## Summary

- The diverse availability of professional-grade masonry hand tools gives masons opportunities to select from those best suited to them and the job at hand.
- Tools should be used only for their intended purpose and only by properly trained individuals.
- Compliance with OSHA standards, manufacturer's recommendations, and employer's safety regulations must be addressed before using any tool.

## Procedures

# Applying Bed Joints to Brick Using the Brickmason's Trowel

**A** For those right-handed, grip the trowel handle in the right hand with the thumb positioned near the end of the handle. Do not permit the thumb to extend beyond the handle and onto the shank.

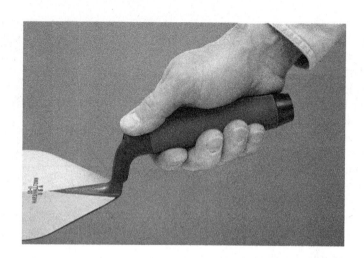

**B** Rotate the trowel counterclockwise so that the blade is *sideways* in a vertical position. Separate a *ribbon* of mortar from the mortar on the board by sliding the trowel towards the right side of the mortar board.

**C** Pick up the ribbon of mortar with the trowel blade.

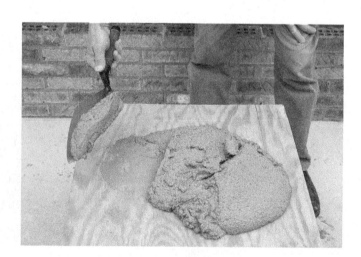

**Procedures**

# Applying Bed Joints to Brick Using the Brickmason's Trowel (continued)

**D** First extend the arm and then bring the arm toward the body. As the arm comes toward the body, lower the toe of the trowel to the level of the brick while rotating the trowel handle counterclockwise, transferring mortar to the brick. Loosely gripping the trowel blade and rotating the blade with the force of the fingers eliminates wrist movement and associated wrist fatigue.

**D**

**E** Lightly *furrow* the mortar by positioning the trowel at approximately a 45-degree angle in relationship to the top of the brick while moving the toe of the trowel along the center of the brick. Turning the trowel blade *topside down* prevents furrowing too deeply and having insufficient mortar for an adequate mortar bond.

**E**

**F** Reclaim the mortar from the face side of the brick displaced by furrowing. Angle the blade of the trowel as shown to prevent collected mortar from staining the face of the brick below the mortar.

**F**

**G** Move the trowel along the backside of the brick as illustrated to reposition the mortar extending beyond the edge of the brick. This procedure prevents excessive mortar from blocking the cavity between the backside of the brick and the adjacent back-up wall.

**G**

## Procedures

# Applying Head Joints to Brick Using the Brickmason's Trowel

 The mortar is applied to the end of the brick before it is bedded in mortar to ensure full coverage of the end of the brick with mortar. Position the thumb on the face of the brick, the index finger extended to the end of its top, and the remaining three fingers on the backside.

 Rotating the wrist counter-clockwise positions the end of the brick for applying a mortar head joint to the end of the brick seen here. This permits moving forward as the brick are laid. Working forward rather than backward is a safer practice and enables you to better observe a uniform bond alignment.

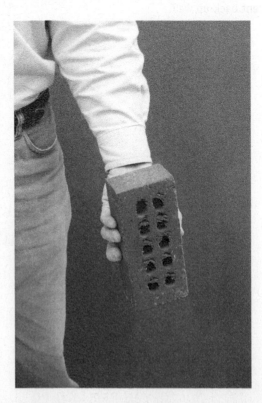

**C** Mortar is applied to the end of the brick after it is *set* on the trowel blade. *Setting* mortar is a technique for preventing the mortar from falling off the trowel when the trowel blade is turned topside down and having the mortar suspended below the blade. Mortar is *set* on the trowel blade from the resulting force of gripping the handle with only the thumb and index finger and quickly shaking the wrist as the remaining fingers clutch the handle in the palm of the hand.

**C**

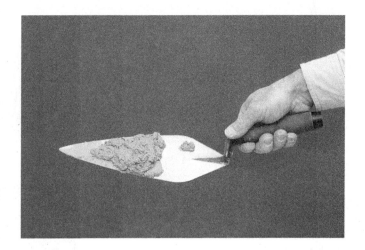

## Procedures

# Applying Head Joints to Brick Using the Brickmason's Trowel (continued)

 Swipe the trowel across the end of the brick while holding the loaded trowel topside down.

 Reload the trowel if necessary. Swipe the trowel blade across the edge of the face of the brick. Angle the blade slightly to prevent mortar from staining the brick's face.

Reload the trowel if necessary. Apply mortar to the back edge of the brick's end with a similar procedure.

# Applying Head Joints to Brick Using the Brickmason's Trowel (continued)

**G** Applying mortar to the opposite end of the brick is accomplished in a similar manner. Position the brick as shown. Follow Procedure C to *set* mortar on the trowel. Hold the brick and loaded trowel as shown, and swipe the mortar over the end of the brick.

**G**

**H** Reload the trowel if necessary. Apply mortar to the front edge of the brick.

**H**

**1** Reload the trowel if necessary. Apply mortar to the back edge of the brick as shown.

**1**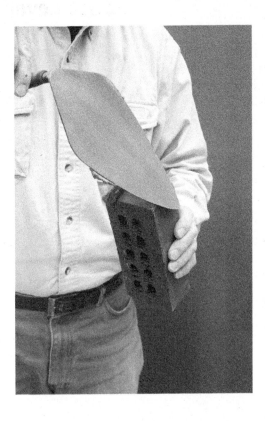

## Procedures

# Checking the Accuracy of a Mason's Spirit Level

**A** Place the level where it can be precisely repositioned, taking note of the *level* vial reading.

**B** Repositioning the level "end-for-end" should give a reading consistent with that first taken.

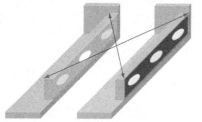

**C** After rotating the edges of the level 180 degrees, the reading should be consistent with those previously taken.

**D** Repositioning the level "end-for-end" on this edge should give a reading consistent with those taken in previous tests. In a similar manner, while holding the level vertically, the vials indicating plumb alignment can be checked.

# Review Questions

*Select the most appropriate answer.*

**1** An example of a mason's trowel pattern is

a. the narrow London

b. the wide London

c. the Philadelphia

d. all of the above

**2** The tool forming a concave joint finish is the

a. colonial jointer

b. concave jointer

c. convex jointer

d. grapevine jointer

**3** The tool forming a recessed bead in the center of the mortar joint is the

a. rake jointer

b. convex jointer

c. grapevine jointer

d. concave jointer

**4** Two lengths of levels masons prefer most are the

a. 24" and 36"

b. 24" and 48"

c. 24" and 72"

d. 36" and 72"

**5** A pointed device suspended from a string to establish a vertical line parallel to the force of gravity is called a

a. plumb bob

b. plumb rule

c. centering point

d. straight-edge

**6** The course spacings on mason's rules are called

a. graduations

b. divisions

c. scales

d. blocks

**7** The device hooked at the end of an outside corner and securing the mason's line is called a

a. line block

b. line pin

c. line twig

d. line trig

**8** A flat, tapered piece of metal that is driven into a head joint to secure a mason's line is called a

a. line block

b. line dog

c. line pin

d. line twig

**9** The mason's line is made of

a. braided nylon

b. bonded-braided nylon

c. twisted nylon

d. all of the above

**10** The chisel used to cut brick and block is called

a. the brick chisel

b. the brick set

c. the plugging chisel

d. both a and b

**11** A chisel used to cut a round hole in masonry is called a

a. concrete chisel

b. brick set

c. star drill

d. tuckpointer's chisel

12  A small, triangular-shaped trowel used to fill mortar joints is called a

a. margin trowel
b. pointing trowel
c. caulking trowel
d. tuckpointer

13  A narrow-bladed tool used to compact mortar into joints is called a

a. slicker
b. tuckpointer
c. margin trowel
d. both a and b

14  A trowel having a small, rectangular-shaped blade is called a

a. margin trowel
b. tuckpointing trowel
c. parging trowel
d. finishing trowel

15  A square having an adjustable head is called a

a. combination square
b. framing square
c. speed square
d. sliding T-bevel

# Chapter 4

# Masonry Construction Equipment

**S**uccessful masonry contractors rely on construction tools and equipment that maximize production, enhance quality workmanship, and meet governing safety standards. This chapter addresses several types of manually operated and power masonry construction tools and equipment, including materials handling equipment, sawing and cutting equipment, scaffolds and accessories, and layout instruments. This chapter only identifies many of the types of tools and equipment used for masonry construction. It is essential for a masonry worker to receive training from a competent, experienced person for each type of equipment before using it.

## OBJECTIVES

After completing this chapter, the student should be able to:

- Identify manually operated and power equipment used in the masonry construction industry.
- Discuss factors to consider when selecting specific types of equipment.
- List safety precautions and care for specific equipment.

# Glossary of Terms

**Dense Industrial 65** the industry standard recognition for sawn planking.

**point up** the process of filling mortar joints.

**sawn planking** scaffold boards sawn from timber wood.

**scaffold buck** another term for a scaffold end frame.

**supported scaffolds** platforms supported by legs, beams, or other approved rigid supports.

**suspension scaffolds** platforms suspended by ropes or other nonrigid means from overhead.

# Mortarboards and Pans

Using a trowel, a mason takes mortar from the *mortarboard* (see Figure 4-1).

Made of wood, fiberglass, steel, or polyethylene, most mortarboards have 24" to 30" square shapes. The steel or polyethylene *mortar pan* measures approximately 30" × 30" and 7" deep. Tubular steel stands fold for easy transportation (see Figure 4-2).

**FROM EXPERIENCE**

Mortarboards can be made using ¾" exterior grade plywood. Eight 24" square boards can be cut from a single piece of 4' × 8' plywood. Round, rimmed steel lids from 55-gallon drums containing nontoxic substances make good mortarboards also.

Mortarboard or mortar pan *stands* are made of tubular steel. Having the mortar at working height reduces fatigue and increases productivity.

# Mortar Box

Mortar is mixed manually in a *mortar box* (see Figure 4-3). Capacities range from approximately 4 to 20 cubic feet. An 8 to 10 cubic foot box is large enough for mixing a single 1 cubic foot bag of cementitious materials with sufficient sand and water. Larger boxes hold mortar mixed in larger machine-powered mixers. Mortar boxes are made of steel or polyethylene.

# Mortar Hoe

A *mortar hoe* is used to manually mix or temper mortar in a mortar box or wheelbarrow (see Figure 4-4). The two large holes in the head reduce the effort of mixing. The handle is either wooden or fiberglass and is approximately 5' long.

**Figure 4-3** The mortar box is used for mixing or transporting mortar. *(Courtesy of Bon Tool Company)*

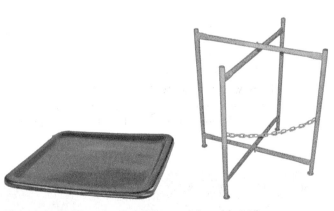

**Figure 4-1** This polyethylene mortar board is 28" square. An optional stand elevates the board. *(Courtesy of Bon Tool Company)*

**Figure 4-4** This 10" wide blade mortar hoe is used to mix and retemper mortar. *(Courtesy of Bon Tool Company)*

**Figure 4-2** A steel stand supports the 30" square metal or polyethylene pan. *(Courtesy of Bon Tool Company)*

**Figure 4-5** From left to right are a "D"-handle round point shovel, long-handle round point shovel, "D"-handle square point shovel, and long-handle square point shovel. (*Courtesy of Bon Tool Company*)

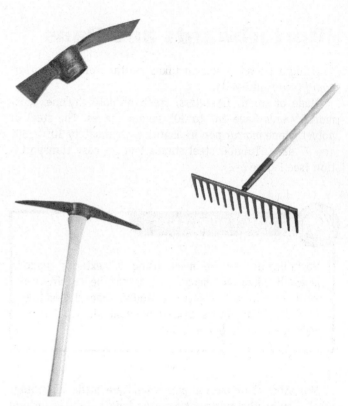

**Figure 4-6** A mattock (upper left), pick (lower left), and rake (right) are used for site preparations. (*Courtesy of Bon Tool Company*)

## Shovels

Both round point and square point shovels are used in masonry (see Figure 4-5). The round point shovel is designed for ground digging. The square point shovel is designed for handling loose materials such as sand, gravel, and soil. Both are used for shoveling mortar and sand. Either style is available with long handles or short, "D" handles.

**FROM EXPERIENCE**

The *shovel count* for adding sand to cementitious materials for making mortar results in inconsistencies between batches of mortar. Sand is more accurately proportioned when measured in a container and added to the mix as a specific volume.

## Rakes, Mattocks, Picks, and Scrapers

An assortment of grounds tools (see Figure 4-6) are used for excavating, leveling, and performing other site-preparation activities. A heavy-duty rake is used to level or clean up the work area, and a pick or mattock is used for digging. A frequent use of scrapers is to remove hardened mortar from foundation footings, concrete surfaces, and concrete masonry walls.

## Wheelbarrows

Wheelbarrows are used to transport brick, block, and mortar. A brick and block wheelbarrow is made of seasoned hardwood (see Figure 4-7). The flat bottom and open

**Figure 4-7** The flat-bottom, open side design makes loading and unloading masonry units easier.

Figure 4-8 **Quality wooden handles and heavy-duty steel undercarriages are found on these steel tray contractor-grade wheelbarrows.** *(Courtesy of Bon Tool Company)*

Figure 4-9 **Two types of brick tongs are steel brick tongs (left) and malleable iron brick tongs with a self-latching handle (right).** *(Courtesy of Bon Tool Company)*

## Brick Tongs

Made from welded steel or malleable iron, *brick tongs* permit quick, single-handed carrying of several brick (see Figure 4-9). Most brick tongs can be adjusted to carry 6 to 11 standard brick. Brick tongs minimize brick handling, preventing hand and finger injuries associated with gripping their rough edges.

sides makes it easy to stack brick or block while the front rail prevents material from falling as the materials are transported.

Tray wheelbarrows (see Figure 4-8) are used to transport mortar, sand, and other aggregates. Their 6 cubic feet trays can be either steel or polyethylene. They have either single-wheel or double-wheel designs.

**CAUTION**

### Brick Tongs Safety and Care

- Adjust the tongs to carry no more units at one time than is comfortable.
- Avoid lifting brick in tongs above chest height.
- Because brick unintentionally released from the tongs can cause bodily injuries, including skin abrasions and foot injuries, wear shirts, long trousers, and safety-toe footwear.
- Carry brick in tongs only where there is adequate foot traction preventing slipping or falling.
- Inspect all welds and fasteners, replacing defective components or damaged tongs.

**CAUTION**

### Wheelbarrow Safety and Care

- Observe load capacities and weight recommendations and limitations.
- Only transport wheelbarrows over unobstructed surfaces permitting good foot traction and over ramps and walkways that have been designed by a competent person qualified in structural design and meeting OSHA standards.
- Daily inspect all hardware for loose, damaged, or missing parts.
- Replace cracked or splintered wood handles and bracing or bent metal handles and bracing.
- Maintain proper air pressure for pneumatic tires and lubricate wheel bearings.
- Avoid poorly distributed loads and shifting loads to keep from upsetting the wheelbarrow.
- Have legs sitting on a level surface capable of supporting the load.
- Transport wheelbarrows only where there is adequate foot traction preventing slipping or falling.

## Mortar Mixers

Either gasoline engines or electric motors power *mortar mixers* (see Figure 4-10). The slow-moving paddles revolving in a stationary drum mix the cementitious materials, sand, and water in a fraction of the time taken to mix manually with a mortar hoe. A six cubic feet drum has the capacity to mix a maximum of two, 1 cubic foot bags of masonry cement with the required sand and water. For larger jobs, mixers are available with drums as large as 16 cubic feet.

Figure 4-10 Inside the stationary drum, paddles attached to a horizontally mounted revolving shaft mix mortar ingredients. *(Courtesy of Bon Tool Company)*

Figure 4-11 The materials table, tracking along the sides of the saw frame, feeds masonry units into the blade. *(Courtesy of Bon Tool Company)*

**CAUTION**

**Mortar Mixer Safety and Care**

- Study the owner's manual for proper and safe use, especially topics related to health and safety.
- Become trained by a competent person before operating.
- Wear eye and respiratory protection meeting OSHA standards.
- Disconnect spark plug or electrical cord before servicing.
- Inspect daily, and follow the manufacturer's maintenance and lubrication schedules.
- Insure that the electric service for electric powered mixers includes inline ground fault circuit protection.
- Keep drum guard in place and engine/motor hood closed when operating.
- Provide adequate foot traction in the area of the mixer to prevent slipping or falling.

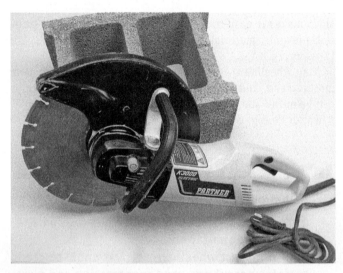

Figure 4-12 Hand-held, gasoline engine and electric motor powered saws are popular among masons.

Hand-held electric or gasoline saws are available also (see Figure 4-12).

## Masonry Saws

Electric or gasoline powered *masonry saws* make quick and accurate cuts on all types of masonry materials. The large, stationary saw is equipped with a 14" or 16" diameter, abrasive or diamond blade (see Figure 4-11). A manually operated foot pedal lowers the blade into the unit being cut. Supplying water to the blade with a submersible water pump cools the blade and helps to contain harmful dusts that can cause occupational diseases when airborne. Similar designs are more portable but do not have the foot pedal for lowering the blade.

**FROM EXPERIENCE**

A 7¼" worm-drive circular saw and 7¼" masonry cutting blade is an economical means for cutting block and brick. Always secure the block or brick against movement to prevent blade binding and dangerous kick-back when cutting with hand-held saws. A fabricated wooden jig having wood cleats fitted to the outside perimeter of the material to be cut and secured to a ¾" plywood stationary base is one means of stabilizing the material for cutting.

Figure 4-13 **Equipped with a special blade, the hand-held tuckpointer's grinder cuts mortar from joints. An attached fence controls the depth of cut.** *(Courtesy of Bon Tool Company)*

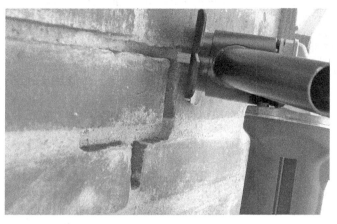

Figure 4-14 **The four-fluted bit rotates at high speeds to remove mortar.** *(Courtesy of Bon Tool Company)*

The compact electric 5" *tuckpointers grinder* (see Figure 4-13) removes mortar joints quicker than hand chiseling.

High-speed diamond-coated router bits are used to cut out mortar joints with a router (see Figure 4-14).

 **FROM EXPERIENCE**

Use extreme care when using power-operated grinders and routers to remove mortar joints. The tool can suddenly leave the joint, taking off across the face of the brick or block as the blade or bit scars their face beyond repair.

**Masonry Saw Safety and Care**

- Study the owner's manual for proper and safe use, especially topics related to health and safety.
- Become trained by a competent person before operating.
- Wear hearing, eye, face, head, and respiratory protection meeting OSHA standards.
- Disconnect spark plug or electrical cord before servicing.
- Inspect daily and follow the manufacturer's maintenance schedule.
- Insure that the electric service for electric-powered equipment includes inline ground fault circuit protection.
- Operate tools only where there is adequate foot traction preventing slipping or falling.
- Use approved masonry blades with imbedded steel reinforcement for sawing masonry materials.

# Paver Splitters

*Paver splitters* operate by mechanical leverage or hydraulic pressure (see Figure 4-15). The force of the blade breaks or *fractures* masonry units rather than cutting them, as do masonry saws.

Figure 4-15 **Shown are a mechanical leverage splitter (left) and a hydraulic pump splitter (right).** *(Courtesy of Bon Tool Company)*

**Figure 4-16** Squeezing the grout bag forces mortar through the tip. *(Courtesy of Bon Tool Company)*

## Grout Bags

A *grout bag* is a heavy vinyl or canvas bag with a metal tip insert that is used to **point up** or fill mortar joints (see Figure 4-16). Squeezing the bag forces mortar through the bag tip directly into mortar joints.

## Grout Guns

Mortar is fed through the tip of a *grout gun* by an auger powered by an electric drill (see Figure 4-17). A grout gun dispenses mortar quicker and with less effort than a grout bag.

 **FROM EXPERIENCE**

Mortar from the grout bag can get on the hands. Wear waterproof gloves to prevent mortar from affecting the skin when using grout bags.

**CAUTION**

**Grout Gun Safety and Care**

- Study the owner's manual for proper and safe use, especially topics related to health and safety.
- Become trained by a competent person before operating.
- Wear eye protection meeting OSHA standards.
- Disconnect electrical cord before servicing.
- Insure that the electric service includes inline ground fault circuit protection.

## Scaffolding

*Scaffolding* is a temporary work platform designed to accommodate both workers and materials. The two classifications of scaffolds are supported scaffolds and suspension scaffolds. **Supported scaffolds** are platforms supported by legs, beams, or other approved rigid supports. **Suspension scaffolds** are platforms suspended by ropes or other nonrigid means from overhead. Masonry construction typically relies upon supported scaffolds rather than suspension scaffolds. There are several types of supported scaffolding. Each design has special features that make one more appropriate than the others for a specific job. The 5' wide *step-type end frame* scaffolding (**scaffold buck**) is made of welded, tubular steel (see Figure 4-18). The working height is adjusted by moving the work platforms or planking from one horizontal rail to another. Typically being 5' in height, structural metal coupling pins enable multitiered assembly of this type of scaffolding.

Two end frames are connected with diagonal braces (see Figure 4-19). Steel braces permitting 5', 7', and 10' spacing are available. Most masons prefer a 7' spacing between end frames.

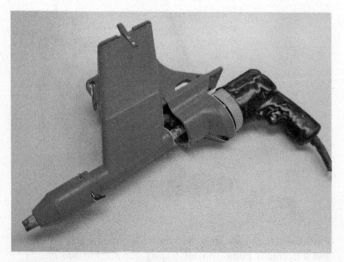

**Figure 4-17 This hand-held grout gun can be used for repointing and filling the joints between mortared brick pavers.**

**Figure 4-18** Shown is a 5' wide by 5' tall step-type end frame, also referred to as a scaffold buck. Diagonal braces are secured to pins by sliding quick locks (exploded view). Coupling pins permit stacking frames atop each other. *(Courtesy of Bon Tool Company)*

**Figure 4-19 Either holes or hooks are used to attach pivoting cross-braces to end frames.** *(Courtesy of Bon Tool Company)*

**Figure 4-22 Shown are a rigid base plate (left), screw jack with base plate (center), and jack extension (right).** *(Courtesy of Bon Tool Company)*

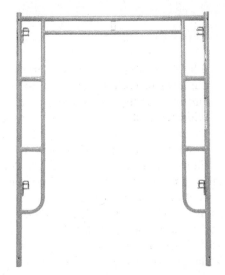

**Figure 4-20 Planking the entire width of walk through end frames gives workers easy access to each tier of multiple stacked scaffolding.** *(Courtesy of Bil-Jax Construction Equipment)*

**Figure 4-23 Independent telescoping legs secured with steel pins permit adjusting the frame height.** *(Courtesy of Bon Tool Company)*

This scaffolding also requires diagonal bracing. Insert pins permit stacking multitiered assemblies.

The design of walk through frames permits alternate height positions for *side brackets* (see Figure 4-21). Side brackets are used to support planking, from which the masons work. OSHA standards require brackets supporting cantilevered loads to support personnel only unless a qualified engineer has designed the scaffold for other loads.

Both *base plates* and *leveling screw jacks* are designed for ladder type and walk through end frame scaffoldings (see Figure 4-22). Rigid base plates prevent the tubular scaffold frames from sinking into the ground. Predrilled holes on the base plates are used to secure the plates to *mud sills*, scaffold grade planks set perpendicular to the wall on which the scaffold frames rest. Leveling screw jacks are used to level scaffold setups.

*Material platform scaffolding* is free-standing scaffolding requiring no bracing between pedestals (see Figure 4-23). They are designed to have a material shelf above the walk

**Figure 4-21 Shown are the adjustable width side bracket (top), 20" side bracket (middle), and side and end bracket (bottom).** *(Courtesy of Bon Tool Company)*

The 6' 4" or 6' 6" tall *walk through end frame* scaffolding allows the entire 5' width of the end frames to be planked for accommodating the materials and workers (see Figure 4-20).

**Figure 4-24 Individual towers are diagonally braced one to another. Elevating the work platforms eliminates stooping and reaching.**

boards. The tubular steel telescoping design can be used to adjust walk board level from 24" to 48". They are intended for single-tier use only.

The *scaffold towers* or *continuous climbing platform scaffolding* are mast-climbing work platforms that can stand freely up to 70' high (see Figure 4-24). The platforms are elevated electrically, hydraulically, or manually.

Compared to tubular steel, diagonally braced scaffolding setup time can be reduced by more than 50%, and the scaffolding can accommodate unusual building shapes. Tower scaffolding is available in both a heavy-duty and light-duty design.

Guardrails (see Figure 4-25) prevent falls. Guardrails consist of vertical posts, horizontal top rails, and mid-rails.

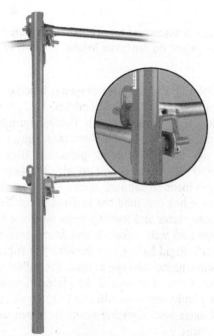

**Figure 4-25 Tubular steel guardrail posts are secured to scaffold frame pins and horizontal top and mid-rails fasten to the guardrail posts.** *(Courtesy of Bon Tool Company)*

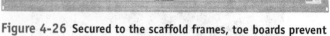

**Figure 4-26 Secured to the scaffold frames, toe boards prevent materials and debris from falling at the edges of platforms.** *(Courtesy of Bil-Jax Construction Equipment)*

Toe boards (see Figure 4-26) extending 3½" vertically above the exposed edges of platforms prevent falls of materials and debris from the edges of planking.

# Scaffolding Planking and Platforms

Supported by the scaffold frames, wooden planks or platforms bear the loads of both workers and materials. **Sawn planking** consists of individual boards 1¾" to 2" thick and 10" to 11½" wide. OSHA standards require sawn wood planks to meet the industry standard known as **Dense Industrial 65.** Select structural Douglas Fir, Southern Yellow pine, or their equivalents are recommended woods. Construction-grade material is not suitable for scaffolding planks because it is engineered for supporting a load on edge rather than when laid flat. Metal plank ties (see Figure 4-27) or drilled tie rods prevent sawn scaffold boards from splitting.

Laminated Veneer Lumber (LVL) is used for scaffolding planks. Gluing together many thin layers of wood makes the *LVL-type scaffold plank.* The LVL plank is usually 1½" or 1¾" thick and 9¼" or 11¾" wide. LVL-type scaffolding planks are heavier then solid wood planks.

A *scaffold platform* consists of a plywood or metal deck supported by an aluminum frame (see Figure 4-28). The ends of the platforms hook over the scaffold frames. Unlike planks, which have to be lapped at the ends to create longer scaffolds, offset hooks are used to arrange scaffold platforms in a smooth and even surface.

# Scaffold Weather Enclosures

Reinforced polyethylene weather enclosures attached to the scaffold framing with special clips prevent precipitation halting work. Portable heaters are used when working in

**Figure 4-27 Driving the *teeth* of plank ties into ends of sawn wood scaffold planks and securing with screws prevents the wood from splitting.** *(Courtesy of Bon Tool Company)*

Figure 4-29 **The hoist arm attaches to the scaffold frame insert pin and is braced by the scaffold frame.** *(Courtesy of Bon Tool Company)*

Figure 4-28 **Half-inch plywood decking supported by aluminum framing give this platform a 75 P.S.F. rating. Offset hooks (inset) are used to create a level line of planking.** *(Courtesy of Bon Tool Company)*

cold weather. To ensure that weather enclosures do not cause tipping of scaffolds in high winds, adequate bracing is necessary. Scaffolds may be braced using guy wires, ties, or other rigid braces meeting the intent of OSHA standards.

> ⚠ **CAUTION**
>
> ### Scaffolding Safety and Care
>
> - Receive training from a person meeting the training requirements of OSHA standard 1926.454 to properly erect, access, or work from scaffolds.
> - Obey OSHA standards, Section 1926.451 and Section 1926.452 of the Construction Industry Regulations, to erect, access, engage in activities upon, and descend scaffolds.
> - Observe manufacturer's recommendations and OSHA standards, Section 1926.451 of the Construction Industry Regulations, for scaffold frames and planking weight load limits.
> - Use only scaffold-grade lumber meeting OSHA standards and never construction-grade lumber for planking.
> - Observe OSHA standards for tying off taller scaffolding to prevent collapsing.
> - Observe OSHA standards, Section 1926.451 of the Construction Industry Regulations for scaffolds near overhead high-voltage power lines.
> - "All employees who work on a scaffold must be trained by a person qualified to recognize the hazards associated with the type of scaffold used and to understand the procedures to control and minimize those hazards." OSHA STANDARDS 1926.454(a)

## Swivel Head Hoist Arm and Wheel

The *gin wheel* or *well wheel* is a 10" to 12" diameter pulley. The attached rope is used to hoist materials to most heights (see Figure 4-29).

Hoisting requires one to be experienced and alert at all times. Because the operator is below the hoist, there is always the danger of falling materials. The scaffolding to which the hoist arm is attached may collapse if the rope is jerked forcibly rather than pulled gently or if the rope is pulled when standing too far from the scaffolding.

> ⚠ **CAUTION**
>
> ### Gin and Well Wheel Hoist Safety and Care
>
> - Obey OSHA standards for overhead hoists, Section 1926.554 of the Construction Industry Regulations.
> - Receive training from a competent person for the proper setup and operation.
> - Do not exceed the safe working limits of the rope, wheel, and hoist arm.
> - Attach hoists to supporting structures with a safe working load equal to that of the hoist.
> - Scaffolds should be braced using guy wires, ties, or other rigid braces meeting the intent of OSHA standards.
> - Hoist materials standing on ground where there is adequate foot traction preventing slipping or falling.

**Figure 4-30** When mounted level to a tripod, this optical instrument provides accurate readings for checking level masonry work. *(Courtesy of Bon Tool Company)*

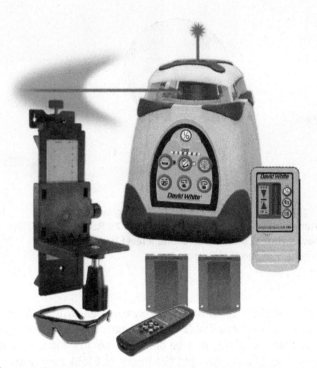

**Figure 4-31** This rotary laser level emits a beam of light detected by the electronic sensor (upper left). *(Courtesy of Southern-Tool)*

# Builders' Levels

Builders' levels are used to lay out and ensure that footings, foundations, and other walls are aligned level. The two classifications of builders' levels are glass optical lens levels and laser light levels. *Optical levels* (see Figure 4-30) require two persons to check results. One person is needed to sight through the optical scope to record the measurement on the ruler or target rod held by another.

*Laser levels* (see Figure 4-31) permit one person alone to verify level footings, masonry corners, or walls. An electronic sensor detects a laser beam of light emitted by the rotating head of the level.

## CAUTION

**Laser Level Safety and Care**

- **Avoid eye exposure to the laser light beam, and wear eye protection, "laser safety goggles which will protect for the specific wavelength of the laser and be of optical density (OD) adequate for the energy involved," meeting the intent of OSHA standards as defined in the OSHA Construction Industry Regulations, Section1926.102(b)(2)(i) and Table E-3.**
- **Do not set up the instrument with the laser light beam at eye level.**
- **Place adequate warnings in the area for others to avoid exposure to the laser beam.**
- **Observe manufacturer's instructions for operation and care.**

# Corner Poles

Corner poles or masonry guides eliminate the need for constructing masonry leads. Made of lightweight aluminum, corner poles are attached to either outside or inside corners of framework using specially designed brackets (see Figure 4-32).

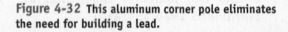

**Figure 4-32** This aluminum corner pole eliminates the need for building a lead.

Telescoping braces are available for bracing corner poles when framework is not present. Approximately 9' long, the poles provide a quick and easy way for laying up walls, because the entire wall is laid to the line.

## Summary

- This chapter identifies many of the types of power tools and equipment used for masonry construction.
- A masonry worker must receive training from a competent, experienced person before using any power tool or type of equipment.
- Not only should training include proper and safe power tool and equipment operation, but it should also include proper inspections procedures and routine maintenance procedures. Inspections and preventive maintenance assure maximum performance and safe operation of power tools and equipment.

- Only masonry power tools and equipment designed and engineered to comply with the intent of OSHA safety standards should be used.
- Altering equipment beyond manufacturers' recommendations may endanger workers' safety. Observe manufacturers' guidelines concerning setups, adjustments, recommended maintenance, and servicing.
- A successful masonry construction business recognizes the need for properly designed and well-maintained equipment operated by well-trained persons. Doing so maximizes tool and equipment life expectancy and performance, minimizes tool and equipment replacement expenses, avoids citations and monetary penalties, saves insurance premiums, and, more importantly, prevents accidents and injuries, boosting employee morale and productivity.

# Review Questions

*Select the most appropriate answer.*

**1** The square size of most mortarboards is between

a. 16" and 20"
b. 20" and 24"
c. 24" and 30"
d. 30" and 36"

**2** A single, 1-cubic foot bag of cementitious material, sand, and water are mixed to make mortar manually in a mortar

a. box
b. drum
c. pan
d. bucket

**3** The two large holes in the blade of a mortar hoe are designed to

a. reduce the physical efforts of mixing mortar
b. blend the ingredients better
c. test the consistency of the mortar
d. do both a and b

**4** The capacity of a contractor's wheelbarrow is

a. 2 cubic feet
b. 4 cubic feet
c. 6 cubic feet
d. 8 cubic feet

**5** Several brick are carried at a time using brick

a. carriers
b. holders
c. tongs
d. straps

**6** Mortar is mixed in power mortar mixers as

a. the drum rotates
b. paddles rotate inside a stationary drum
c. the drum vibrates
d. none of the above

**7** An electric tool used to remove mortar joints in masonry walls is called the

a. cut-out saw
b. tuckpointer's grinder
c. stationary saw
d. oscillating grinder

**8** A tool designed to break masonry units by applying force is called

a. a mechanical leverage splitter
b. a hydraulic pump splitter
c. a chipper
d. both a and b

**9** A canvas or vinyl bag that holds mortar for filling mortar joints is called a

a. grout bag
b. mortar bag
c. tote bag
d. tuckpointer's bag

**10** Temporary work platforms designed to accommodate both workers and materials are called

a. piers
b. scaffolding
c. platforms
d. risers

**11** Wood scaffold planking must meet the industry standard called

a. Dense Industrial 65
b. structural framing lumber
c. hardwood lumber
d. both a and b

⑫ **A rope pulley used to hoist materials is called**

    a. a come-along

    b. a gin and wellwheel

    c. a derrick

    d. both a and b

⑬ **A type of builder's level is**

    a. a laser level

    b. an optical level

    c. a periscope

    d. both a and b

# Chapter 5 | Laying Brick to the Line

Lines provide direction. Children learn to keep crayons within the lines in coloring books. Drivers keep their vehicles between the lines of their lane. Sports fields and courts are defined with lines. And masons rely on lines for accurate alignment of masonry units. Without the mason's line, it would be very difficult to align the masonry walls of buildings level, plumb, and straight.

## OBJECTIVES

After completing this chapter, the student should be able to:

- Lay out a brick wall in the running bond pattern.
- Demonstrate options for placing a cut brick in a wall.
- List the four procedures performed for laying every brick.
- Demonstrate procedures for hanging a line and twigging a line.
- Lay brick to the line in the running bond pattern.

# Glossary of Terms

**closure brick** the last brick laid in each course.

**crowding the line** an expression describing brick that are too close to or possibly touching the line.

**dry bonding** the process of establishing brick arrangement and the width of head joints.

**facing the brick** aligning the bottom edge of each brick's face with the top edge of the brick faces on the course below it.

**hanging the line** attaching a mason's line to the leads at opposite ends of a wall.

**holding bond** maintaining a plumb aligned bond or brick pattern.

**layout** the brick pattern and joint spacing to satisfy the intended wall construction.

**lipping** a condition where the bottom of a brick unintentionally extends beyond the face of the brick below it.

**racking** the process of intentionally setting back or stepping back the brick from the course below

**raising the line** repositioning the line to the next course of the lead upon completing each course of brick.

**set-back** a condition exhibited by a brick when its bottom edge is unintentionally back of the face of the brick below.

**twigging the line** clipping a twig, also known as a trig, onto a mason's line and positioning the twigged line above a brick bedded in mortar that is aligned level and plumb with the leads.

The process of aligning brick in a wall with the aid of a mason's string line attached to and guided by brick leads or guides at either end is known as laying brick to the line. It is not the scope of this chapter to describe the construction of brick leads. Chapter 6, "Constructing Brick Leads," introduces procedures for constructing a variety of brick corners and jambs to which lines can be attached for laying brick to the line. The surface upon which the brick are to be bedded in mortar should be free of soil, masonry dust, mortar droppings, and all other debris obstructing the process.

## Dry Bonding

**Dry bonding** refers to arranging the brick in the intended pattern bond with a desirable head joint width. Adjusting the width of the head joints within an acceptable range changes the overall length of a brick course, sometimes eliminating the need for brick pieces. Joint widths no less than ¼" or greater than ½" are acceptable in most cases. A brick at the end of each course needs cutting if adjusting the head joints within an acceptable range does not permit using full brick. The brick nearest the end of the wall laid in the stretcher position is cut to permit a plumb bond (see Figure 5-1).

*For step-by-step instructions for dry bonding a brick wall in the running bond, see the Procedures section on page 70.*

After the brickwork pattern and head joint width is decided and before removing the brick so that they may be bedded in mortar, clearly mark the intended brick spacing onto the face of the surface to which the brick are to be bedded in mortar, identifying the center of every head joint and also each end of the wall (see Figure 5-2).

*For step-by-step instructions on making allowances for walls where brick need cutting, see the Procedures section on page 71.*

Figure 5-1 Cutting the brick laid in the stretcher position nearest the end of the wall permits a wall to have a uniform bond and a head joint width of acceptable size.

Figure 5-2 Pencil marks clearly identify intended brick spacing.

## Choosing Course Spacing Scale

Where brick leads to which a mason's line is attached have been built at opposite ends of a wall, the mason's line is aligned level with each course as the wall is built. But in cases where the line is attached to a corner pole or masonry guide, a temporarily secured, plumb-aligned, rigid member for attaching a tensioned line, the mason's scales permit uniform course spacing. As with head joints, it is recommended that bed joints be no less than ¼" and no more than ½" wide. Choose a spacing that creates the desired bed joint width. For example, when laying standard size brick, brick whose face height is approximately 2¼", and choosing ⅜"-wide bed joints, the overall course spacing height should be 2⅝". Or brick can be laid on one of the scales intended for brick spacing. Using the mason's scales for brick course spacing is a topic addressed in Chapter 7 "Masonry Spacing Scales."

## Hanging the Line

A mason's line repositioned at the top of the face side for each course of brick enables a mason to align brick courses level and plumb. Attaching a line to the leads or corner poles at opposite ends of a wall is referred to as **hanging the line.** Line blocks are usually used to hang or secure tensioned lines at the ends of brick leads or the edges of corner poles, whereas line pins secure tensioned lines to inside brick corner leads that have nowhere to hook line blocks.

**CAUTION**

**Always wear eye protection when stretching mason's lines to tension, working near or in the presence of tensioned lines, or removing tensioned lines from walls.**

## Twigging the Line

Although not always needed, a twig, also called a trig, is sometimes used to position a mason's line. Adequately tensioned shorter lines seldom need a twig. However, it may be necessary to twig longer lines to eliminate sagging lines between leads farther apart or to eliminate line movement caused by high winds.

**Twigging the line** to eliminate a mason's tensioned line from sagging or moving involves clipping the twig around a mason's line and weighting the twig above a brick near the middle of the length of the course (see Figure 5-3).

**Figure 5-3 The weight of a brick above the twig secures the twigged line.**

The brick on which the twig is placed must be bedded in mortar to align level and plumb with the leads at the opposite ends of the wall.

> ◢◢◢◢ **CAUTION** ◤◤◤◤
>
> **Remove twigs from lines before repositioning lines to successive courses, tensioning, or removing lines from walls.**

Using a builder's level or rotary laser level ensures that the twigged brick is level with the leads at opposite ends of long walls.

> ◢◢◢◢ **CAUTION** ◤◤◤◤
>
> **Do not stare into the laser light. Post warnings in the area for the safety of others.**

Confirming that the twigged brick is aligned plumb with the face of the wall below it helps to build a plumb wall.

> ◢◢◢◢ **CAUTION** ◤◤◤◤
>
> **Do not use your eyes to sight the length of the twigged line from one end to the other to confirm a straight and level line. Having your eye aligned with the mason's line imposes risk of serious eye injuries and the possibility of vision loss should the tensioned line break or become dislodged while sighting it.**

A twig serves four purposes:

1. Although not always needed, a twig is sometimes used to properly position the mason's line at the lead.
2. A twig prevents a longer line from sagging lower near the middle than intended at the leads to which it is attached.
3. A twig minimizes line interference caused by more than one mason laying brick simultaneously along the same line.
4. A twig prevents line movement caused by high winds.

# Laying Brick to the Line

Spreading a mortar bed joint and applying head joints is described in the "Procedures" section of Chapter 3. For a better bond strength, the Brick Industry Association recommends bedding brick "within one minute or so" after spreading the mortar. Four operations are observed for each brick as it is laid. The order of operations observed by the author is 1) aligning the pattern bond, 2) facing the brick, 3) aligning the top of the brick level, and 4) aligning the face of the brick plumb.

***For step-by-step instructions for laying brick to the line, see the Procedures section on page 72.***

For step-by-step instructions for laying brick to the line, see the Procedures section on page 72.

>  **FROM EXPERIENCE**
>
> "Wetter is better!" Having the consistency of mortar as wet as possible without being too soft as to cause the mortar to bleed or the brick to "float" makes bedding brick in mortar easier than when using stiff mortar. In most cases, wetter mortar allows a better bond between mortar joints and brick. Because of freezing temperatures, wetter mortar may not be desirable during cold weather construction.

## Aligning the Running or Stretcher Brick Pattern Bond

For the first course bedded on a concrete slab, the author aligns the bottom edge of each standard size brick's face to a chalk line or pencil line representing the face side of the wall. The head joints are aligned with the marks made to establish the pattern bond when dry bonding. On the second course, each brick is centered above the two brick below it. Brick on the third course and all odd numbered courses thereafter are positioned so that their head joints are aligned plumb with those of the first course. Brick on the fourth course and all even numbered courses thereafter are positioned so that their head joints are aligned plumb with those of the second course (see Figure 5-4). With variations

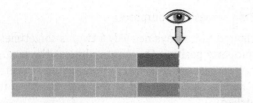

**Figure 5-4** A plumb pattern bond is maintained by aligning head joints of alternating courses in a vertical, plumb alignment.

in the length of common building brick, the widths of the head joints may vary slightly if the brick pattern is to remain plumb.

Prevent the thumb used to grip the face of the brick from unnecessarily moving the line and interfering with the work of other masons.

*For step-by-step instructions for checking for a plumb bond, see the Procedures section on page 73.*

**FROM EXPERIENCE**

Using the mason's 4' level to draw plumb lines aligned with the head joints of the first and second courses onto a backup wall when present makes aligning the pattern bond easier. The mason simply needs to align the head joints of each course with the plumb line markings.

## Facing the Brick

Facing the brick refers to aligning the bottom edge of each brick's face with the top edge of the brick faces on the course below it. Use the edge of the trowel blade near the toe of the trowel to feel the alignment of the bottom of each brick with those below it as protruding mortar is removed from the bed joint (see Figure 5-5). The alignment of the two courses is visually examined after the protruding mortar is removed.

Brick that are not faced exhibit conditions known either as lipping or setback. Lipping is a condition where the bottom of a brick unintentionally extends beyond the face of the brick below it (see Figure 5-6).

Set-back is a condition exhibited by a brick when its bottom edge is unintentionally back of the face of the brick below it (see Figure 5-7).

**Figure 5-5** The bottom edge of the brick's face is *faced* or aligned even with the top edge of the two brick below it.

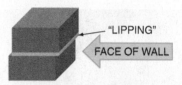

**Figure 5-6** The face of the top brick is *lipping* the brick below and needs to be moved back from the face of the wall for proper alignment.

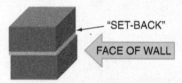

**Figure 5-7** The top brick exhibits *set-back* and must be moved *outward* to align with the face of the brick below it.

**Figure 5-8** Applying light pressure with the hand aligns the brick level.

## Aligning the Brick Level

A brick is aligned level by slightly pressing down, bedding the brick in mortar until the top edge of the brick's face is aligned level with the mason's line. Rather than tapping on each brick with the trowel, the technique of pressing against the top of the brick increases productivity and permits the trowel blade to remove mortar protrusions while reconfirming that the brick remains faced at its bottom. A brick is considered aligned level when the top edge of its face is level with the tensioned mason's line (see Figure 5-8).

## Aligning the Brick Plumb

A brick is aligned plumb by applying more pressure with the hand to the top of the brick either near the brick's back or front edge, tilting the top edge of the brick's face closer to or farther from the line as it is being aligned level. A space no less than $\frac{1}{16}$" or more than $\frac{1}{8}$" should be maintained between the line and the brick's face (see Figure 5-9). Some brick are intentionally made with crooked and uneven faces.

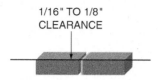

Figure 5-9 **Keeping the top edge of the brick level with the line while maintaining a spacing of ¹⁄₁₆" to ⅛" between the line and the brick's face enables aligning brick level, plumb, and straight.**

While parts of such brick may be more than this distance from the line, no part of the brick should be permitted closer to the line than ¹⁄₁₆". It is important that a brick's face be tilted at the top to align a brick plumb rather than moving the entire brick closer to or farther from the line. Moving a brick closer to or farther from the line will result in lipping or set-back.

Crowding the line is an expression describing a brick whose face is too close to or possibly touching the line. Crowding can result in brick pushing against the line, resulting in a wall neither plumb nor straight. Brick crowding the line are referred to as being "hard to the line".

 **FROM EXPERIENCE**

Beginning masonry students should gauge the space between the line and the brick using a U.S. nickel. Brick closer than this can easily crowd or touch the line, which are mistakes that result in walls not being aligned straight or plumb.

## Laying the Closure Brick

The last brick laid in each course is called the closure brick. To ensure full head joints between the closure brick and those adjacent, mortar can be applied to the ends of the two adjacent brick as well as each end of the closure brick before installing it. Another method is to apply a full head joint as usual to one end of the closure brick and fill the remaining joint with mortar after laying the closure brick.

The mason's line is repositioned to the next course of the lead upon completing each course of brick. This procedure is often referred to as raising the line. It is recommended to raise the line before spreading mortar for the next course bed joints. Spreading mortar bed joints before raising the line allows mortar to accumulate on the line.

## Holding Bond at Openings

There are times that openings such as a doors, windows, or vents break up a brick course. When an opening interrupts the brick course, it becomes necessary to cut the brick

to fit at both sides of the opening. However, the bond pattern should not be altered because of the opening. Maintaining a plumb aligned bond or brick pattern is known as holding bond. The mason must keep the joints of every other course aligned plumb.

*For step-by-step instructions on holding bond at openings in a wall, see the Procedures section on page 74.*

## Holding Bond When Racking a Wall

The process of intentionally stepping back at the end of a course of brick is known as racking. The brick stepped back are referred to as racked brick. Maintaining a plumb bond, the brick at the end of a course is cut shorter when racking reduces the length of the wall. Cutting the brick laid in the stretcher position nearest the wall's end preserves the desired vertically plumb head joint alignment of alternate courses throughout the wall.

*For step-by-step instructions for maintaining a plumb bond when racking a wall, see the Procedures section on page 75.*

## Summary

- Laying brick to the line is one of the first wall-building tasks that an apprentice mason has the opportunity to perform. The feelings of success and pride from such an accomplishment are gratifying to both the apprentice and experienced mason. It should become a top priority for every apprentice mason to master the skills of laying brick to the line.
- Effective procedures must be followed for consistent results. Four operations are observed each time a brick is laid to the line: 1) align the bond, 2) face the brick, 3) align the brick level, and 4) align the brick plumb. Mastering these skills enables a supervised apprentice to take part in wall construction.
- Not only should a well-laid wall be level, plumb, and straight, but also the bed joints should have uniform widths, and the head joints of alternate courses should be aligned plumb, while both the sizes of bed joints and head joints comply with recommended industry standards.

## Procedures

# Dry Bonding a Brick Wall in the Running Bond

**A** Prepare the concrete surface on which brick are to be bedded in mortar by removing dust, soil, hard mortar, and other debris effecting the placement of brick. Use a broom to remove dust, wash soil off of the surface, and chip hard mortar using the mason's hammer.

> **CAUTION**
>
> Wear eye protection and respiratory protection in the presence of dust and debris.

**B** Starting with a full brick, dry bond each brick of the first course. Space brick approximately ⅜" apart. Draw arrows indicating the center of each joint and both ends of the wall.

| Procedures | Making Allowances for Walls Where Brick Needs Cutting |

## First Solution

**A** Starting with a full brick, dry bond the first course keeping head joints ⅜" wide and cutting the last brick to fit. All odd numbered courses align in the wall as appearing in the first course.

**B** Evenly space each full brick of the second course above the two full brick below it.

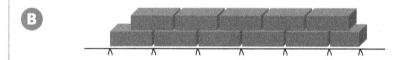

**C** Cut the brick at each end to length, allowing for a ⅜"-wide head joint. All even-numbered courses align in the wall as appearing in the second course.

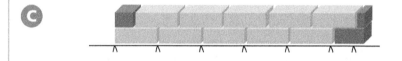

## Alternate Procedure

**A** Starting with a full brick, dry bond the first course keeping head joints ⅜" wide and cutting the last brick to fit. All odd-numbered courses align in the wall as appearing in the first course.

**B** Evenly space each full brick of the second course above the two full brick below it.

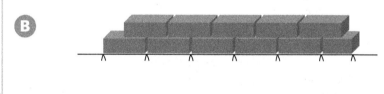

**C** Eliminate having pieces smaller than a half-brick or *bat* at the end of the wall by substituting the smaller piece with a bat and shortening the adjacent brick also. All even-numbered courses align in the wall as appearing in the second course.

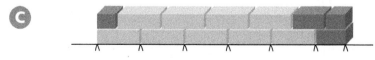

## Procedures

# Laying Brick to the Line in the Running or Stretcher Bond Pattern

**A** Align the top edge of each brick's face along the first course level with the mason's line while maintaining a space no less than ¹⁄₁₆" and no more than ⅛" between the line and the brick.

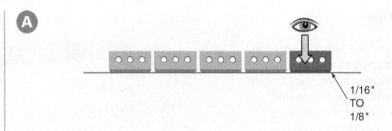

1/16"
TO
1/8"

**B** Begin successive courses with each full brick centered above the head joint below it and also aligned vertically plumb with the brick two courses below it.

**C** *Face*, or align, the bottom edge of each brick as it is laid with the top edges of the brick faces below it.

**D** Align the top edge of each brick as it is laid level with the line by applying slight pressure to the top of the brick. Tapping the top of the brick with the edge of the trowel blade is sometimes necessary for level alignment.

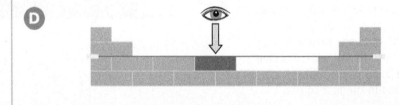

**E** Tilt the top of each brick as it is laid until a space no less than ¹⁄₁₆" and no greater than ⅛" exists between the line and the brick. Tapping with the edge of the trowel blade or applying slight pressure with the hand to either the back edge or front edge of a brick's top tilts the face of the brick, establishing a plumb wall line.

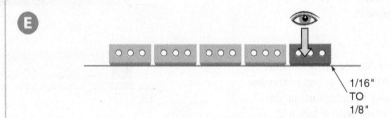

1/16"
TO
1/8"

## Procedures

# Checking For a Plumb Bond

*The **plumb bond** exists when the head joints on every other course align vertically in a straight or plumb line. A plumb bond provides the most uniform distribution of weight for masonry units within a masonry wall. Skilled masons maintain a plumb bond to insure a stronger wall that is more pleasing to the eye.*

**A** Align the edge of the mason's level in plumb alignment with a head joint on the first course.

**A**

**B** A plumb bond exists if head joints on alternate courses are in vertical alignment.

**B**

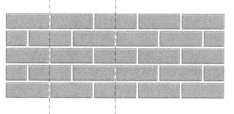

**C** This wall ***does not*** exhibit a plumb bond. Neither the head joints on courses 1, 3, and 5 nor courses 2 and 4 align vertically, as the broken lines indicate.

**C**

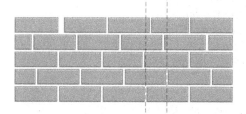

## Procedures

# Holding Bond at Openings in a Wall

**A** Brick walls often contain openings such as doors, windows, and vents. When an opening interrupts the brick course it becomes necessary to cut the brick to fit. However, the bond pattern should not be altered because of the opening. Maintaining a plumb bond and brick pattern is known as *holding bond*. This is shown in the illustration. The darker brick are cut to permit the opening. The cut brick to the left and right of an opening are not necessarily the same length. Notice that the cut brick to each side of the vent are different lengths. This permits a vertically plumb head joint alignment for every other course.

### Alternate Procedure for Holding Bond at Openings in a Wall

**A** "A Bat and a Three-Quarter Piece"

- The shorter brick piece to the right of the vent shown in the preceding procedure can be more difficult to cut and lay than a longer brick. For this reason, many masons prefer to eliminate this smaller piece. A half-brick is laid in its place, and the full brick beside it is cut shorter. This method is referred to as using *a bat and a three-quarter piece*. However, this method creates a new line of head joints that is neither below nor above the opening.

**A**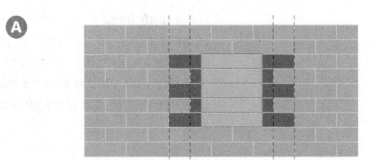

**A**

## Procedures

# Maintaining a Plumb Bond When *Racking* a Wall

**A** The process of intentionally setting back the face of brick from the wall line below is known as racking. The brick set back are referred to as *racked* brick. Maintaining a uniform bond, end brick are cut shorter when racking reduces the length of a wall. Cutting the brick laid in the stretcher position nearest the wall's end preserves the vertically plumb head joint alignment of alternate courses throughout the wall. To protect the brick's *integrity* or strength, do not reduce the widths of brick whose ends are oriented to the face of the racked wall. The darker brick represent those cut to permit holding bond and preserving the integrity of the brick.

**B** The *set back* is gauged with a block of wood whose thickness is equivalent to that which is intended. Dry bonding the brick at each end allows the mason to mark the brick needing to be cut. A plumb line aligned with the end of the stretcher nearest the end of the first course ensures accurate lengths of brick requiring shortening.

**C** The fourth course and all other even-numbered courses are aligned with the second course.

**A**

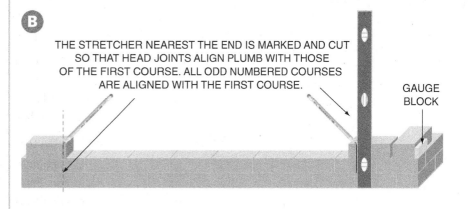

**B**

THE STRETCHER NEAREST THE END IS MARKED AND CUT SO THAT HEAD JOINTS ALIGN PLUMB WITH THOSE OF THE FIRST COURSE. ALL ODD NUMBERED COURSES ARE ALIGNED WITH THE FIRST COURSE.

GAUGE BLOCK

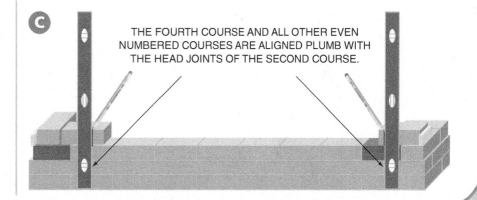

**C**

THE FOURTH COURSE AND ALL OTHER EVEN NUMBERED COURSES ARE ALIGNED PLUMB WITH THE HEAD JOINTS OF THE SECOND COURSE.

# Review Questions

*Select the most appropriate answer.*

**1** The process of establishing brick arrangement and head joint size is called

a. coursing
b. dry bonding
c. scaling
d. none of the above

**2** It is recommended that the size of mortar joints be between

a. ⅛" and ¼"
b. ¼" and ⅜"
c. ¼" and ½"
d. ⅜" and ¾"

**3** Attaching a mason's line to the leads is called

a. hanging the line
b. scaling the line
c. spotting the line
d. twigging the line

**4** A line is twigged to

a. eliminate the line from sagging
b. eliminate the effects of movement caused by high winds
c. position the line accurately at leads
d. do all of the above

**5** Aligning the bottom edge of each brick's face with the top edges of the brick faces below it is called

a. bonding the brick
b. facing the brick
c. leveling the brick
d. straight-edging the brick

**6** A condition observed when a brick unintentionally extends beyond the face of the brick below it is called

a. corbelled
b. lipping
c. racked
d. set-back

**7** Brick too close to the mason's line are considered to be

a. crowding the line
b. hard to the line
c. irritating the line
d. topping out the line

**8** The last brick to be laid in a course is called the

a. closure brick
b. final brick
c. last brick
d. queen brick

**9** Laying brick to the line requires

a. aligning the pattern bond plumb
b. aligning the brick level
c. aligning the wall plumb
d. all of the above

# Chapter 6 | Constructing Brick Leads

asonry leads are similar to good leaders, guiding and giving direction to others. Leads guide the construction of the walls between them that are laid to the line, a procedure already covered in Chapter 5, "Laying Brick to the Line." The accuracy of masonry wall construction depends upon the accuracy of the leads. Masons who build leads should be able to lay out the wall, establish the pattern bond, and select the scale for bed joint width.

Frequently, experienced masons build the leads, and those less experienced, including apprentices, lay to the line. The experienced mason is more capable of making decisions involving the layout of the wall. It requires much experience to become proficient at using the mason's level to align leads level and plumb.

## OBJECTIVES

After completing this chapter, the student should be able to:

- Lay out and construct an outside corner.
- Lay out and construct an inside corner.
- Lay out and construct 4", 8", and 12" brick jambs.
- List precautions taken when brick toothing.
- Demonstrate procedures for setting a corner pole.

# Glossary of Terms

**brick jamb** a lead permitting brickwork along a wall in only one direction.

**checking the range** confirming the straight alignment of a wall.

**corner** two walls connected at ends, normally forming a right angle.

**corner of the lead** that part of the corner where the two walls connect.

**double-wythe wall** a masonry wall comprised of two masonry units back-to-back.

**lead** that part of the wall first constructed, becoming a guide for building the remaining wall.

**mortar bridgings** mortar protrusions so large that they contact the adjoining wall.

**mortar protrusions** mortar coming from the joints as the brick are bedded in mortar.

**quoined corners** patterns of brick projecting beyond the wall line, creating block designs.

**rack of the lead** the brick alignment at the tail end of the courses on the lead.

**single-wythe** a masonry wall of only a single width masonry unit.

**tail of the lead** the stepped-back end of corners and jambs.

**toothing** continuing with courses on a lead where the tail of the lead normally limits additional courses by temporarily laying a half-brick or bat on the stretcher below.

# Leads

In masonry, lead refers to that part of the wall first constructed and used as a guide for building the remaining wall. A lead is built at each end of a wall. As covered in Chapter 5, a tensioned mason's line is then attached to each lead, and the remaining brick are laid to the line. Leads require more knowledge of masonry construction than laying brick to the line. Mastering the procedures for using the mason's level to align leads level, plumb, and straight are required. Understanding the mason's scales and using them to maintain uniform course spacing as walls are raised to a specified height is necessary. Two of the most frequently built leads are 4" corners and 4" jambs. A 4" lead is a single-wythe lead, meaning a wall whose width is comprised of a single masonry unit. A corner, two walls connected at ends normally forming a right angle, can be faced either on the outside or the inside of the angle (see Figure 6-1).

A brick jamb is a lead permitting brickwork along a wall in only one direction (see Figure 6-2). 4" jambs are constructed at a wall's end, where there is no adjacent wall requiring brickwork. 4" brick jambs are evident next to doors and windows.

Although the techniques for building leads sometimes differ between masons, their resulting leads must be aligned accurately. The procedures described in this chapter are those followed by the author. An apprentice mason should observe and attempt other techniques practiced by competent masons. Eventually, the apprentice mason adapts those techniques and improves their mastery of skills.

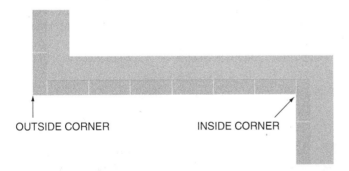

OUTSIDE CORNER          INSIDE CORNER

**Figure 6-1 The *outside corner* is aligned on the outside of the angle, and the *inside corner* is aligned on the inside of the angle.**

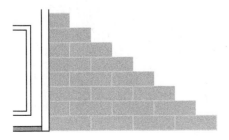

**Figure 6-2 A vertical line of brickwork forms a brick *jamb* next to the opening.**

# Building an Outside Brick Corner

The entire length of the wall is dry bonded before beginning the corner. A small length wall may require a shortened brick at its corner to accommodate its length.

***For step-by-step instructions on placing a brick requiring cutting at the end of a lead, see the Procedures section on page 88.***

The number of courses attainable when building a corner depends on the corner's first course layout. With no more than a difference of one brick on one side of the corner than the other, the number of courses possible equals the total number of brick laid on the corner's first course (see Figure 6-3). For example, a total of six courses can be attained for a corner with three brick on each side of the first course.

A corner with three brick on one side and four on the other side of the first course can be built seven courses in height. But a corner with three brick on one side and five brick on the opposite side, having a difference of more than one brick on one side than the opposite side, can only reach seven courses in height. Using a 4' mason's level, limit either side of the first course to no more than six brick, or approximately the length of the 4' level. This permits building a corner to a height of 12 courses. With the height of the leads limited to 12 courses, the walls between the leads must be constructed before going higher using the procedures for laying brick to the line, unless corner poles or brick toothing procedures are used. Corner poles and brick toothing procedures are addressed later in this chapter.

The accurate alignment of the lead's first course is critical. Extreme care should be taken to align it level, plumb, and straight. All other courses of the lead are in plumb alignment with the first course. Each course of a brick lead is aligned level, plumb, and straight before proceeding to the next course. For a better bond of the mortar with the brick, the Brick Industry Association recommends bedding brick in the mortar within "one minute or so" of spreading the mortar.

**Figure 6-3 Six brick on the first course of this corner are used to build six courses.**

To prevent bedding brick in mortar where water evaporation has begun to occur, lay and align brick for one side of a corner before spreading mortar for the opposite side. This will prevent weakening the bond strength. Remembering the order of operations, "level, plumb, and straight" is easy if one does them alphabetically, that is, level (l) before plumb (p) and plumb (p) before straight (s).

That part of the corner where the two walls connect is sometimes referred to as the **corner of the lead**, and the stepped-back opposite end is called the **tail of the lead** (see Figure 6-4). Starting at the corner of the lead and working toward the tail of the lead helps to accurately align plumb the brick at the corner with the wall below it.

Laying brick from the corner of the lead towards the tail of the lead necessitates walking forward on one side of the corner and stepping backward on the opposite side.

**CAUTION**

Keep the walking area or scaffold planking unobstructed to prevent tripping, especially when stepping backward to build leads. When working on scaffolds, guardrails and mid-rails meeting the intent of OSHA standards must be in place to prevent falls should the mason back up to the outer edge of the scaffold while building leads.

Beginning with the brick at the corner of the lead, lay all of the brick for that side of the course, and align them level, plumb, and straight before continuing to the opposite side. Not only must the brick for the corner be aligned straight, but they must also be aligned straight with the opposite end of the wall (see Figure 6-5).

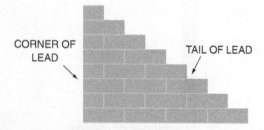

**Figure 6-4** The *corner of the lead* establishes the end of the wall. Brick eventually *laid to the line* complete the wall at the *tail* of the lead.

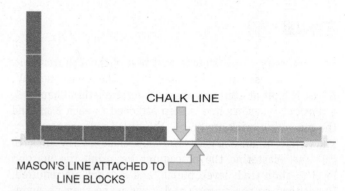

**Figure 6-5** The three brick to the right are aligned straight and also aligned straight with the corner at the opposite end of the wall.

Confirming the straight alignment of a wall is known as **checking the range.**

A chalk line snapped on the footing or concrete slab is a good guide for first aligning the wall, but you can also check alignment with a mason's line held at each end of the wall while aligning the face of each brick along the course.

The course spacing is verified at the outer edge corner of the lead (see Figure 6-6). Maintaining equal course spacing assures uniform bed joints, a desirable trait for quality brickwork.

The remaining brick of each course of the lead are aligned level with the brick that is scaled at the end of the corner (see Figure 6-7).

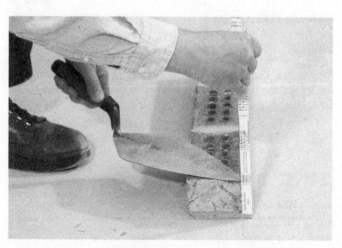

**Figure 6-6** The scale is checked at the outermost edge of the corner.

**Figure 6-7** **The mason's level aligns the brick** *level.*

After leveling the first side of the first course, the end of the brick at the corner of the lead, its face, and the face of all other brick on the first course are aligned plumb (see Figure 6-8).

 **FROM EXPERIENCE**

Using the shorter two-foot level rather than the 4' level to align plumb the first few courses is recommended. The shorter length 2' level is easier to hold steady against the face of a brick.

After each brick is aligned plumb, the 4' mason's level is used to align straight the course of brick (see Figure 6-9).

The range is then checked to confirm straight alignment of the first course of the corner with the lead or end of the wall at the opposite end.

For the first course of the adjacent side of the corner, brick are bedded in mortar, and then aligned level, plumb, and straight as is done for the first side.

One side at a time is laid and aligned for the remaining courses. Beginning on the side of each new course with the brick at the corner of the lead assures proper alignment of the first brick.

Building a properly aligned corner also requires facing the brick, or aligning the bottom edge of each brick as it is laid with the top edge of the brick below it. Facing each brick as it is laid requires minimal repositioning of the brick as it is aligned plumb, helping to assure a good mortar bond with the brick and improving productivity. Once the bottom edge of all brick along the course are faced with those below, a mason's level is used to align plumb each brick at the

**Figure 6-8** **The mason's level is held vertically against the face of each brick on the first course for a** *plumb* **alignment.**

**Figure 6-9** **Aligning the top edge of each brick against the edge of the level aligns the brick in a straight line.**

**Figure 6-10** Aligning all brick straight with the brick at each end of the course previously aligned plumb creates a plumb course.

**Figure 6-12** The mason compares the alignment of the top corner edge of the last brick of every course. This is known as *checking the rack of the lead*.

corner of the lead and the tail of the lead. With each brick at the corner of the lead and the tail of the lead aligned plumb, the edge of the mason's level is used to align straight the brick between them, tilting the top of each brick until the faces of all are aligned straight (see Figure 6-10). It is important to remember that the two end points aligned plumb are the reference points for aligning the remaining brick straight. These two end points are no to be moved once aligned plumb.

Joints are tooled as they become thumb-print hard. Joints becoming harder require excessive pressure applied to the jointer. Applying too much pressure on the joints can misalign the brick lead.

Upon completion, inspect the corner's accuracy, giving close attention to the plumb lines both at the corner of the lead and tail of the lead of each course. Because of the uneven profile of some brick, it is important to remember that the entire face of a brick may not be touching the edge of the mason's level after aligning the brick in a straight line. A straight wall line is considered established once as much of the face of the brick as possible is aligned straight (see Figure 6-11).

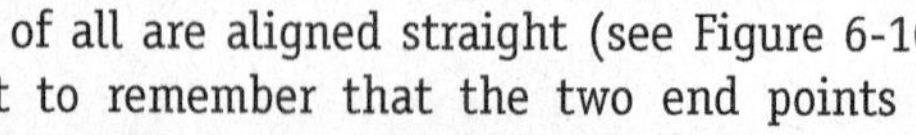

Observing the rack of the lead can indicate inaccuracies of lead construction. The **rack of the lead** is the brick alignment at the tail of the lead for the combined courses. The rack of the lead is observed by holding the edge of the level diagonally with the top edge of each brick at the tail of the lead for each course (see Figure 6-12).

If the rack of the lead is not uniform, meaning that not all brick align against the straight edge of the mason's level, reexamine the tail of the lead of each course to be sure the brick are aligned plumb with the straight-aligned first course. Correct all brick not in plumb alignment with the first course, and realign the course straight. It should be noted that brick with uneven or irregular profiles may not necessarily have a uniform rack of the lead.

*For step-by-step instructions on building a 4" outside brick corner, see the Procedures section on pages 89–95.*

## Building an Inside Brick Corner

The procedures for building an inside corner are similar to those for building an outside corner. However, the mason may feel restricted in the limited area created by the adjacent walls. The brick at the corner of the lead is laid first, followed by the other brick on that side of the lead. Referring to Figure 6-13, this is illustrated as brick #1.

![Brick profile diagram]

**Figure 6-11** Although the face of each brick exhibits an irregular profile, the brick are considered to be aligned straight.

Figure 6-13 **Brick #1, #2, and #3 are laid and checked before laying brick #4, #5, and #6.**

FACE OF LEAD

Brick #2 and brick #3 are laid in that order. The scale or height is checked on the face side of the wall at the vertex of the angle, the point where brick #1 and brick #4 adjoin. Brick #2 and #3 are aligned level with course scaled brick #1. Brick #1, #2, and #3 are aligned plumb as the level is positioned against the face of each brick. Next, the brick course is aligned straight. The range of the wall is checked to confirm straight alignment of the first course of the corner with the lead or end of the wall at the opposite end. For the first course of the adjacent side of the corner, brick #4, #5, and #6 are bedded in mortar, and then aligned level, plumb, and straight as is done for the first side. The same procedures used to align brick #1, #2, and #3 are used for brick #4, #5, and #6. For all remaining courses, one side of the corner is completed at a time, beginning with the brick at the corner of the lead. By facing the brick as they are laid, the mason's level is used at the corner of the lead and the tail of the lead to verify plumb alignment of these two brick at the extreme ends of the course (see Figure 6-14).

Figure 6-14 **In addition to the *tail end*, the brick at the intersection of the two sides of the corner is aligned *plumb*.**

Figure 6-15 **The one head joint in the very inside corner of every course is *trowel struck*, becoming a *troweled* joint.**

Once they are aligned plumb, the edge of the mason's level is used to align straight the brick between them, tilting the top of each brick until the faces of all are aligned straight. Remember not to move the brick at the ends of the lead that have been aligned plumb. This will result in a misaligned lead.

For walls specifying a concave joint finish, it is common practice to use the toe of the trowel blade to make a flush-cut head joint where the two sides intersect (see Figure 6-15). With concave joints, tooling the one head joint in the very corner of every course creates a wavy-looking vertical line and the illusion of a crooked or misaligned wall.

Check the accuracy of scale, level alignment, plumb alignment, straight alignment, and the rack of the lead upon completing the corner and make necessary corrections before attaching the mason's line and laying brick to the line between leads.

## Building a 4" Brick Jamb

A 4" jamb is built similar to an outside corner. Examples of 4" jambs are found next to window or door openings or at the end of walls where there is no brickwork to the adjacent side. Where the pattern bond permits, alternating the brick at the corner of the lead between courses as being a full- or half-brick simplifies construction (see Figure 6-16).

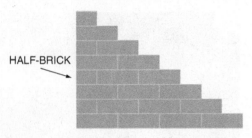

**Figure 6-16** A *bat*, or half-brick, begins every other course of this 4" jamb.

**Figure 6-17** Each jamb on both sides of the window requires cutting the brick differently to maintain a plumb bond.

The same order of procedures is used for building a 4" jamb as are used for either side of an outside corner. The operations and their order are 1) establish scale at the end of the jamb, 2) align the brick course level, 3) align both the end of the jamb and the face of the jamb plumb, and 4) align the brick course straight.

At locations requiring jambs and where brickwork exists below or continues above the opening, a uniform plumb bond should be continued at each jamb. This can necessitate cutting the end brick for each course (see Figure 6-17).

Jambs such as those illustrated in Figure 6-17 are usually constructed as brick are laid to the line on a longer wall during wall construction. The mason's line is attached to leads at opposite ends of the wall, and these jambs are a result of an opening in the wall.

It is easy to unintentionally misalign a 4" jamb. The weight of the mason's level can unintentionally misalign brick when aligning the brick plumb and so can the jointer when tooling joints.

# Building an 8" Brick Jamb

An 8" jamb is built at the end of a **double-wythe wall**, which is a wall comprised of two masonry units back-to-back (see Figure 6-18).

**Figure 6-18** The actual width of an 8" jamb is equal to the length of a brick, which is approximately 7⅝" long.

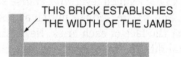

THIS BRICK ESTABLISHES THE WIDTH OF THE JAMB

**Figure 6-19** Laying the first brick across both wythes ensures accurate wall width.

The brick laid across the two wythes at the end of an 8" jamb on every other course must be the same length if the wall is to align plumb on both sides. Slight variations in the length of common building brick can affect plumb alignment at the end of the wall unless end brick having equal lengths are selected. Wall design permitting, laying the first brick across the end of the lead establishes the width of the wall (see Figure 6-19).

As when building a brick corner, one side of each course is completed before spreading mortar for the opposite side. Observing the following guidelines for the brick laid across the end of an 8" jamb prevents errors:

- Each brick laid across the end of the jamb is aligned level.
- All brick laid across the end of the jamb have equal lengths.
- Only one of the two ends of a brick laid across the end of the jamb is aligned vertically plumb using the mason's level. Having the opposite end aligned plumb with the face of the wall below requires the brick to have a length equal to those brick spanning the two wythes below it.
- Aligning plumb the end of the brick on the side of the jamb that is to be more noticeable permits uncontrollable slight variations in brick length to be less noticed on the opposite side.

*For step-by-step instructions on building 8" brick jambs, see the Procedures section on pages 96–103.*

The face of each brick at the end of the jamb on the courses where two brick are separated by a head joint are aligned plumb with the wall below. Exposed as the face of the wall at its end, the tops of the two brick should be aligned level with one another (see Figure 6-20).

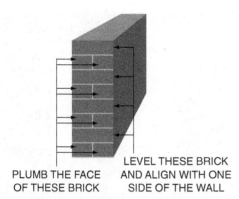

PLUMB THE FACE
OF THESE BRICK

LEVEL THESE BRICK
AND ALIGN WITH ONE
SIDE OF THE WALL

**Figure 6-20** Brick at the end of an 8" jamb should be level. Brick across the end of the jamb are aligned *plumb* on one end only. Detailed illustrations for the plumb alignment of 8" brick jambs are found on page 104.

Eliminating excessive mortar protrusions (mortar coming from the joints as the brick are bedded in mortar) and mortar bridgings (mortar protrusions so large that they contact the adjoining wall) between the two wythes prevents mortar from pushing the wythes apart. Spreading no more mortar than sufficient for bedding the brick minimizes mortar protrusions and bridgings.

For each side of every course, align level, align plumb, and align straight the brick. Be certain that the two wythes are aligned level at the end of the jamb. Laying two courses on one side of the jamb before continuing with the opposite side assists in the removal of mortar protrusions (see Figure 6-21).

*For step-by-step instructions on plumb alignments of 8" brick jambs, see the Procedures section on page 104.*

When the first course brick pattern requires the ends of two brick forming the 8" jamb, then both wythes of the first course must be built before continuing to the second course (see Figure 6-22).

**Figure 6-21** Two courses are laid consecutively on each side.

**Figure 6-22** This layout necessitates building the opposite side of the brick jamb before continuing with the next course.

**Figure 6-23** Regardless of variations in brick lengths and widths, changing the head joint width permits a plumb wall line on each side of every course at the end of a 12" jamb.

# Building a 12" Brick Jamb

The width of a 12" jamb is viewed as being a brick and a half. Including the head joint between the two brick, its actual width varies from approximately 11½" inches to 12", depending on actual length and width of the selected brick as well as the width of the head joint.

Procedures for building a 12" brick jamb are similar to those for an 8" jamb. Because every course of the 12" jamb's end requires a head joint, each side at the jamb's end can be aligned plumb using the mason's level without affecting the opposite side (see Figure 6-23). The presence of a head joint at the end of the wall on each course permits slight variations in brick lengths without affecting plumb alignment of either side.

Examples of 12" brick jambs and 12" brick walls can be found around patios and courtyards. Properly designed and reinforced with concrete, grout, and steel, 12" walls are sometimes used as reinforced brick masonry (RBM) walls.

Wider jambs such as 16" and 20" are built using the same procedures as for a 12" jamb.

# Toothing a Corner

The racking back of brickwork at the tail of a lead limits the courses to which a lead can be built. Toothing is used to complete a lead to any height before the adjoining wall is built. Toothing is a procedure for continuing with courses on a lead where the tail of the lead normally limits additional courses by temporarily laying a half-brick or bat on the stretcher below, typically bedding it in mortar (see Figure 6-24).

Several items of concern must be monitored when toothing a corner. First, the half-brick or bats must be carefully removed before the mortar hardens. Accidentally moving the brick in the lead while removing the bats not only alters the plumb and level alignment of the brick, but it also reduces the mortar bond strength. In addition, the mortar joints between the toothed lead and wall should be filled solid. A tuckpointer is used to fill the joints solid. Also, both the brick and the mortar for the toothed corner and the adjoining wall must match in color. Using brick from

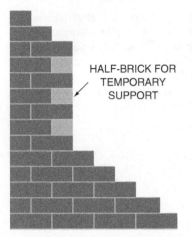

Figure 6-24 Half-brick temporarily support the stretchers. The half-brick are removed before the mortar hardens.

HALF-BRICK FOR TEMPORARY SUPPORT

Figure 6-26 A corner pole eliminates the need for building a lead.

Figure 6-25 The corner is toothed to permit completing one wall before beginning the adjacent wall.

Figure 6-27 Straight, 2" framing lumber is used as a corner pole.

the same kiln firing and proportioning the mortar ingredients identically for the toothed lead and the wall prevents noticeable color variations between the toothed corner and the adjoining wall.

Toothing is used to complete one section of brickwork or one side of a structure before an adjoining wall is built (see Figure 6-25). Specifications may sometimes limit or prohibit toothing of brick leads.

## Using Corner Poles

Attaching a metal or wooden corner pole or masonry guide at a corner eliminates constructing a lead. One mason can quickly and easily erect a corner pole, increasing productivity by eliminating the greater time taken for building a lead. Commercially available, 2" × 2" aluminum poles with adjustable top and bottom fittings are used for quick and accurate setups on both outside and inside corners (see Figure 6-26).

Straight, 2" framing lumber is often used for making corner poles on the job site (see Figure 6-27). The corner pole must remain straight and resist the forces of a tensioned mason's line.

*For step-by-step instructions on adjusting the top of a corner pole and for adjusting a corner pole at its base, see the Procedures section on pages 105–106.*

## Quoined Corners

Quoined corners are architectural details dating back to previous centuries. Quoined corners are constructed by projecting the brick at corners to create block designs (see Figure 6-28).

**Figure 6-28** **The brickwork at the corner projects to form brick quoins.**

Quoin patterns vary, usually two to three brick in length and five to six courses tall. All quoins may be identical patterns of brickwork, or alternating similar patterns is common. The number of courses between the quoins is a design element that can vary between jobs. Quoins should project no more than ⅝" when using cored brick or more than ¾" for solid brick.

## Summary

- Constructing leads requires the knowledge of layout, dry bonding, scaling, and basic alignment operations performed with the mason's level, including level, plumb, and straight alignments.
- As the term *lead* implies, accurate wall construction is lead or guided by accurate leads.
- A competent mason should closely supervise an apprentice mason while the apprentice develops the skills required for lead construction.
- Not until required proficiency is demonstrated should an apprentice attempt constructing leads without supervision.

## Procedures

# Placing a Brick Requiring Cutting at the End of a Lead

**A** A brick may require shortening to create a wall of specific length. Place the shortened brick as the first brick laid in the *stretcher* position at the end of the wall.

**A**

**B** Align the cut brick in a plumb line for the entire height of the wall.

**B**

## Procedures

# Building a 4" Outside Brick Corner

 Beginning with the brick at the corner, lay the brick for that side of the lead. Laying no more than six brick permits aligning the brick using the 4' level.

 Align the course *level* at the intended course height. The course height is confirmed at the end of the corner using the intended mason's scale.

| **Procedures** | **Building a 4" Outside Brick Corner (continued)** |

 Align each brick of the first course *plumb*.

 Align the brick straight using the edge of the level. Not only must the brick be aligned straight, but also the corner must be aligned with the lead at the opposite end when building walls.

Repeat procedures A, B, C, and D for the opposite side of the corner. Confirm the corner to be square using a framing square.

 For the second course, begin with the brick at the corner, and lay all brick for only the side to which the first brick's face is oriented, carefully *facing* the bottom edge of each brick with those brick below it.

 Align the course level and at the intended height. The course height is confirmed at the end of the corner using the intended mason's scale.

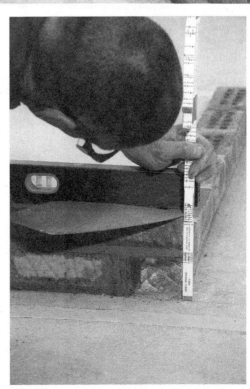

## Procedures

# Building a 4" Outside Brick Corner (continued)

 Using the mason's level, vertically align plumb both the end and face of the first brick. Vertically align plumb the brick at the tail end of the lead.

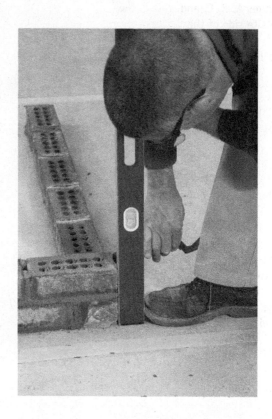

**I** Using the level as a straight edge, position the edge of it at the top edge of the course. If necessary, reposition the top edge of each brick in straight alignment with the ends.

With the bottom of the brick faced, tilt the top of each brick by tapping on it with the edge of the trowel blade until all are aligned.

**J** Lay the brick on the opposite side of the corner, carefully facing each brick as it is laid.

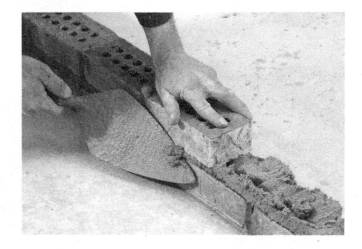

# Building a 4" Outside Brick Corner (continued)

 **K** Align the brick level with the brick at the corner.

 **L** Vertically align plumb the brick at the tail end of the lead.

Using the level as a straight edge, position the edge of it at the top edge of the course. If necessary, reposition the top edge of each brick in straight alignment with the brick at the corner and the tail of the lead. With the bottom of the brick faced, tilt the top of each brick by tapping on it with the edge of the trowel blade until all are aligned.

 Continue the procedures F through M for each successive course.

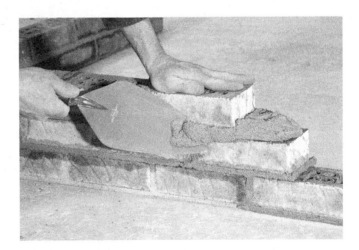

## Procedures

# Building an 8" Brick Jamb

**A** Lay the brick whose face is oriented toward the end of the jamb and the brick for one side of the lead. Laying no more than six brick permits aligning the brick using the 4' level. The brick whose face is oriented toward the end of the jamb must be aligned square with the front side of the jamb.

**A**

**B** Align *level* the brick at the end of the jamb at the intended course height. Align the course *level* with the brick at the end.

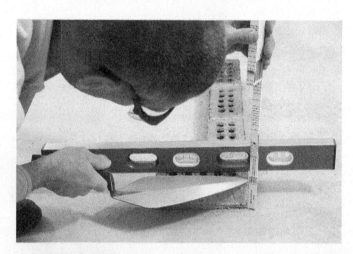

**B**

**C** Align the face of each brick *plumb*.

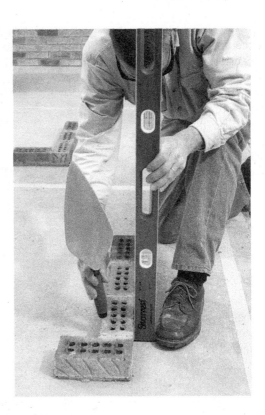

**D** Align the brick *straight* along the front of the jamb using the edge of the level. Not only must the brick be aligned straight, but also the jamb must be aligned with the lead at the opposite end when building walls.

## Procedures

# Building an 8" Brick Jamb (continued)

**E** Choose to lay either the first course for the opposite side of the jamb or the second course above the brick just laid. The author prefers to proceed with the second course of the first side. This procedure helps prevent mortar protrusions from misaligning the opposite side, the course just laid.

**F** Align the course level at the intended course height. The course height is confirmed at the end of the jamb using the intended mason's scale.

**G** Using the mason's level, vertically align *plumb* both the end and face of the first brick. Vertically align *plumb* the brick at the tail end of the lead.

Using the level as a *straight edge*, position the edge of it at the top edge of the course. If necessary, reposition the top edge of each brick in straight alignment with the ends.

Tilt the top of each brick by tapping on it with the edge of the trowel blade until all are aligned.

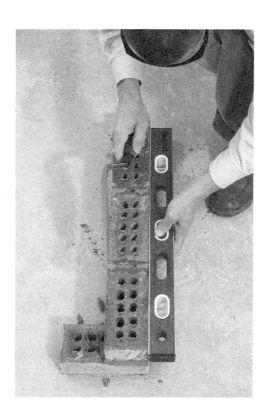

 Lay the first course of brick on the opposite side of the jamb. Align the brick *level* with the brick at the end of the jamb.

## Procedures

## Building an 8" Brick Jamb (continued)

 Align each brick *plumb*.

 Align the brick *straight* using the edge of the level. Not only must the brick be aligned straight, but also the double-wythe wall created by the jamb must be equivalent to the length of a brick.

**L** Lay the second course, beginning with the brick at the end of the jamb, carefully *facing* the bottom edge of each brick with those brick below it.

**M** Align the brick at the end of the jamb *level* with the end brick on the opposite side of the jamb.

**N** Align the remaining brick of this side level with the brick at the end of the jamb.

## Procedures

# Building an 8" Brick Jamb (continued)

 Using the mason's level, vertically align *plumb* both the end and face of the first brick. Vertically align *plumb* the brick at the tail end of the lead.

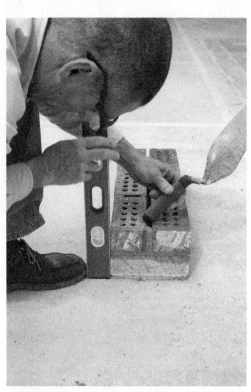

**P** Using the level as a *straight edge*, position the edge of it at the top edge of the course. If necessary, reposition the top edge of each brick in straight alignment with the ends.

Tilt the top of each brick by tapping on it with the edge of the trowel blade until all are aligned.

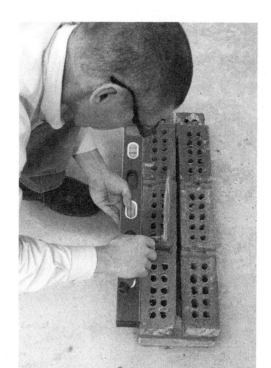

**Q** Continuing to raise the jamb, repeat the preceding procedures for each side of the jamb.

## Procedures

# Plumb Alignments of 8" Brick Jambs

**A** For an 8" jamb to be aligned plumb on both sides of a wall, all brick laid across both wythes at the end of the wall (darker brick) must have the same lengths. The single-brick design at the end of such courses permits aligning plumb only one side of the wall with the mason's level.

**A**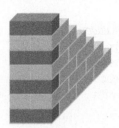

**B** Attempting to plumb the opposite side of brick with different lengths misaligns the side initially aligned plumb. Laying brick of different lengths (darker brick) at the end of an 8" jamb causes misaligned jambs as illustrated here.

**B**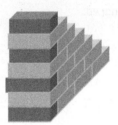

**C** Courses having a single brick at the end of the wall (dark brick) restrict plumb alignments to one side only. Accurate alignment of the opposite side requires selecting brick with equivalent lengths.

**C**

**D** Courses with two brick at the end of the jamb bonded with a head joint require the use of the mason's level to establish plumb alignment of both sides of the wall.

**D**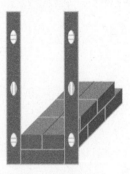

**E** Aligning the ends of 8" jambs plumb at the edges of both the front and back sides for each course ensures that the end of the jambs remain plumb and square.

**E**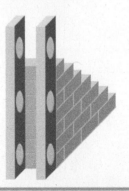

## Procedures

# Adjusting the Top of a Corner Pole

**A** With the corner pole's base bracket supported on the masonry shelf below it, fasten the top bracket to the wall framing, and adjust the pole to provide the desired spacing between the face of the brick wall and the wall framing.

1. The *top bracket* is anchored to the wall framing. Screws or double-headed form nails permit easy removal.

2. Before tightening the wing nut securing the two-piece top bracket, the corner pole is adjusted for the desired spacing. In addition to providing required spacing between the brick and wall framing, an additional ⅛" is allowed for the combined width of the mason's line (¹⁄₁₆") and spacing between the line and the brick face (¹⁄₁₆") (arrow).

**A**

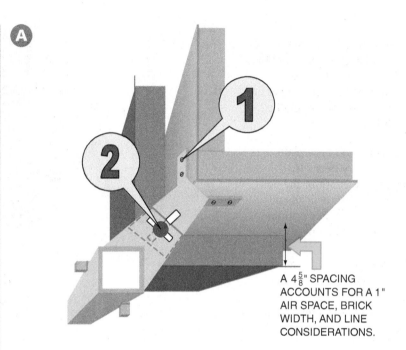

A $4\frac{5}{8}$" SPACING ACCOUNTS FOR A 1" AIR SPACE, BRICK WIDTH, AND LINE CONSIDERATIONS.

## Procedures

# Adjusting a Corner Pole at Its Base

**A** After adjusting the corner pole at its top, adjust the bottom of the corner pole until its adjacent sides are plumb. Mason's lines are secured by the corner pole to guide brickwork.

1. The corner pole's base plate rests on the masonry below it.
2. Adjusting the depth of the bolts between the wall and support brackets on each side positions the corner pole in a plumb position.
3. Self-threading masonry screws secure the corner pole to the masonry wall.
4. The corner pole secures line blocks holding the mason's lines.

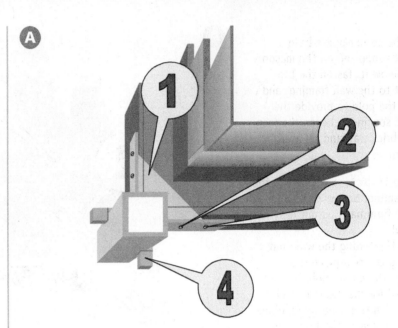

## Review Questions

*Select the most appropriate answer.*

**1** The part of a wall first constructed to which a line is attached for laying brick to the line is called

a. a guide

b. a lead

c. a pier

d. none of the above

**2** A type of lead is

a. a corner

b. a jamb

c. a sill

d. both a and b

**3** The time recommended for bedding a brick in mortar after the mortar is spread is within

a. "1 minute or so"

b. 3 minutes

c. 5 minutes

d. 10 minutes

**4** The end of the lead where brick coursing steps back is called

a. the jamb of the lead

b. the base of the lead

c. the tail of the lead

d. none of the above

**5** Examining the straight alignment of leads at opposite ends of a wall is called

a. checking the range

b. coursing

c. dry bonding

d. spacing the wall

**6** Examining the brick alignment at the tail end of a lead is called

a. straight-edging

b. checking the rack of the lead

c. scaling the lead

d. plumbing the lead

**7** A procedure permitting the continuation of a lead to any height is called

a. jambing

b. racking

c. scaling

d. toothing

**8** Constructing leads may sometimes be eliminated by using

a. corner poles

b. masonry guides

c. plumb lines

d. both a and b

# Chapter 7

# Masonry Spacing Scales

The masons' scales are used to maintain uniform course spacing while building a wall to a predetermined height. Scales simplify equal spacing of courses and eliminate mathematical operations requiring the division and multiplication of whole numbers and fractions. Masons rely upon the scales when laying out walls for choosing desired course heights. Those building masonry leads confirm proper course spacing using the scales. Using brick spacing scales also ensures uniform spacing of brick laid in the rowlock and soldier positions.

## OBJECTIVES

After completing this chapter, the student should be able to:

- Identify and give the application for the three sets of spacing scales.
- Lay out a wall to a specific height using the spacing scales.
- Lay out a brick rowlock using masonry scales.
- Discuss factors influencing the selection of scales.

# Glossary of Terms

**engineered brick** also called oversize brick, brick whose approximate dimensions are 7⅝" long, 3½" wide, and a 2¾" face height.

**mason's scales** graduated markings on mason's rules and tapes permitting different course spacing of masonry units.

**modular masonry construction** a construction design permitting different types of masonry units being used to build walls to the same module or repetitive dimension.

**modular masonry unit** a masonry unit permitting modular masonry construction.

**oversize brick** brick whose approximate dimensions are 7⅝" long, 3½" wide, with a 2¾" face height.

**standard size brick** brick whose approximate dimensions are 7⅝" long, 3½" wide, with a 2¼" face height.

# Mason's Scales

The **mason's scales** are graduated markings on mason's rules and tapes permitting different course spacing of masonry units. There are three different sets of masons' scales:

1. Standard-size brick spacing scales
2. Oversize brick spacing scales
3. Modular scales

These scales are included on mason's folding rules and measuring tapes. Typically only one of the three sets of scales is on a single rule or tape.

## Standard-Size Brick Spacing Scales

The standard-size brick spacing scales are used for spacing standard size brick. Ten numbers, 1-9 and 0, represent the scales. Scale 1 represents a brick course that has the smallest spacing. Each consecutively larger number represents wider course spacing. Scale 0 represents the largest course spacing. The scale numbers appear black on the mason's brick spacing rule. Associated with each scale number is a corresponding red number (see Figure 7-1). This red number indicates the number of brick courses from the origin of the rule to the given scale. The overall height for each course of brick increases as the scale number increases due to wider bed joints. Scale 1 allocates $2\frac{3}{8}$" per brick course. Scale 2 is equivalent to laying a course of brick every $2\frac{1}{2}$". Scale 3 is equivalent to a course every $2\frac{9}{16}$". Scale 2 is $\frac{1}{8}$" greater than scale 1, but there is only a $\frac{1}{16}$" difference between any two consecutively numbered scales 3, 4, 5, 6, 7, 8, 9, and 0. Using the scales eliminates the need for spacing brick courses using fractions of inches, a process requiring mathematical calculations at the work site.

At the opposite end of the 6' folding rule, the scales and coursing appear as illustrated in Figure 7-2. Starting at the left, it is read as "27 courses laid on scale 3," "26 courses laid on scale 5," "28 courses laid on scale 2," "25 courses laid on scale 7," "24 courses laid on scale 9," "27 courses laid on scale 4," "30 courses laid on scale 1," "26 courses laid on scale 6," "28 courses laid on scale 3," and "25 courses laid on scale 8."

Factors to consider for selecting an appropriate scale are preferred bed joint width and specified wall height.

### Preferred Bed Joint Width

Most agree that mortar joint width should be no less than $\frac{1}{4}$" or more than $\frac{1}{2}$". Improper joint width can lead to water migration from the exterior face of the wall to the air cavity behind the wall, a situation causing many moisture-related problems. Acceptable joint widths limit the selection of scales for any given brick. For example, a brick with a height of $2\frac{1}{4}$" (**standard-size brick**) and laid on scale 1 results in a bed joint less than $\frac{1}{4}$", and scale 2 allows a bed joint of only $\frac{1}{4}$". For such a brick, scale 1 should not be considered, and scale 2 may be questionable if the face height of the brick vary. Scales 7, 8, 9, and 0 require joints wider than $\frac{1}{2}$" for the same brick. Likewise, these scales should not be considered. Brick coursing should be limited to scale 3, 4, 5, or 6 for this particular brick, allowing bed joints to be no less than $\frac{1}{4}$" or more than $\frac{1}{2}$". Although called standard-size brick, manufacturer's brick can vary slightly in height, some more and some less than $2\frac{1}{4}$". The average brick height should be considered before selecting a particular scale.

### Specified Wall Height

With scale selection, a wall can be constructed to a specific height. Proper spacing permits brick rowlocks below windowsills. Likewise, when scaling permits, it is desirable to have the brick coursing able to cross just above the top of windows and doorframes, supported by steel angle lintels (see Figure 7-3).

**Figure 7-1** The standard-size brick spacing rule includes 10 scales (black numbers) for spacing standard-size brick. Red numbers indicate the quantity of intervals from the end of the rule.

| 3 | 5 | 2 | 7 | 9 | 4 | 1 | 6 | 3 | 8 |
|---|---|---|---|---|---|---|---|---|---|
| 27 | 26 | 28 | 25 | 24 | 27 | 30 | 26 | 28 | 25 |

**Figure 7-2** The scales (black numbers) at the opposite end of the brick mason's rule appear as in this illustration. Red numbers indicate the quantity of intervals from the end of the rule for the attached scale number.

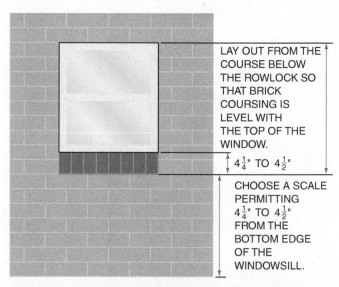

LAY OUT FROM THE COURSE BELOW THE ROWLOCK SO THAT BRICK COURSING IS LEVEL WITH THE TOP OF THE WINDOW.

$4\frac{1}{4}$" TO $4\frac{1}{2}$"

CHOOSE A SCALE PERMITTING $4\frac{1}{4}$" TO $4\frac{1}{2}$" FROM THE BOTTOM EDGE OF THE WINDOWSILL.

**Figure 7-3** Scales can permit brickwork to accommodate openings without cutting the brick below or above the openings.

*For step-by-step instructions for using brick spacing scales to accommodate doors and windows, see the Procedures section on pages 113–114.*

The brick spacing scales are also used for the horizontal equal spacing of brick laid in the rowlock and soldier positions. Although it is a common procedure to align three brick in the rowlock or soldier positions with one brick in the stretcher position (see Figure 7-4), this three-to-one alignment is not always applied.

In many cases, the brick spacing rule is used to insure uniform joints for the rowlocks or soldiers. The brick spacing rule can be used to lay out the horizontal spacing for brick rowlocks and brick soldiers similar to the method of vertically spacing a wall to a specified height (see Figure 7-5).

In Figure 7-5, scale 5 is chosen because it permits a head joint to be laid between the last brick and the existing brick. As indicated at the arrow above the spacing scale, 26 brick (red number) are to be laid on scale 5 (black number) to complete the rowlock.

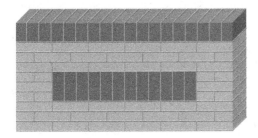

**Figure 7-4** Three standard-size brick laid in either the rowlock or soldier positions are aligned with each brick below them.

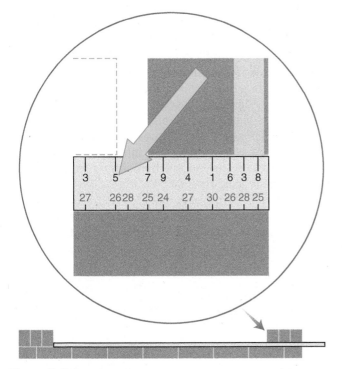

**Figure 7-5** The brick mason's scales are used to provide uniform joints and brick spacing.

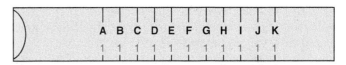

**Figure 7-6** Oversize or engineered brick spacing scales are labeled A through K. Red numbers indicate the quantity of intervals from the end of the rule for the attached scale letter. The oversize brick spacing scales can also be used for spacing queen size brick.

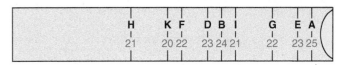

**Figure 7-7** The brick spacing scales at the opposite end of the oversize brick spacing rule appear as in this illustration. Red numbers indicate the quantity of intervals from the end of the rule for the attached scale letter.

## Oversize Brick Spacing Scales

The oversize brick spacing scales are used for laying oversize or engineered brick. The 11 scales are lettered A through K with the course spacing increasing $\frac{1}{16}$" for each scale from A through K. As with the standard-size brick spacing rule, red numbers indicate the number of brick courses from the origin of the rule to any given scale (see Figure 7-6).

At the opposite end of the 6' folding rule, the scales and coursing appear as illustrated in Figure 7-7. Starting at the left, it is read as "21 courses laid on scale H," "20 courses laid on scale K," "22 courses laid on scale F," "23 courses laid on scale D," "24 courses laid on scale B," "21 courses laid on scale I," "22 courses laid on scale G," "23 courses laid on scale E," and "25 courses laid on scale A."

The height of most **engineered brick** or **oversize brick** is approximately 2¾". Scale D or E results in bed joints with widths of approximately ⅜" thick with such brick. As with standard-size brick, scale selection should limit bed joint width to no less than ¼" or no more than ½".

As with the standard-size brick mason's rule, the oversize brick spacing scales may be used to uniformly space brick horizontally in the rowlock or soldier positions.

## Modular Spacing Scales

The modular spacing scales permit the construction of double-wythe walls where each wythe is a different size and type of masonry unit. All stock sizes and types of masonry materials permit course spacing at 16" intervals. A construction design permitting different types of masonry units to be used to build walls to the same module or repetitive dimension is called **modular masonry construction**. The masonry units for modular construction are called **modular masonry units**.

With two walls level with one another at 16" intervals, wire joint reinforcement is used to bond or tie the two walls together. There is one modular scale representing each of six different masonry units. The scale number refers to the number of courses for every 16" interval of wall height. For example, concrete masonry units are laid on modular scale 2, meaning two courses each at a 16" interval. Standard-size brick are laid on modular scale 6, meaning six courses at

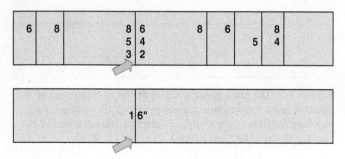

**Figure 7-8a** The 6 modular scales merge at 16" intervals.

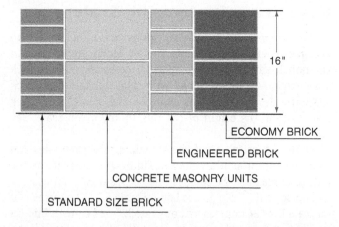

ECONOMY BRICK

ENGINEERED BRICK

CONCRETE MASONRY UNITS

STANDARD SIZE BRICK

| SCALE | MASONRY UNIT | # COURSES IN 16" |
|-------|--------------|------------------|
| 2 | CONCRETE MASONRY UNIT | 2 |
| 3 | FACING TILE | 3 |
| 4 | ECONOMY BRICK | 4 |
| 5 | ENGINEERED/OVERSIZE BRICK | 5 |
| 6 | STANDARD BRICK | 6 |
| 8 | ROMAN BRICK | 8 |

**Figure 7-8b** Modular course spacing is illustrated here. Although not identified on the modular rule, modular scale 5 can also be used for spacing queen size brick.

every 16" interval. Similarly, oversize or engineered brick are laid on modular scale 5, representing five courses every 16". All modular scales merge at 16" intervals to accommodate different types and sizes of masonry units while allowing dissimilar units to be joined with horizontal wire joint reinforcement (see Figure 7-8). Using horizontal wire joint reinforcement to join multiwythe masonry walls is explained in Chapter 16, Composite and Cavity Walls.

In addition to modular scales for standard size brick, oversize brick, and concrete masonry units, there is a modular scale for laying each of these units: economy brick, Roman brick, and facing tile. Whereas the mason has choices for the course height when using the standard-size brick scales or oversize brick scales, modular scaling does not permit the mason to vary course height.

Modular scales must not be confused with the standard-size brick spacing scales. Both sets of scales refer to the numbers 2, 3, 4, 5, 6, and 8 as scales. For the standard-size brick spacing scales, these numbers represent different course heights attainable for standard-size brick courses as the bed joint width is changed. But when using the modular scales, these same numbers represent different masonry units laid in height intervals of 16". To be clear and to prevent confusion, one should address the scales for example as "modular rule scale 5" or "brick spacing rule scale 5."

## Summary

- The masonry scales are intended to provide uniform course spacing and permit constructing masonry walls to desired heights.
- Both the brick spacing scales and the oversize scales permit optional bed joint widths. It is important to remember that not all scales on the brick spacing or oversize rules are applicable for any particular brick.
- Keeping bed joint width within acceptable width range and complying with job specifications or governing regulations limits scale options.
- The modular scaling system gives the mason no options for varying bed joint widths. Its intention is to permit 16" intervals for the construction of two adjacent walls of composite and cavity walls.
- Regardless of the masonry units selected, modular scales permit building two adjacent walls at 16" intervals, facilitating the installation of horizontal joint reinforcement.

## Procedures

# Using Brick Spacing Scales to Accommodate Doors and Windows

**A** A 4¼" to 4½" clearance below windows permits brick sills with brick in the rowlock position. Wood-framed structures necessitate clearances between the brick masonry and the bottom edges of windows or doors to allow for possible shrinkage of framing materials. Follow window and door manufacturers' recommendations for clearances between brickwork and their products. The scales on the brick spacing rule or the oversize brick spacing rule permit building brick walls to the desired height below windows while having the required spacing for brick sills. The same or different scale may permit continuing the wall to the top of an opening, so that brick coursing aligns level with the door or window. Otherwise, brick below and above openings must be cut as shown in the illustration for Procedures B. There are no options for brick coursing heights if using the modular spacing rule, and brick are cut below and above openings where necessary.

**B** The brick below the window are cut to permit a sill with brick laid in a rowlock position. 4¼" to 4½" is a suggested spacing for the rowlock sill. Depending on window and door manufacturers' instructions, additional clearances may be required. The brick above the window are cut to maintain level coursing with the remaining wall.

**A**

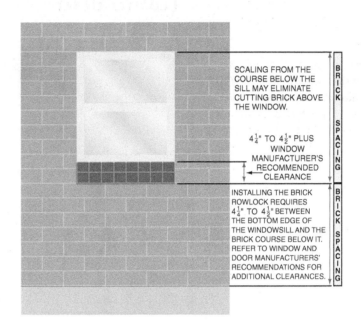

SCALING FROM THE COURSE BELOW THE SILL MAY ELIMINATE CUTTING BRICK ABOVE THE WINDOW.

4¼" TO 4½" PLUS WINDOW MANUFACTURER'S RECOMMENDED CLEARANCE

INSTALLING THE BRICK ROWLOCK REQUIRES 4¼" TO 4½" BETWEEN THE BOTTOM EDGE OF THE WINDOWSILL AND THE BRICK COURSE BELOW IT. REFER TO WINDOW AND DOOR MANUFACTURERS' RECOMMENDATIONS FOR ADDITIONAL CLEARANCES.

BRICK SPACING

BRICK SPACING

**B**

# Using Brick Spacing Scales to Accommodate Doors and Windows (continued)

**C** The brick above the window are laid in the rowlock position. This is an alternative for the method shown in Procedure B. Shortened brick laid in the soldier position give similar results.

**C**

## Review Questions

*Select the most appropriate answer.*

**1** The difference between any two standard-size brick scales except scales 1 and 2 is

a. ¹⁄₁₆"
b. ⅛"
c. ¼"
d. ⅜"

**2** Oversize brick scales are identified by the letters

a. A through H
b. A through J
c. A through K
d. A through L

**3** The spacing rule permitting the construction of multiwythe walls so that adjacent wythes having different types of masonry materials can be tied together at repetitive heights is the

a. engineered brick spacing rule
b. modular spacing rule
c. oversize brick spacing rule
d. standard-size brick spacing rule

**4** All modular scales merge at every

a. 4"
b. 8"
c. 12"
d. 16"

**5** The scaling system used for residential brick veneer construction permitting choices for bed joint width is

a. the oversize brick scales
b. the standard-size brick scales
c. the modular spacing scales
d. both a and b

**6** A factor considered when selecting a specific scale on the standard-size brick spacing rule is

a. the desired joint width
b. the desired height of the wall
c. the restrictions for bed joint width
d. all of the above

**7** Scales are used to lay out

a. vertical brick coursing
b. brick rowlocks
c. multiple brick laid in the soldier position
d. all of the above

**8** The oversize brick spacing scales are also called the

a. colonial brick spacing scales
b. engineered brick spacing scales
c. modular brick spacing scales
d. tudor brick spacing scales

**9** A masonry unit permitting modular masonry construction is

a. the standard concrete masonry unit
b. the economy brick
c. the engineered brick
d. all of the above

**10** Which of the following statements is true for modular scales?

a. 2 courses of CMUs equal 16"
b. 5 courses of oversize brick equal 16"
c. 6 courses of standard-size brick equal 16"
d. All of the above

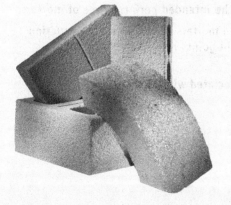

# Chapter 8 | Masonry Mortars

*T*he value of mortar to the integrity of a wall cannot be under-estimated. Mortar bonds individual masonry units into walls that are required to support extreme loads and resist strong lateral forces. For exterior walls, mortar must be water resistant and not adversely affected by the freeze-thaw cycles of nature. The mason must be able to effortlessly place it on masonry units, yet it must resist falling off the units as they are placed in the wall. It is often said that mortar is perhaps the only construction material produced at the job site, and the responsibility for mixing mortar is often entrusted to those with little understanding of a wall's reliance upon properly proportioned and mixed ingredients. Additionally, masons must understand factors influencing the performance of mortar in wall systems so that the structural integrity of the walls is not sacrificed.

## OBJECTIVES

After completing this chapter, the student should be able to:

- List the ingredients of masonry mortars.
- Identify the types of cementitious materials used to make mortar.
- List additives contained in some cementitious materials.
- Describe the procedures for mixing mortar manually and with a power mixer.
- List procedures for maximizing the intended performances of mortars.
- Describe the differences between mortars used for new construction and mortars used for repairing the joints of older and historical brick walls.
- Describe potential problems associated with mortars.

# Glossary of Terms

**accelerators** admixtures that increase the masonry cement's rate of hydration.

**admixtures** materials added to masonry cements for specific purposes.

**air-entraining agents** additives increasing the air content in cement.

**autogeneous healing** a unique characteristic of lime-based mortars that have the capability by chemical reaction to fill small voids that may occur in the joints.

**bond strength** a measure of the resistance to separation of mortar and masonry units.

**cold weather construction** construction executed when either air or material temperatures are below 40°F.

**elasticity** a property of mortar that enables it to retain its original size and shape after compaction.

**flexural strength** the measure of a material's capability to withstand bending.

**grout** a mixture of cement and aggregates to which sufficient water is added to produce a pouring consistency without separation of the ingredients used to strengthen masonry walls.

**hot weather construction** construction executed when either air or material temperatures are above 100°F.

**masonry cements** a mixture of Portland or blended hydraulic cement (a cement that hardens under water) and plasticizing materials (materials such as limestone, hydrated or hydraulic lime intended to enhance workability) whose air content varies widely between manufacturers and are recognized for their good workability.

**mortar** a mixture of cementitious materials, sand, and water bonding individual brick or concrete masonry units to create masonry walls.

**mortar cements** similar to propriety masonry cements except they have lower air content and include a minimum flexural bond strength requirement.

**pigments** insoluble fine powders added to masonry cements to obtain a desired mortar color.

**plasticizers** admixtures that result in mortars that are easier to spread and that adhere to the trowel and masonry units better.

**retarders** admixtures used to delay the mortar's rate of hydration.

**retempering** also called tempering, the process of adding water to the mortar to replace water lost by evaporation.

**tensile strength** the measure of a material's capability to withstand stretching.

**water retention** the ability of mortar to hold water when placed in contact with absorbent masonry units.

**workability** a measure of the ease with which the mortar can be mixed, placed, and finished.

# The Development of Mortars

Ancient Egyptians used gypsum and sand to make mortar. Lime and sand were used for making mortar before the development of Portland cement in 1824. Nineteenth-century lime and sand mortars show evidence of much deterioration, as illustrated in Figure 8-1. Horsehair and oyster shells were sometimes added to strengthen the mortar. The development of Portland cement accounts for the durability and high strength of today's mortars. Since the late 1800s, Portland cement has been the main ingredient of most mortars. Ingredients of today's mortars can include Portland cement, mortar cement, masonry cement, hydrated lime, sand, and water. The three classifications of mortars are 1) Portland cement-lime mortars, 2) masonry cement mortars, and 3) mortar cement mortars.

## Cement and Mortar

The terms *cement* and *mortar* are sometimes misused when referring to masonry construction. Cement is an ingredient of mortar, a bonding agent used in masonry construction. Mortar is a mixture of cementitious materials, sand, and water. Mortar is the bonding agent that allows brick and concrete masonry units (CMUs) to be assembled as masonry walls. Mortar must provide sufficient bond strength, be water-resistant, and permit differential movement of materials.

## Types of Mortars

Mortars are categorized based upon the cementitious materials included in the mix. Three categories of cementitious materials include Portland cement-lime mixes, masonry cements, and mortar cements. Each has different properties and characteristics.

# Portland Cement-Lime Mixtures

Portland cement-lime mixtures contain Portland cement and hydrated lime proportioned by the manufacturer under controlled conditions. The Portland cement accounts for the durability and strength of the resulting mixed mortar. Portland cement also contributes to the quick setting of the mortar. Although mortars with higher air content are more resistant to potential weathering caused by freeze-thaw cycles, Portland cements containing air-entraining agents, additives increasing the air content in cement, can reduce the bond strength between the mortar and the masonry units. Bond strength is a measure of the resistance to separation of the mortar and the masonry units. Although reducing bond strength, air-entraining agents improve mortar's workability, a measure of the ease with which the mortar can be mixed, placed, and finished. Bedding mortar must be easily spread onto brick using the mason's trowel (see Figure 8-2).

Lime contributes to bond strength, workability, water retention, and elasticity. Manufacturers use Type S hydrated lime. Lime-based masonry cement mortars permit slight

**Figure 8-1** The lime-sand mortar joints of this nineteenth-century building are slowly eroding.

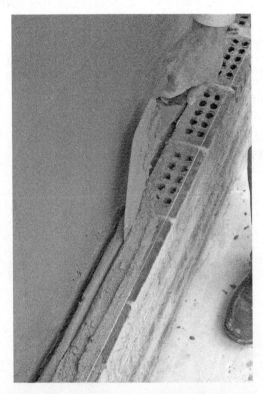

**Figure 8-2** Mortar flows in a continuous ribbon from the trowel blade onto the brick, exhibiting good workability.

movement of the wall due to expansion and contraction without experiencing bond failure. Type S lime improves the water retention of mortar, which in turn improves workability. **Water retention** is the capability of mortar to hold water when placed in contact with absorbent masonry units. If the masonry unit absorbs water too quickly, then the bond strength is reduced. **Elasticity** is that property of mortar permitting it to retain its original size and shape after compaction. A unique characteristic of mortars containing lime is their capability to fill small voids that may occur in the joints. If hairline cracks develop, water and carbon dioxide react with the lime to form calcium carbonate. The hardened calcium carbonate seals the crack from further water penetration. This process is called **autogeneous healing.**

## Masonry Cements

**Masonry cements** are a mixture of Portland or blended hydraulic cement (a cement that hardens under water) and plasticizing materials (materials such as limestone, hydrated or hydraulic lime intended to enhance workability). The air content of these mortars varies widely between manufacturers. The Brick Industry Association Technical Notes 8 reports that "In the model building codes, allowable flexural tensile stress values for masonry built with masonry cement mortars are lower than those for masonry built with non air-entrained Portland cement-lime mortars." Mortars using masonry cements as the cementitious ingredient are recognized for their good workability. The three types of masonry cements are M, S, and N. The Brick Industry Association Technical Notes 8 states that air content for all three types must be a minimum of 8%. The maximum air content for Types M and S is 19% and 21% for Type N.

The Brick Industry Association Technical Notes 8 states that "provided other parameters are held constant, as air content is increased, compressive strength and bond strength are reduced, while workability and resistance to freeze-thaw deterioration are increased."

## Mortar Cements

**Mortar cements** are similar to masonry cements in that they are a mixture of Portland or blended hydraulic cement and plasticizing materials. However, mortar cements have lower air content and include a minimum flexural bond strength requirement. The Brick Industry Association reports that "However ASTM C 1329, Specification for Mortar Cement, includes requirements for maximum air content and minimum flexural bond strength that are not found in the masonry cement specification. Because of the strict controls on air content and the minimum strength requirement, mortar cement and Portland cement mortars are treated similarly in the Masonry Standards Joint Committee (MSJC) Code

and Specification." The three types of mortar cements are M, S, and N. The Brick Industry Association Technical Notes 8, states that air content for all three types must be a minimum of 8%. The maximum air content for Types M and S is 14% and 16% for Type N.

## Strength Classifications of Mortar

Three types of mortar typically used for new construction are Types M, S, and N. A fourth type, Type O, has very limited applications. Two notable differences between them are air content and bond strength. Mortar strength is given in terms of compressive strength and is measured in psi (pounds per square inch) that the mortar is capable of supporting. The Brick Industry Association Technical Notes 8B, gives these recommendations for mortar selection.

- *Type N mortar.* General all-purpose mortar with good bonding capabilities and workability.
- *Type S mortar.* General all purpose mortar with high flexural bond strength.
- *Type M mortar.* High compressive strength mortar but not very workable.
- *Type O mortar.* Low strength mortar, used mostly for interior applications and restoration.

In addition to compressive strength, water retention, workability, and elasticity, the tensile strength and flexural strength for walls depend partly on the type of mortar used. **Tensile strength** is the measure of a material's capability to withstand stretching. **Flexural strength** is the measure of a material's capability to withstand bending.

According to The Brick Industry Association Technical Notes 8B, "no single mortar is best for all purposes. Select a mortar Type with the lowest compressive strength meeting the project requirements."

The selection of a type of cementitious material varies from one job to another. In most cases, the type of cementitious materials, whether categorized as masonry cements, mortar cements, or cement-lime mixes, depends on the use of the masonry materials rather than the type of masonry materials being used. Exceptions include glass block mortar, specifically manufactured for the laying of glass block, and fire clay, used for the laying of firebrick. Consult governing regulations and project specifications when in doubt as to the type of cementitious material required.

## Admixtures

**Admixtures** are materials added to masonry cements for specific purposes. Some admixtures can be harmful to the mortar and the finished wall. Nothing should be added to the masonry cement unless the effect on the mortar, the masonry units, and embedded items such as wall ties and joint reinforcement are known. For these reasons, the manufacturers should determine admixtures under controlled plant conditions. Types of admixtures include accelerators, retarders, pigments, colored aggregates, and plasticizers.

## Accelerators

Accelerators are admixtures that increase the cementitious material's rate of hydration. Accelerators shorten the time required for the mortar to set or harden. Manufacturers may include accelerators in cementitious materials intended for use in cold weather. The use of some accelerators may be reason for concern. Admixtures containing calcium chloride should not be used as accelerators where there is a potential for corrosion of metal wall ties and joint reinforcement. Air-entraining admixtures improve the resistance of the mortar to freeze-thaw deterioration but reduce the compressive strength and the bond strength of the mortar.

## Retarders

Retarders are admixtures used to delay the mortar's rate of hydration. Manufacturers may include retarders in cementitious materials intended for use in hot weather. Admixtures allow some mortars properly stored to be used several hours after mixing. This permits the mixing of mortar away from a jobsite when construction space is at a minimum.

## Pigments

Pigments are insoluble fine powders added to cementitious materials to obtain a desired mortar color. One group of pigments includes metallic oxide powders. The two types of metallic oxides are natural iron oxides and synthetic iron oxides. Natural iron oxides occur naturally in nature and are mined. Natural iron oxides used as pigments are more likely to result in slight variations in color. Synthetic iron oxides are produced from controlled laboratory processes. Pigments made from synthetic iron oxides are used in masonry cements because they provide more consistent colors. Synthetic iron oxide pigments include iron, manganese, and chromium oxides. Carbon black and ultramarine blue are used as pigments to create various mortar colors. Organic pigments weaken the mortar and are not used. Manufacturers limit the concentration of each pigment so that the strength and durability of the mortar is not sacrificed. For example, the maximum quantity of most metallic oxide pigments is limited to 10% of the cement content by weight, whereas carbon black is limited to 2%. Color uniformity of the finished mortar joints is maintained by 1) keeping the amount of mixing water consistent, 2) using the same amount and source of sand when mixing, 3) maintaining the same moisture content of the brick when laid, 4) striking or tooling the joints when the mortar is at the same consistency, and 5) using approved and uniform cleaning techniques.

Colored cementitious materials are available in Types M, S, and N. Colored cementitious materials should be mill mixed and prepackaged before shipment under controlled plant conditions to assure that the pigments and their proportions conform to industry requirements of ASTM C979.

## Plasticizers

Plasticizers are admixtures that result in mortars that are easier to spread and that adhere to the trowel and masonry units better.

## Sand

Sand is added to cementitious materials for making masonry mortar. Sand helps to control shrinkage. Most often, quartz (silicon oxide), or silica (silicon dioxide) are the chosen sands. Sand may be used in its natural state or manufactured by crushing. Particles of natural sand are round. Crushing sandstone, limestone, or other aggregates results in particles that feel sharp or angular. Ground limestone, ground granite, or ground marble are used as aggregates for white mortar. Colored aggregates, including white sand, ground granite, marble, or stone, are preferred when the desired mortar color can be obtained because the aggregates do not weaken the mortar, and the color is permanent. Acquire sand from a reputable supplier of masonry sand. Sand temperatures are to be between 40 and 100°F for making mortar.

**FROM EXPERIENCE**

The sand source should not be too dry. There should be sufficient moisture in the sand to clump when squeezed in the hand. Inaccurate volume measurements result when sand is too dry or too wet.

## Water

Water is added to the cementitious material-sand mixture to make mortar. The water must be clean. Water containing acid, salt, silt, or other organic matter weakens the mortar. Water temperature is to be between 40 and 120°F.

## Standard Proportions for Mortar Mixtures

Volumes or weights proportion masonry mortar ingredients. According to the Brick Industry Association Technical Notes 8, Table 1, "Proportion Specification Requirements," mortar for exterior and interior walls should be composed of 1 part cementitious material and no less than 2¼ or more than 3 parts of clean, well-graded aggregates, that is, sand. Graduation limits for sand are given in ASTM C 144, "Specification for Aggregates for Masonry Mortar." The maximum bond strength of mortar is achieved by mixing the ingredients with the maximum amount of water while maintaining workability. Water should be clean and free of acids, alkalies, or organic material. Using clean drinking water is a safe solution for the fitness of the water.

**CAUTION**

**Wear eye protection, hand protection, and a particulate respirator mask when exposed to sand and cements. Cement dust and sand can injure the eyes and skin, and breathing air contaminated with sand and cement can lead to silicosis, an occupational disease impairing the lungs.**

The procedures differ for manual mixing and machine mixing. A mortar box and a mortar hoe are needed for manually mixing mortar. A contractor's 6-cubic feet wheelbarrow can be used when mixing mortar. However, there is room for mixing only ½ of a cubic foot bag of cementitious material, sand, and water when using the wheelbarrow. Flush the mortar box or wheelbarrow with cool water before adding the ingredients. Place the sand in the mortar box or wheelbarrow, followed by the cementitious material. Use a chopping action with the mortar hoe, immediately begin mixing the cement and sand so that the light, powdery cement does not become airborne. Mix the sand and cement until the two ingredients are well blended. Add water to the cement and sand mixture, and mix thoroughly using a chopping action with the hoe (see Figure 8-3).

**Figure 8-3** Mortar is mixed in a mortar box using a long-handled hoe.

**Figure 8-4** This mixer has the capacity to hold a 1-cubic foot bag of masonry cement, 3 cubic feet of sand, and sufficient water to thoroughly mix the ingredients to make mortar.

If using a powered mortar mixer, put water in the mixing drum first. The water acts as a lubricant and reduces the friction between the rotating paddles and the stationary drum. With the mixer paddles rotating, add approximately one-half of the total volume of sand and then the cementitious materials. A typical mixer used for residential construction has the capacity to mix a 1-cubic foot bag of cementitious materials and the required sand and water (see Figure 8-4). Finally, add the remaining volume of sand and more water if needed. Allow the ingredients to mix for a minimum of 3 minutes but not more than 5 minutes after adding all ingredients.

The silo mixer stores, blends, and mixes mortar, ensuring that specified proportions of mortar materials can be accurately maintained. The multi-compartment silo stores and protects the cementitious materials and sand separately. Silo mixers are delivered to the construction site filled with sand and cementitious materials specified by the customer. The silo is unloaded from the truck, raised upright, and ready to operate just by connecting the water supply and electricity to power the motor (see Figure 8-5). A simple push of a button begins the blending and mixing of the mortar, fed by an auger into a wheelbarrow or mortar box. The controls permit the operator to specify the strength type of mortar. There are many advantages to the silo mixing method.

- There is continuous quality control because materials are blended in specified proportions regardless of the quantity of mortar being mixed.
- Materials stored in the protective silo require a minimum of space when working in restricted or congested workspaces.
- The volume of sand is uniform because the moisture content is consistent.
- Optional heating elements permit cold weather operation.
- Colored mortars can be produced with excellent consistency.

**Figure 8-5** Sand and cement materials stored in the silo are mixed at the touch of a button, discharging mortar as it is mixed in any amount desired. *(Courtesy of Ennstone, Inc.)*

**Figure 8-6** Mortar is mixed with a minimum amount of water, just enough to permit it to retain its shape when formed into a ball, and allowed to prehydrate.

In the Brick Industry Association's Technical Notes on Brick Construction 8A, page 5, and Technical Notes on Brick Construction 8B, page 1, four rules are given for mortars:

1. Never use a mortar having a higher compressive strength than is required.
2. Retemper to restore consistency but use within 2.5 hours after initial mixing or within 1.5 hours when temperatures exceed 80°F.
3. Lay units within 1 minute of spreading mortar.
4. Comply with cold weather masonry construction for temperatures below 40°F.

# Retempering Mortar

**Retempering** or tempering mortar is the process of adding water to the mortar to replace water lost by evaporation. Putting less mortar on the mortarboard and covering the wheelbarrow or mortar box containing the mortar to reduce evaporation can reduce retempering. Retempering should be kept at a minimum because it reduces the compressive strength of the mortar; however, it is desirable to sacrifice some compressive strength in favor of improved bond strength resulting from retempering. Provided the mortar is used within the manufacturer's recommended timeframe after initial mixing, ranging from 90 minutes to 2.5 hours depending on manufacturer, retempering is an acceptable practice.

# Mortars for Historical Buildings

Mortar used for repairing masonry joints should be mixed differently from mortar used for new construction. Mix the cementitious material and sand before adding water. Then add just enough water to make a dough-like mixture that retains its shape when formed as a ball (see Figure 8-6). Let the mortar set for 1 to 2 hours. This allows the mortar to prehydrate, or chemically combine with the water. Finally, add enough water to produce a workable mortar. The mortar should have strength characteristics similar to that of the original mortar. Mortars having strength characteristics of those used for new construction, Types M, S, and N, are not recommended for repairing masonry joints. It is not the intention of this chapter to address masonry restoration, only new construction.

# Specialty Mortars

Specialty mortars are designed for specific materials or purposes, such as laying glass block, and structural tile, and setting granite or marble stone. The characteristics of these mortars may include high bond strength, excellent flexibility, excellent water and frost resistance, and excellent resistance to chemicals or impacts. Epoxies and latex additives account for the unique characteristics of most specialty mortars.

## Surface Bonding Cements

Surface bonding cements are a mixture of Portland cement, glass fibers, and specific chemicals mixed with water. They are applied manually with a cement finishing trowel or with power spray equipment to both sides of dry-stacked block walls (see Figure 8-7). Surface bonding cements give walls greater flexural and tensile strengths than masonry walls with mortared joints. These cements are also water-resistant, reducing the rain penetration of CMU walls.

### Building Surface Bonded CMU Walls

The first course of block for a wall is bedded in mortar or surface bonding cement. The remaining blocks are stacked

**Figure 8-7** A cement finishing trowel is used to manually apply surface bonding cement to both sides of this block wall.

dry. Either applying manually with a finishing trowel or with power spray equipment, both sides of the CMU wall are covered with a $\frac{1}{8}$"-thick structural coating. Finish decorative coatings as little as $\frac{1}{16}$" thick can be applied over the structural coatings. Entire sections of walls should be surfaced at one time to lessen the possibilities of cold joints, which are separation cracks resulting from applications at different times.

## CAUTION

**Wear eye protection, hand protection, and a particulate respirator mask when exposed to surface bonding cements. Cement dust, sand, and fiber glass can injure the eyes and skin, and breathing air contaminated with sand, cement, and fiber glass can lead to occupational diseases impairing the lungs.**

# Cold Weather Construction

Requirements for cold weather procedures must be followed to prevent damaging effects of freezing temperatures to the mortar. **Cold weather construction** is defined as construction executed when either air or material temperatures are below 40°F. Cold weather construction requires heating the water and sand to produce mortar temperatures between 40°F and 120°F. Walls under construction must be covered, and a heated enclosure is required at temperatures below 20°F. These and other cold weather construction recommendations are given in the Brick Industry Association Technical Notes 1.

# Hot Weather Construction

**Hot weather construction** is defined as construction executed when either air or material temperatures are above 100°F. At these temperatures, mortar sets rapidly due to water evaporation. Both the extent of bond and bond strength is reduced because a lack of sufficient water prevents cement hydration. To prevent compromising the strength of the masonry work, the Brick Industry Association Technical Notes 1 makes these recommendations:

1. Keep the materials cool during hot weather, storing them in a shaded area.
2. Wet the brick 3 to 24 hours before use if the absorption rate is too high.
3. Consider using mortars with a high lime content and high water retention.
4. Use mortar within 2 hours of initial mixing, and shade mortar from sunlight.
5. Sprinkle sand piles to keep sand damp.
6. Use cool water for mixing.
7. Flush mortar pans, wheelbarrows, and mixers with cool water before using.
8. Fog spray newly constructed walls three times a day for three days.
9. Use wind breaks and cover walls with weather-resistant covers at the end of the work day.
10. Use admixtures only if their full effect on the mortar is known.

# Applications and Types of Grout

**Grout** is a mixture of cement and aggregates to which sufficient water is added to produce a pouring consistency without separation of the ingredients. Grout is used to fill hollow masonry units or the cavities between walls of two or more wythes for the purpose of increasing the compressive, shear, and flexural strength of masonry walls.

# Standard Proportions for Grout Mixtures

There are two classifications of grout: fine grout and coarse grout. Noted by some standards, fine grout has fine aggregates up to $\frac{3}{8}$", and coarse grout has aggregates between $\frac{3}{8}$" and $\frac{1}{2}$". According to the Brick Industry Association Technical Notes 3A, the standard proportions of ingredients for fine grout is 1 part Portland cement or blended cement, 0 to $\frac{1}{10}$ part hydrated lime or lime putty, and $2\frac{1}{4}$ to 3 parts fine aggregate. Coarse grout has these same proportions, but in addition, it has 1 to 2 parts additional coarse aggregates.

## Mixing Grout

The ingredients are mixed with a maximum amount of water to allow the grout to flow without separating the ingredients. The recommended slump should be between 8" and 11". Grout can be mixed manually, but ready-mix trucks and powered mixers are mostly used. On some jobs, workers place grout in walls manually, pouring it from buckets. Grout pumps speed up the process, as grout is discharged through a hose.

## Placing Grout

Areas to be grouted must be free of mortar droppings. Dampening the space to be grouted prior to grouting can prevent grout shrinkage and cracking, two conditions resulting in weaker walls. Vertical bar reinforcement placed in the wall during construction should be free of mortar droppings if the grout is expected to bond to it.

Grout is placed either as low-lift or high-lift applications. Low-lift grouting is done as the wall is built, usually for narrow grout spaces such as cavities between two or more wythes. High-lift grouting is done after a wall is built to its story height or final height, whichever is less, usually for walls constructed of units with open cells such as CMU walls. High-lift grouting of CMU walls is popular because ready-mix trucks or powered grout mixers and pumps speed up the grouting process (see Figure 8-8).

When multiple lifts are poured, industry standards recommend stopping each grout lift halfway between the

**Figure 8-8** Lifted by a crane or forklift, this hydraulic-powered equipment has an extended auger tube and dispensing hose, making grout-pouring operations easier.

courses. This ensures that a "cold joint," possible separations between pours, does not align with courses. High-lift grouts should not be placed within three days of wall construction. Placing grout too soon can result in "blowout," in which mortar joints or even entire wall sections come apart.

The grout is consolidated to remove voids within the wall or cavity. Powered mechanical pencil vibrators, powered rebar shakers, or manually using sticks are three methods used to consolidate the grout.

## Summary

- It is said that mortar is perhaps the only building material manufactured at the construction site. The structural integrity of the walls depends on its quality. Only trained personnel should be entrusted with proportioning the ingredients and mixing the mortar.
- Mortar choices including PCL mortars, masonry cements, and mortar cements, complicated with mortar types M, S, N, O, and K, are issues requiring one to understand job specific industry recommendations and governing regulations before choosing any cement.

- Protecting materials from contaminates and the weather is vital. Covering sand to prevent excessive drying and contamination, elevating bagged cement on palettes and protecting it with waterproof covering, and providing clean water permits mixing mortars having the intended strengths and characteristics.
- Bedding the materials in mortar in a timely manner and assuring that both the materials and the air are within an acceptable temperature range are key factors for maximum bond strength.

- There is no one mortar type suitable for all jobs. The selected type of cement depends upon its intended use, as well as governing regulations.
- Never use a stronger mortar, or a mortar having a higher compressive strength, than is recommended. Mortars having greater Portland cement content have higher compressive strength ratings but are more likely to cause separation cracking between the masonry units and the mortar, resulting in wind-driven rain penetration of the wall.

# Review Questions

*Select the most appropriate answer.*

**1   Mortar is**

a. a bonding agent containing cementitious material, sand, and water

b. a structural element containing cement, gravel, sand, and water

c. a major ingredient of Portland cement

d. both a and b

**2   The ingredient accounting for the durability and high strength of today's mortar is**

a. calcium

b. lime

c. magnesium

d. Portland cement

**3   The ingredient accounting for the workability and water retention of mortar is**

a. calcium

b. lime

c. magnesium

d. Portland cement

**4   Small voids and hairline cracks in masonry mortar joints are sealed by a process called**

a. autogeneous healing

b. automatic sealing

c. joint fixation

d. mortar netting

**5   The resistance to separation of mortar and the masonry units is called**

a. bond strength

b. compressive strength

c. mortar strength

d. all of the above

**6   A measure of the ease with which mortar can be placed is called**

a. flowability

b. spreadability

c. workability

d. all of the above

**7   A measure of a mortar's capability to withstand stretching is called**

a. bond strength

b. compressive strength

c. flexural strength

d. tensile strength

**8   A measure of a mortar's capability to withstand bending is called**

a. bond strength

b. compressive strength

c. flexural strength

d. tensile strength

**9   The type of masonry cement recommended for construction below grade level such as foundation walls is**

a. Type S

b. Type N

c. Type O

d. both a and b

**10   A type of masonry cement often recommended for exterior walls above grade, including those exposed to severe weather is**

a. Type K

b. Type O

c. Type N

d. both a and b

11. **Higher levels of air in masonry cements**

    a. improve mortar's resistance to freeze-thaw deterioration

    b. reduce mortar's compressive strength

    c. reduce mortar's bond strength

    d. do all of the above

12. **Ingredients added to masonry cement for specific purposes are called**

    a. admixtures

    b. adjusting compounds

    c. optional compounds

    d. both a and b

13. **Masonry cements intended for use in cold weather contain ingredients to reduce the time required for the mortar to set called**

    a. accelerators

    b. pigments

    c. plasticizers

    d. retarders

14. **Masonry cements intended for use in hot weather contain ingredients to delay the time required for the mortar to set called**

    a. accelerators

    b. pigments

    c. plasticizers

    d. retarders

15. **Pigments used in masonry cements providing more consistent colors are made from**

    a. natural iron oxides

    b. organic pigments

    c. synthetic iron oxides

    d. both a and b

16. **Ingredients intended to make masonry cements easier to spread and better adhere to the trowel are called**

    a. accelerators

    b. pigments

    c. plasticizers

    d. retarders

17. **Sand is added to masonry cement to help control**

    a. color

    b. freezing

    c. expansion

    d. shrinkage

18. **A finely-ground aggregate used to produce white mortar is**

    a. granite

    b. limestone

    c. marble

    d. all of the above

19. **Mortar quality is reduced when the mixing water contains**

    a. acids

    b. salt

    c. silt

    d. all of the above

20. **An acceptable ratio for mixing mortar is**

    a. 1 part sand to 3 parts cementitious material

    b. 1 part sand to 2 parts cementitious material

    c. 3 parts sand to 1 part cementitious material

    d. 4 parts sand to 1 part cementitious material

21. **Retempering mortar can**

    a. reduce its compressive strength

    b. increase its bond strength

    c. decrease its bond strength

    d. do both a and b

22. **Mortar used for repointing or repairing mortar joints is**

    a. called prehydrated mortar

    b. mixed initially with a minimum of water

    c. allowed to set for 1 to 2 hours before adding all needed water

    d. all of the above

**23** Specialty mortars designed for laying glass block, tile, and marble contain

a. epoxies

b. latex

c. silicates

d. both a and b

**24** When referring to mixing or using mortar, cold temperatures are defined as either air or material temperatures below

a. 40°F

b. 32°F

c. 20°F

d. 0°F

**25** The chemical action of Portland cement combining with water to form a hardened, bonding agent for masonry units is called

a. calcification

b. hydration

c. tempering

d. all of the above

# Chapter 9 | Concrete Masonry Units

oncrete masonry units are often referred to as CMUs by architects and engineers. Masons and home builders frequently refer to them as cinder block, or simply block. The term block is used both as a singular and plural description. Concrete masonry units provide an economical structural masonry wall for residential, commercial, and industrial applications. Whether for constructing a single-home foundation, an anchored masonry wall facade, a multistory hotel, a shopping mall, or an industrial manufacturing facility, CMUs are a popular building material.

## OBJECTIVES

After completing this chapter, the student should be able to:

- Identify types of concrete masonry units.
- Identify the sizes of concrete masonry units.
- List the ingredients of concrete masonry units.

## Glossary of Terms

**anchored veneer** a single-wythe masonry wall, requiring vertical support and anchored to either a load-bearing or non-load-bearing wall, with an air space between it and the backup wall.

**architectural CMUs** structural concrete masonry units whose one or more faces are produced in a multitude of texture and color combinations, suitable for both interior and exterior walls.

**autoclaved CMUs** block cured in sealed steam kilns at steam pressures above 120 psi and a temperature of 350°F, which is harder and less effected by changes in weather than block cured in steam kilns at atmospheric pressures at lower temperatures.

**exposed aggregate CMUs** also called aggregate faced units, block having the natural texture of crushed stone or small pebbles embedded in one or more faces.

**fluted CMUs** also called ribbed or scored block, CMUs whose one or more faces have vertical, machine-molded grooves.

**glazed CMUs** block with a glazing compound permanently molded to one or more faces, becoming an integral part of the unit and providing an impervious finish highly resistant to staining, impact, and abrasion.

**ground face CMUs** concrete masonry units with faces ground smooth and highly polished, exposing the natural colors of the aggregates and reflecting light.

**heavyweight CMUs** block containing primarily Portland cement, crushed limestone, and sand.

**hollow unit** a CMU in which the cross-sectional area of the cells is more than 25% of the overall cross-sectional area of the unit.

**lightweight CMUs** block containing expanded shale, clay, and slate or pumice to reduce the weight of the block and improve its performance.

**solid unit** a CMU either having no cells or the cross-sectional area of the cells are 25% or less than the overall cross-sectional area of the unit.

**sound-absorbing CMUs** also called acoustical masonry units, structural concrete masonry units designed to reduce sound transmission.

**split-face CMUs** block produced by mechanically splitting multiple-molded units, having one or more rough, exposed aggregate, three-dimensional faces.

**stone-face CMUs** block having the natural look of hand-chiseled stone.

**structural load** the weight or pressure exerted on a structure, including the weight of the building materials, occupants, furnishings, wind, snow, and rain.

# Hollow and Solid CMUs

Concrete masonry units are a combination of Portland cement, water, and various fine aggregates, including sand, crushed limestone, and lightweight aggregates. Concrete masonry units, also called CMUs or block, are machine molded and cured in steam kilns with atmospheric pressures at temperatures ranging from 120°F to 180°F for periods up to 18 hours. An optional curing method is with autoclave kilns. Typical autoclave kiln-cured CMUs are cured in sealed steam kilns at steam pressures above 120 psi and a temperature of 350°F. Manufacturers claim **autoclaved block**, as they are called, to be harder and less affected by changes in weather. CMUs can be either hollow or solid units. All hollow units and some solid units contain two or three cavities known as cores or cells. A **hollow unit** is one in which the cross-sectional area of the cells is more than 25% of the overall cross-sectional area of the unit. A **solid unit** either has no cells, or the cross-sectional area of the cells are 25% or less than the overall cross-sectional area of the unit (see Figure 9-1).

CMUs are designed to support the structural loads of a building. A **structural load** is the weight or pressure exerted on a structure, including the weight of the building materials, occupants, furnishings, wind, snow, and rain.

Concrete masonry units are classified as either heavyweight CMUs or lightweight CMUs. CMUs are used for both **anchored veneer**, a single-wythe masonry wall, requiring vertical support and anchored to either a load-bearing or non-load-bearing wall, with an air space between it and the backup wall and through-wall applications.

---

**CAUTION**

To prevent back injuries, consider having help lifting CMUs weighing over 30 pounds when such units are repeatedly lifted for several hours during a work day. Wear eye protection to protect the eyes from small particles and hand protection to protect the skin.

---

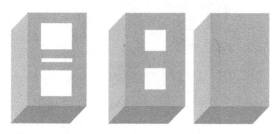

**Figure 9-1** The cross-sectional area of the unit on the left is less than 75% solid and is considered a *hollow* unit, but the center and right units are each considered *solid* units.

## Heavyweight CMUs

**Heavyweight** concrete masonry units contain Portland cement, crushed limestone, and sand. The ingredients are mixed with water and cast in molds. Aggregates used in some of the earlier blocks include cinders (a byproduct from burning coal) and slag (a byproduct of the steel industry). The term *cinder block*, which is still used today, was derived from the fact that cinders were used in block manufacturing. As the name implies, these block are heavier than lightweight block.

## Lightweight CMUs

In addition to Portland cement, crushed limestone, and sand, **lightweight** CMUs contain expanded shale, clay, and slate or pumice. Firing shale, clay, and slate in a rotary kiln at a temperature of approximately 2300°F will produce expanded shale, clay, and slate as the natural moisture within the materials reaches temperatures causing them to expand. The resulting lightweight aggregate permits the manufacture of CMUs that are superior to heavyweight block. Pumice is a naturally occurring lightweight aggregate used in the manufacture of lightweight CMUs.

There are several advantages to using lightweight units. First, walls built using lightweight CMUs are more economical than walls constructed with heavyweight units. Although lightweight CMUs cost more, construction costs are lower. Those using lightweight CMUs report increases in productivity as much as 30%. Faster construction means faster completion dates and therefore a savings on total construction costs. Furthermore, it is reported that lightweight block act as thermal insulators. The lightweight aggregates have high insulating properties. Gases are formed within the materials as they are fired in the rotary kiln, causing them to expand and creating millions of small air cells that act as an insulating medium. Finally, although lighter in weight, tests show that lightweight block are stronger than heavyweights and the expansion and contraction of walls caused by varying temperatures are less with walls built using lightweight block. Therefore, there is less stress on the wall and adjoining materials.

Block manufacturers develop their own blend of aggregates used in lightweight block. Crushed limestone is used in the manufacture of some lightweight block. Other manufactures report using only lightweight aggregates. Depending on the blend of ingredients, some units are referred to as medium weight block because they contain both crushed limestone and lightweight aggregates.

---

**CAUTION**

In addition to eye, face, and hearing protections, wear respiratory protection when sawing CMUs, for protection from breathing airborne dusts.

---

FROM EXPERIENCE

Have the suppliers of CMUs give printed specifications, including percentages of lightweight aggregates and block weight.

**Figure 9-3 Fluted block laid in the running bond create a wall with continuous vertical lines.**

# Architectural CMUs

Architectural CMUs are structural concrete masonry units whose one or more faces are produced in a multitude of texture and color combinations, suitable for both interior and exterior walls. With a variety of looks from natural quarried stone to that of polished granite, architectural CMUs are appreciated for enhancing the design and appearance of masonry structures. There are several categories of architectural CMUs. Unlike other masonry units, architectural CMUs can be manufactured with an integral water repellant system.

## Split Face CMUs

Split face CMUs are block having one or more rough, exposed aggregate, three-dimensional faces (see Figure 9-2). Split face CMUs are created by mechanically splitting multiple-molded units. Split face block are available in a variety of colors.

## Fluted CMUs

Fluted CMUs, also called ribbed or scored block, are block whose one or more faces have vertical, machine-molded grooves, permitting the faces of walls to exhibit continuous,

three-dimensional, vertical grooves (see Figure 9-3). The ribs or flutes can vary in number and can be rounded or squared. Fluted block are available in a variety of sizes and colors.

## Exposed Aggregate CMUs

Exposed aggregate CMUs, also called aggregate faced units, have the natural texture of crushed stone or small pebbles embedded in the facing of lightweight CMUs (see Figure 9-4). Different aggregates account for the large selection of colors.

## Ground Face CMUs

Ground face CMUs are concrete masonry units with faces ground and polished to expose the natural colors of the aggregates (see Figure 9-5). The surface can be finished with a coating to highlight the color. Although not an impervious surface, the low-maintenance, smooth finish is ideal for walls requiring to be sanitized.

**Figure 9-2 Split face block add dimension and texture while providing economy of construction.**

**Figure 9-4 Crushed stone are visible in these exposed aggregate CMUs.** *(Courtesy of Trenwyth Industries, Inc.)*

**Figure 9-5** The ground face units present a colorful appearance and smooth texture for this wall.

## Glazed CMUs

CMUs with a glazing compound permanently molded to one or more faces are called glazed CMUs. The glazing compound becomes an integral part of the unit and provides an impervious finish highly resistant to staining, impact, and abrasion (see Figure 9-6). The glazed faces can add ⅛" to the unit's dimensions, which allows for a ¼" joint using modular coursing.

**CAUTION**

**Protect the faces of both ground-faced and glazed CMUs from scratches and chips, damages that leave the units in disrepair.**

**Figure 9-6** A wide variety of colors make the glazed block a versatile masonry unit. *(Courtesy of Trenwyth Industries, Inc.)*

**Figure 9-7** The excellent sound-absorbing properties of these acoustical block improve the acoustical properties of rooms such as auditoriums and gymnasiums. *(Courtesy of Trenwyth Industries, Inc.)*

## Sound-Absorbing CMUs

Sound-absorbing CMUs, or acoustical masonry units, are structural concrete masonry units with closed tops and vertical slots in front of the cavities (see Figure 9-7). The cells can have factory-inserted insulating fillers to reduce sound transmission. Indoor applications include mechanical equipment rooms, noisy plant areas, gymnasiums, and auditoriums. As an example for exterior use, noninsulated units can be used to screen residential areas from expressway noise.

Sound-absorbing masonry units are also available either as glazed units or as ground face units.

## Stone face CMUs

Stone face masonry units are concrete masonry units that have the natural look of stone (see Figure 9-8). They are available in a variety of colors, shapes, and sizes. These attractive architectural CMUs are used either as anchored veneer or through-wall application.

 **FROM EXPERIENCE**

Consider using architectural CMUs for foundation walls. There are a variety of styles from which to choose to create a desired appearance when used as the interior or exterior finished face of foundation walls.

**Figure 9-8** Stone veneer masonry units give the appearance of natural stone. *(Courtesy of Trenwyth Industries, Inc.)*

# CMU Sizes and Dimensions

Standard CMUs measure 15⅝" long and 7⅝" tall. Including ⅜"-wide bed and head joints, each full unit laid in the wall measures 16" in length and 8" in height. The *size* of a block refers to the width of the block. The five standard, nominal or approximate sizes for referring to CMUs are 4", 6", 8", 10", and 12". The actual size or width of these CMUs is ⅜" less than the nominal or referred sizes. For example, an 8" CMU is actually 7⅝" wide.

Job specifications and local building codes dictate the block size to be used. For example, a 12" block may be required for below-ground foundation walls, whereas an 8" block may be specified for a one-story garage. In addition, there are many specialty block with dimensions other than the five mentioned.

The chart in Figure 9-9 is a guide for the application of block sizes. This data is for general guidance only. Prior to

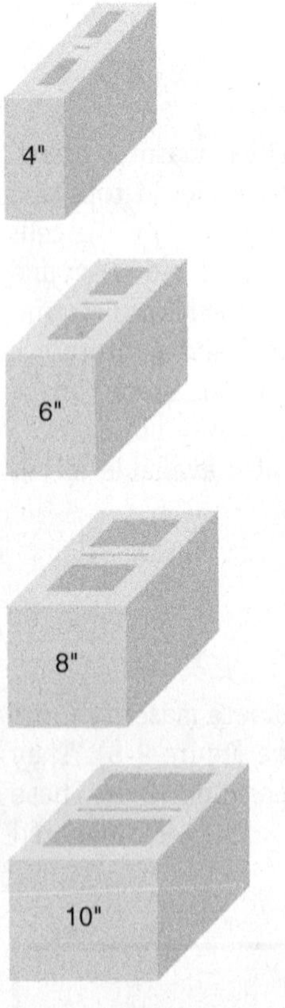

| | RESIDENTIAL APPLICATIONS | COMMERCIAL APPLICATIONS |
|---|---|---|
| 4" | COMPOSITE WALLS | COMPOSITE WALLS, COMPOSITE-CAVITY WALLS |
| 6" | NON-LOAD-BEARING PARTITION WALLS | NON-LOAD-BEARING PARTITION WALLS |
| 8" | GARAGES AND WORKSHOPS<br><br>FOUNDATION WALLS NOT EXCEEDING 4 FEET BELOW FINISHED GRADE | PARTITION WALLS<br><br>LOAD-BEARING WALLS |
| 10" | FOUNDATION WALLS NOT EXCEEDING 5 FEET BELOW FINISHED GRADE | PARTITION WALLS<br><br>LOAD-BEARING WALLS |
| 12" | FOUNDATION WALLS NOT EXCEEDING 6 FEET BELOW FINISHED GRADE | PARTITION WALLS<br><br>LOAD-BEARING WALLS |

**Figure 9-9** This data is for general guidance only. Prior to using this information, a registered professional engineer should be consulted to ensure appropriate sizes for a particular project. No warranties (expressed, implied, or statutory) are made in connection with this data.

using this information, a registered professional engineer should be consulted to ensure appropriate sizes for a particular project.

**FROM EXPERIENCE**

Bigger is better. Consider using a CMU that is a size larger than minimum requirements to give additional strength to a wall.

## Pattern Bonds

CMU walls are mostly laid in the running bond pattern. This half-lap arrangement provides the most even distribution of weight. However, a decorative stack bond pattern is sometimes used for walls. Stack bond pattern walls need to be reinforced with wire joint reinforcement between the courses. Wire joint reinforcement is also used to strengthen walls laid in the running bond pattern.

## Summary

- A wide variety of sizes, shapes, textures, and colors account for the popularity of using CMUs for both residential and commercial construction.
- Properly designed and reinforced CMU walls provide for economical construction of durable, low maintenance, and fireproof walls.
- A registered professional engineer and governing codes should be consulted to ensure that plans for using a particular CMU is appropriate for a given project.

# Review Questions

*Select the most appropriate answer.*

1. The hollow spaces in CMUs are called

   a. cavities
   b. cells
   c. cores
   d. holes

2. The percentage that the openings account for in a hollow CMU is

   a. more than 10%
   b. more than 25%
   c. less than 20%
   d. less than 10%

3. CMUs containing expanded shale and slate are called

   a. cinder block
   b. heavyweight block
   c. lightweight block
   d. both a and b

4. The weight of the building materials, occupants, furnishings, and snow are examples of

   a. dead loads
   b. live loads
   c. structural loads
   d. weight loads

5. The abbreviation for referring to masonry building block is

   a. CMU
   b. HWB
   c. LTB
   d. RBM

6. Structural CMUs that have color and/or texture are called

   a. appealing masonry units
   b. architectural masonry units
   c. designer masonry units
   d. engineered masonry units

7. CMUs that have polished, smooth faces to expose the natural colors of the aggregates are called

   a. designer faced
   b. ground faced
   c. rubble faced
   d. stone faced

8. The size of a CMU refers to its

   a. height
   b. length
   c. width
   d. volume

9. The length of a standard CMU is

   a. 12"          c. 15⅝"
   b. 15"          d. 16"

10. The height of a standard CMU is

    a. 7"          c. 8"
    b. 7⅝"         d. 8⅜"

11. The approximate size of a CMU is called its

    a. average size
    b. intended size
    c. nominal size
    d. proportioned size

12. The five standard sizes of CMUs are

    a. 2", 4", 6", 8", and 10"
    b. 3", 5", 7", 9", and 11"
    c. 4", 6", 8", 10", and 12"
    d. 5", 7", 9", 11", and 13"

13. The mortar joint width recommended for CMUs is

    a. ¼"          c. ⁷⁄₁₆"
    b. ⅜"          d. ½"

# Chapter 10

# Laying Block to the Line

Laying block to the line is similar to laying brick to the line. A tensioned mason's line attached to leads at opposite ends of the wall guides the alignment of block for constructing a wall. Much practice is required to become proficient using the line as a guide. Line work accounts for the majority of block being laid for foundation walls. Line work is also usually the first opportunity given apprentice masons to partake in actual wall construction.

## OBJECTIVES

After completing this chapter, the student should be able to:

- Lay out a block wall in the running bond pattern.
- Explain procedures for placing a cut block in a wall.
- List the four procedures performed for laying each block to the line.
- Demonstrate procedures for hanging a line and twigging a line.
- Lay block to the line in the running bond pattern.

# Glossary of Terms

**CMU** the abbreviation for concrete masonry unit, also called a block.

**face-shell spreading** a term describing the bedding of mortar along the tops of both face sides of a block.

**facing the block** aligning the bottom edge of each block's face with the top edge of the block faces on the course below it.

**hanging the line** attaching a tensioned line to the leads.

**twigging the line** clipping a twig on a mason's line and securing the twigged line to a block properly aligned.

This chapter addresses building CMU (concrete masonry unit) walls using the procedures known as laying block to the line. Before beginning, it is necessary to prepare the supporting surface, typically a concrete footing or concrete slab, and lay out the wall. Before bedding the first course of block in mortar, the surface should be free of soil, masonry dust, mortar droppings, and all other debris obstructing the process and having the potential to reduce the bond strength.

## Laying Out the Wall

The wall should be laid out before attaching a line to the leads and laying block to the line. Wall lay out involves the following:

- Marking the block spacing for the first course
- Determining the length of the cut block at each end of the wall where no lead exists

Unlike common building brick, there is little or no difference in the length of CMUs. The manufacturing processes and the materials account for their uniformity. Therefore, dry bonding as with common brick, spacing several brick to determine the bond, is not necessary, because the block do not vary in length. Spacing each block at 16" intervals permits a head joint width of $^3/_8$" between each block whose full length is $15^5/_8$". A course spacing height of 8" permits a $^3/_8$" bed joint width, since block height is $7^5/_8$".

Some walls have door openings beginning with the first course, requiring the block wall to eventually be bonded above the opening. For this reason, the block spacing for the first course is laid out in 16" increments as if there are no openings. Block are cut to length to fit each side of an opening without altering the 16" spacing established as the layout pattern.

***For step-by-step instructions on installing steel doorframes in CMU walls, see the Procedures section on pages 147–150.***

Designing CMU wall lengths in 8" increments minimizes cutting shorter length block at the end of walls. Fewer cuts mean greater productivity. However, shortening a block may be necessary for building a wall to its specified length. Where corners are built at the ends of walls, the block laid in the stretcher position nearest the end of the corner is cut shorter to permit a plumb bond (see Figure 10-1). Where a

**Figure 10-1 Cutting the block nearest the end of the wall accommodates wall lengths.**

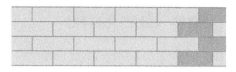

**Figure 10-2 The darker color block represent block that are shortened to accommodate wall length.**

jamb is built at a wall's end, the block at the end of the jamb is cut to length. Corners and jambs are topics of Chapter 11, "Constructing Block Leads."

***For step-by-step instructions on the placement and alignment of block cut to length, see the Procedures section on page 145.***

For each course, the block nearest the end is cut so that head joints align plumb on alternate courses to permit a half-lap running bond (see Figure 10-2).

Block require cutting to form vertical jambs adjacent to wall openings. The mason cuts block to allow a plumb bond alignment A masonry saw rather than a mason's hammer is recommended to shorten the length of a full-length block. Impacts from the hammer can reduce the block's structural integrity.

> **CAUTION**
>
> **Only those trained by a competent person should operate a masonry saw. Comply with all governing regulations. Wear eye protection, face protection, hearing protection, and respiratory protection when using a saw or when in the presence of sawing operations.**

## Hanging the Line

Attaching a tensioned mason's line to the leads is referred to as **hanging the line.** Attaching a mason's line to a lead with a line block, line pin, or line dog secures the line. These tools are identified in Chapter 3, "Masonry Hand Tools." The mason's line is observed for aligning each block level, plumb, and straight. Properly tensioning the line prevents it from sagging between leads further apart. Twigging the line on longer walls may be necessary to prevent the line from sagging between leads further apart or line movement caused by high winds.

> **CAUTION**
>
> **Always wear eye protection when stretching mason's lines to tension, working near or in the presence of tensioned lines, or removing tensioned lines from walls.**

# Twigging the Line

A twig, also called a trig, serves four purposes.

1. Although not always needed, a twig is sometimes used to properly position the mason's line at the lead.
2. A twig prevents a longer line from sagging lower near the middle than intended at the leads to which it is attached.
3. A twig minimizes line interference caused by more than one mason laying block simultaneously along the same line.
4. A twig prevents line movement caused by high winds.

Twigging the line involves clipping a twig on a mason's line and securing the twigged line above a block properly bedded in mortar. Weighting the twig with another block maintains the twig's position (see Figure 10-3).

**Remove twigs from lines before repositioning lines to successive courses, tensioning, or removing lines from walls.**

Using a builder's level or rotary laser level ensures that the twigged block is level with the leads at opposite ends of the wall. Confirming the twigged block to be aligned plumb

Figure 10-3 **The twigged line is secured above the block.**

with the block below it enables building a plumb wall. The author does not recommend sighting the length of the twigged line from one end to the other to confirm a straight and level line, a practice some masons are observed doing. Having your eye aligned with the mason's line is a risk for serious eye injuries and possible vision loss should the tensioned line break while sighting it.

**Only those trained by a competent person are to operate and adjust laser equipment. Where there is potential exposure to direct or reflected laser light greater than 5 milliwatts, antilaser eye protection meeting the intent of OSHA Standard 1926.54 Subpart E is required. Do not stare into the laser light. Post warnings in the area for the safety of others.**

# Laying Block to the Line in the Running Bond Pattern

Six procedures are recommended when laying block to the line, in the following order:

1. Spread the mortar bed joints.
2. Apply the mortar head joints.
3. Establish bond.
4. Face the block.
5. Align the block level.
6. Align the block plumb.

 **FROM EXPERIENCE**

Wearing rubber-coated, cotton-lined industrial work gloves or non-allergen synthetic rubber coated work gloves when handling block provides hand protection without sacrificing finger grip and dexterity, which is often associated with leather gloves.

## Spreading the Bed Joints

Ribbons of mortar are spread on the concrete footing or slabwork for the bed joints supporting the face shells of the first course (see Figure 10-4). The mortar should not be furrowed with the point of the trowel as is often done with the bed joints of brick walls. Furrowing the mortar displaces too

**Figure 10-4** *Ribbons* **of mortar support the first course.**

**Figure 10-5** Known as *face-shell spreading*, mortar is applied above the face shells.

much mortar below the face shells of the block. A better bond between mortar and block is achieved when the block are bedded in the mortar within one minute or so of spreading the mortar.

Sufficient mortar must be spread to completely bond to the shells of each block while providing for a ³/₈" thick bed joint. Job specifications may require a solid bed joint below the first course, necessitating placing mortar entirely below the block for its entire width. A solid bed joint can also be used where the footing is slightly lower and a wider bed joint is needed to align the block level. However, mortar bed joints should not be wider than recommended, typically ¹/₂". For all courses after the first course, mortar is spread on the top of the two face shells of each block. This is called **face-shell spreading.** Fully covering the outer shells' top surfaces with mortar improves bond strength (see Figure 10-5).

## Applying the Head Joints

Applying mortar to the end of a block forms head joints. Typically, a ribbon of mortar is applied to each side of the end of a block rather than a fully mortared joint between two block (see Figure 10-6).

## Establishing the Bond

Observing the 16"-interval spacing lines made when laying out the wall assures proper placement of each block along the first course. As each block is shoved into position adjoining

**Figure 10-6** Head joints are applied to the block's end in the same manner as spreading the bed joints.

the end of the previous one, the mortar is compacted between the units, forming well-bonded, water-resistant head joints. Since the last head joint of the course cannot be compacted otherwise, it should be filled solid from the top to insure a strong joint (see Figure 10-7).

For all courses after the first course, each block is centered above the two blocks on which it is laid. This procedure forms a vertical, plumb alignment with head joints and block two courses below it (see Figure 10-8).

**Figure 10-7** Filling the last head joint assures a strong joint.

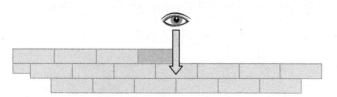

**Figure 10-8** Each block is centered above the two blocks on which it beds and aligned vertically plumb (arrow) with the block two courses below it.

**Figure 10-9** The block's webs are evenly spaced above the cells of the blocks below it.

 **FROM EXPERIENCE**

Observing the alignment of the block's cells and cross webs with the block on which it beds helps maintain a uniform bond. A uniform bond can be confirmed by having a block's middle cross web centered above the open cell formed by the ends of the two supporting block below it (see Figure 10-9).

## Facing the Block

**Facing the block** refers to aligning the bottom edge of each block's face flush with the top edge of the block faces on the course below it. Use the edge of the trowel blade to feel the alignment of the bottom of each block with those below it as protruding mortar is removed from the bed joint (see Figure 10-10). The alignment of the two courses is visually examined once the protruding mortar is removed.

**Figure 10-10** *Looking* at the alignment and *feeling* the alignment with the edge of the trowel as the excess mortar is removed accomplishes proper facing.

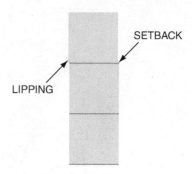

**Figure 10-11** Lipping and setback occur between the two courses when the block is not faced.

Because there is nothing below the first course to face these block, each block on the first course can be aligned plumb, using the mason's level.

Failure to face the block results in lipping and setback (see Figure 10-11).

## Aligning the Block Level

Align a block with the line by tapping the top of the block's length with the trowel (see Figure 10-12). A block is considered properly aligned when its top edge is level with the line. Reexamine the bottom edge of the block to ensure that it remains properly faced.

**Figure 10-12** Tapping the block along its top centerline with either the trowel blade or trowel handle *levels* the block.

**Figure 10-13 As indicated by the arrow, block are aligned uniformly within ¹⁄₁₆" and ¹⁄₈" to the line.**

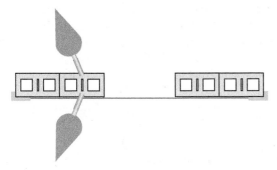

**Figure 10-14 The block is aligned *plumb* by tapping the top of the block near either edge (trowels), tilting the top of the block until a ¹⁄₁₆" to ¹⁄₈" spacing between the line and the block exists.**

## Aligning the Block Plumb

Maintain a uniform space no less than ¹⁄₁₆" or more than ¹⁄₈" between the line and the block's face. A width between the line and block that is equivalent to the thickness of the nickel coin is another suggestion (see Figure 10-13). The uniform spacing insures a wall aligned straight and plumb with the leads to which the line is attached.

Tapping the top of one of the two face shells of the block tilts the block closer to or farther from the mason's line, aligning it close to but not touching the line (see Figure 10-14). It is important to tap only near the middle of the block's length to maintain level alignment of the block with the mason's line.

 **FROM EXPERIENCE**

The edge of the trowel blade may chip the face of a block if it is used to tap along a block's face shells to align it plumb. Using the trowel handle, cushioned with a rubber chair tip, prevents chipping the block.

Facing the bottom of each block and aligning its top edge with the tensioned mason's line attached to accurate leads results in a level and plumb wall.

These procedures, face, level, and plumb, are conducted for each block as it is laid. These operations are necessary for building a plumb and level wall as block are laid to the line.

***For step-by-step instructions for laying block to the line, see the Procedures section on page 146.***

# Beginning Walls on Surfaces Not Level

There are times that concrete footings or slabs are found not to be level and the walls they support are required to be level. Yet the mason feels compelled to build a wall with structural performance that complies with engineered specifications and building code regulations. Unless the width of the bed joint can be adjusted within acceptable limits, the block should be cut with a saw to the height needed. Mortar joint widths no less than ¹⁄₄" or more than ¹⁄₂" are generally acceptable; however, one must comply with governing regulations and project specifications.

Where the footing is slightly lower than intended, a solid bed joint below the first course can help build up the height of the first course. Another procedure used is to slightly increase or decrease the bed joint width within acceptable limits for several courses. This alternative can eliminate cutting the blocks. For example, after six courses a wall constructed on a footing ³⁄₄" higher than intended can be at the required height without cutting any block by having bed joint widths of ¹⁄₄" rather than the usual ³⁄₈" width.

A masonry saw should be used to amend block height. Modifying the block for this purpose using the mason's hammer is not recommended. Using a hammer to cut the block's outer shells and cross webs can weaken the block. Sawing, rather than chipping with a hammer, preserves the structural reliability of the block.

**CAUTION**

**Only those trained by a competent person are to operate a masonry saw. Comply with all governing regulations. Wear eye protection, face protection, hearing protection, and respiratory protection when using a saw or in the presence of sawing operations. Electric saws must have inline ground fault protection to protect operators from electric shock and electrocution.**

# Wire Joint Reinforcement

Wire joint reinforcement is embedded in the horizontal bed joints of masonry to increase the compressive, lateral, and tensile strength of walls. There are two designs of prefabricated 10' length reinforcement: ladder-type and truss-type (see Figure 10-15).

Most all reinforcement is galvanized or zinc coated to protect it from corrosion. However, only hot-dipped galvanized reinforcement should be used on exterior walls where

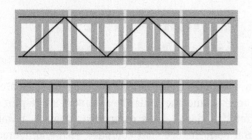

**Figure 10-15 Either truss-type (upper) or ladder-type (lower) wire joint reinforcement is embedded in the bed joints of CMU walls.**

maximum protection from corrosion is needed. Using uncoated or mill galvanized reinforcement for exterior walls results in excessive wire corrosion, leading to mortar joint failures.

One recommendation is to bed joint reinforcement above the first course and every other joint thereafter. Design considerations or code regulations may require a closer spacing. The width of the wire should be approximately 2" less than the nominal width of the masonry units. The ends of sections should be lapped 6".

**CAUTION**

**The exposed ends of wire reinforcement can impale workers, causing serious injuries. Wear safety glasses and gloves when working with it. Metal wire reinforcement is a conductor of electricity. Make certain that wire reinforcement can come no closer than 10' to overhead high voltage electrical lines as it is handled.**

## Dry-Stacked Surface-Bonded CMU Walls

Dry-stacked, surface-bonded CMU walls are an alternative design for building block walls where block courses are bed in mortar and mortar head joints bond adjacent block along each course. Dry stacked, surface-bonded CMU walls are built by stacking all but the first course of block dry, without mortar joints. The first course of block is bedded in mortar or surface bonding cement. Block are butted tightly against one another without applying mortared head joints. Either manually with a finishing trowel or with power spray equipment, both face sides of the CMU wall are covered with a 1/8"-thick structural coating. Finish decorative coatings as little as 1/16" thick can be applied over the structural coatings. Entire sections of walls should be surfaced at one time to lessen the possibilities of cold joints, which are separation cracks resulting from applications at different times.

**CAUTION**

**Wear eye protection, hand protection, respiratory protection, and protective clothing when working with surface-bonding cements.**

## Summary

- It should become a top priority for every apprentice mason to master the skills of laying block to the line.
- With each block laid to the line, an apprentice should 1) observe the pattern bond, 2) face the bottom of the block, 3) align it level with the line, and 4) align it plumb. Mastering these skills enables an apprentice mason to partake in wall construction, supervised by experienced masons building the leads and performing more difficult tasks.

## Procedures

# The Placement and Alignment of Block Cut to Length

**A** For illustration purposes, a wall is laid out to be 76" long. This requires cutting a block at the end of each course of the wall to permit the desired length. Each of the cut block (darker block) on the first and third courses measures 12", and the cut block on the second and fourth courses measures 4".

**A**

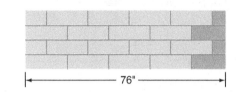

76"

**B** Courses 1 and 3 for the illustrated wall require a block length of 12" (arrow) at one end of the wall.

**B**

**C** Courses 2 and 4 for the illustrated wall require a block length of 4" (arrow) at one end of the wall.

**C**

## Procedures

## Laying Block to the Line

**A** Center the block above the two blocks below it.

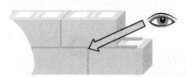

**B** **Face** or align the bottom edge of the block flush with the top of the wall below.

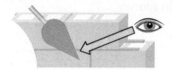

**C** **Level** the top of the block with the line by tapping either or both ends.

**D** **Plumb** the block by tapping near the center of the top of one of the two face shells, tilting the top until a ¹⁄₁₆" to ¹⁄₈" space, the thickness of a nickel, exists between the line and the face of the block. It is important to tap only near the middle of the block's length to maintain level alignment of the block with the mason's line.

## Procedures

# Installing Steel Doorframes for CMU Walls

*The use of steel doorframes anchored to masonry walls is a widespread practice in commercial construction. Although it may not be the masons' responsibility to set frames, masons are responsible for accurately anchoring the frames to the walls they construct. Masons must ensure that frames are plumb, square, and aligned with the walls. The two types of steel frames are the wrap-around frame and the butt-type frame.*

## Wrap-Around Frame

**A** The throat opening of the wrap-around frame is slightly wider than the width of the block size used for the wall. The block is fitted between the backbends of the frame and into the frame's throat, which is grouted with mortar to strengthen the frame.

**A**

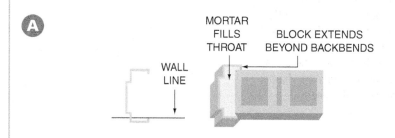

## Butt-Type Frame

**B** The width of the butt-type frame is equivalent to the block size. The block butts the backbends of the frame. The frame's throat is grouted with mortar for added strength.

**B**

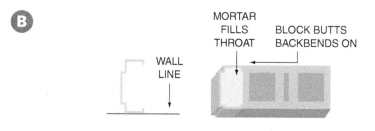

## Procedures    Installing Steel Doorframes for CMU Walls (continued)

**C** Base anchor clips fit tightly in the frame throat between the backbend returns and the front rabbets. Hardware secures the base clips to the doorframe and surface below it. Either wire anchors or T-anchors are installed at specified intervals between block courses on both sides to secure doorframes to masonry walls. Grouting the frame throat and block cells with mortar ensures good anchorage. CAUTION! Temporary stiffening braces must be placed in the opening of doorframes to prevent the pressure of the grout from pushing frames out of straight alignment.

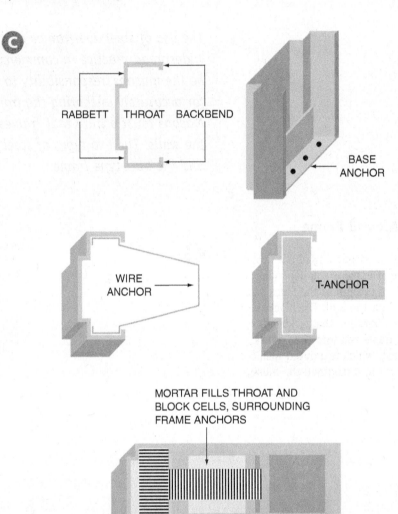

RABBETT   THROAT   BACKBEND

BASE ANCHOR

WIRE ANCHOR

T-ANCHOR

MORTAR FILLS THROAT AND BLOCK CELLS, SURROUNDING FRAME ANCHORS

**D** Wrap-around doorframe backbends are aligned to chalk line.
Base clips anchor the bottom of the frame on each side.
Well-fastened bracing from the door head to the ground ensures that the frame remains aligned plumb until anchored to a masonry wall.
Bracing at the bottom and middle keep the doorframe aligned straight.

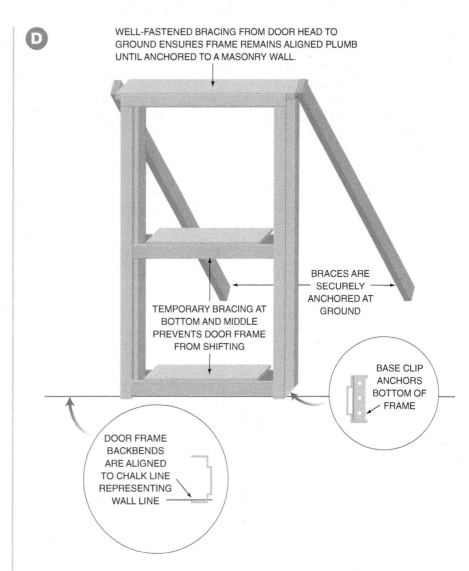

**D**

WELL-FASTENED BRACING FROM DOOR HEAD TO GROUND ENSURES FRAME REMAINS ALIGNED PLUMB UNTIL ANCHORED TO A MASONRY WALL.

TEMPORARY BRACING AT BOTTOM AND MIDDLE PREVENTS DOOR FRAME FROM SHIFTING

BRACES ARE SECURELY ANCHORED AT GROUND

BASE CLIP ANCHORS BOTTOM OF FRAME

DOOR FRAME BACKBENDS ARE ALIGNED TO CHALK LINE REPRESENTING WALL LINE

## Procedures

# Installing Steel Doorframes for CMU Walls (continued)

**E** Blocks are laid out on 16" spacing to include a ³⁄₈" head joint. The length of the cut block at each side of the doorframe depends on the width and location of the frame.

**E**

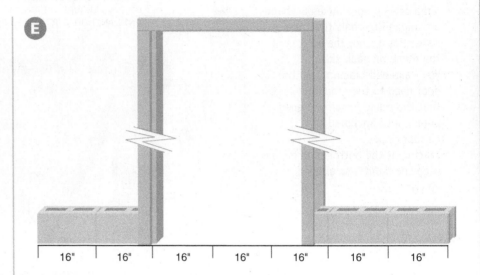

| 16" | 16" | 16" | 16" | 16" | 16" | 16" |

**F** Once above the frame, blocks create a half-lap bond with those on either side of the frame below. Bond-beam CMUs filled with concrete and rebar or a masonry lintel supports masonry above the wrap-around doorframe. A temporary center support is placed below the head until grout, mortar, and concrete harden.

**F**

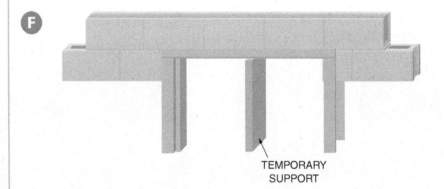

TEMPORARY
SUPPORT

# Review Questions

*Select the most appropriate answer.*

**1** The combined length of a block and a head joint is

a. 15 3/8"

b. 15 5/8"

c. 16"

d. 16 3/8"

**2** The combined height of a block and a bed joint is

a. 7 3/8"

b. 7 5/8"

c. 8"

d. 8 3/8"

**3** Applying mortar along each face side of the top of block without applying mortar to the cross webs is called

a. edge bedding

b. face-shell spreading

c. surface bonding

d. web coating

**4** Aligning the bottom edge of each block as it is laid with the top edges of the block below it is called

a. centering the block

b. facing the block

c. gauging the block

d. lipping the block

**5** Block should be aligned level by tapping the top near its

a. back edge

b. center line

c. front edge

d. face

**6** For building a plumb wall, a uniform spacing should be maintained between each block and the mason's line equivalent to

a. 1/16" to 1/8"

b. the thickness of a U.S. nickel coin

c. 1/2"

d. both a and b

**7** The size for mortar joints should be

a. no less than 1/4" or more than 1/2"

b. no less than 1/4" or more than 3/4"

c. no less than 3/8" or more than 3/4"

d. no less than 1/2" or more than 3/4"

**8** To preserve their structural integrity, block should be cut with

a. a chisel and hammer

b. a mason's hammer

c. a masonry saw

d. all of the above

**9** The mason's line is used to align block in a wall to be

a. level

b. plumb

c. straight

d. all of the above

**10** A device clipped to the line to position it as needed is called

a. a trig

b. a twig

c. a line clip

d. both a and b

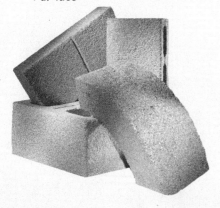

# Chapter 11 | Constructing Block Leads

As with brick leads, block leads are constructed at the ends of walls, establishing both the length and height for the walls. The accuracy of the finished walls is dependent upon the construction of the leads. Only those demonstrating the skills for aligning masonry units with the mason's level can be entrusted to build the leads. Learning the necessary procedures early in your training and practicing the procedures frequently will enable you to become competent at building leads.

## OBJECTIVES

After completing this chapter, the student should be able to:

- Lay out and construct block corners and jambs.
- Identify the special offset corner blocks and demonstrate their installations.
- Demonstrate the proper alignment for block cut to length at the end of a lead.

## Glossary of Terms

**block jamb** a lead permitting block work along a wall in only one direction.

**block size** the nominal or approximate width of a block.

**checking the range** confirming the straight alignment of a wall.

**corner** two walls connected at ends, typically forming a right angle.

**corner of the lead** that part of the corner where the two walls connect.

**lead** that part of the wall first constructed, becoming a guide for building the remaining wall.

**nominal size** the expressed size of material, the approximate width of a block.

**rack of the lead** the block alignment at the tail end of the courses on the lead.

**tail of the lead** the stepped-back end of corners and jambs.

**toothing** continuing with courses on a lead where the tail of the lead normally limits additional courses by temporarily laying a half-block on the stretcher below.

# Laying Out the Wall

Constructing leads requires mastering the procedures for using the mason's level to align masonry units level, plumb, and straight while building to a specific height. Wall layout involves the following:

- Marking the block spacing for the first course
- Determining the length of a cut block if needed at the end of the wall
- Scaling for maintaining uniform course spacing

Unlike common building brick, there is little or no difference in the length of concrete masonry units. The manufacturing processes and the materials account for their uniformity. Therefore, dry bonding, as with brick, to acquire a representative spacing is not necessary because the block do not vary in length. Spacing each block at 16" intervals permits a head joint width of ³/₈". A course spacing of 8" permits a ³/₈" bed joint width.

Designing CMU wall lengths in 8" increments minimizes cutting block at the end of walls. Fewer cuts mean greater productivity. Shortening a block may be necessary for building a wall to its intended length. The block laid in the stretcher position nearest the end of the wall is cut to permit a plumb bond.

*For step-by-step instructions on laying out a block wall and corner requiring a block to be cut, see the Procedures section on page 162.*

A lead can be either a corner or a jamb. A **corner** consists of two adjoining walls, typically at a 90-degree angle (see Figure 11-1).

A **jamb** is a lead that allows block to be laid along a wall in only one direction. Examples of jambs are found at door and window openings (see Figure 11-2).

There is more than one procedure for constructing leads. The procedures presented in this chapter are just one of many that are used to accurately construct leads.

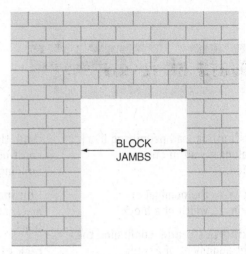

Figure 11-2 **The block jamb creates the opening in the wall.**

# Building a Block Corner

The entire lengths of the walls are laid out before building the corners. The number of block laid for a corner's first course determines the maximum number of courses for the corner. Having no more than a difference of one block on one side of a corner than the other, the maximum number of courses equals the total number of block on the corner's first course (see Figure 11-3). For example, having three block on each side of the corner's first course, a total of six block, creates a corner with six courses.

*For step-by-step instructions for laying out a block wall, see the Procedures section on page 162.*

A corner with three block on one side and four on the other side of the first course can be built seven courses in height. But a corner with three block on one side and five block on the opposite side is still limited to seven courses in height. Using a 4' mason's level to align the block all together limits either side of the first course to no more than three block. This permits a six-course corner with a 48" height. After the leads are constructed to a 48" height, a mason's line is attached to them, and the wall between opposite corners is laid to the line. These two procedures,

Figure 11-1 **A block corner establishes the end of the wall.**

Figure 11-3 **Six block on the first course of this corner permits building six courses.**

constructing leads and laying to the line, are repeated until the finished wall height is achieved.

**FROM EXPERIENCE**

Wearing rubber-coated, cotton-lined industrial work gloves or nonallergen synthetic rubber-coated work gloves when handling block provides hand protection without sacrificing finger grip and dexterity, which is often associated with leather gloves.

A lead's first course becomes a reference for constructing the remaining courses. It must be laid out, aligned, and constructed accurately. A better bond between mortar and block is achieved when the block are bedded in the mortar within one minute or so of spreading the mortar. Therefore, it is better to lay and align block for one side of a corner before spreading mortar for the opposite side.

The vertically aligned edge of the corner is called the **corner of the lead,** and the stepped back opposite end is called the **tail of the lead** (see Figure 11-4). Working from the corner of the lead toward the tail of the lead assures exact placement of the first block at the corner's edge.

Laying the block at the corner of the lead first necessitates walking forward on one side of the corner and stepping backward on the opposite side. Keeping the work area unobstructed prevents tripping when stepping backward.

Beginning with the block at the corner of the lead, all of the block for one side of the corner are laid and aligned level, plumb, and straight before beginning the opposite side.

**FROM EXPERIENCE**

Remembering the order of operations, "level, plumb, and straight" is easy if you do them alphabetically, that is, level (l) before plumb (p) and plumb (p) before straight (s).

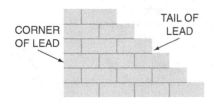

**Figure 11-4 The corner of the lead establishes the end of the wall. Block eventually laid to the line complete the wall at the tail of the lead.**

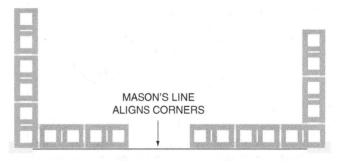

MASON'S LINE
ALIGNS CORNERS

**Figure 11-5 The corners are aligned straight using the mason's line.**

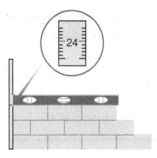

**Figure 11-6 Block are aligned level and a height of 24" is observed for a three-course lead with ³/₈"-wide bed joints.**

Not only must the block be laid in a straight line, but also they must be aligned with the opposite end of the wall. A chalk line or mason's line is used to align the wall straight (see Figure 11-5).

Aligning a wall accurately using the mason's line is known as **checking the range.**

The standard course spacing for CMU walls is 8". The block are aligned level, and the course height is checked at the corner of the lead (see Figure 11-6).

It may be necessary to increase or decrease the course spacing for a lead. This permits eventually aligning walls level from end to end where the concrete footing or slab upon which the leads are supported are not level. However, bed joint widths should remain within the tolerances specified, typically no less than ¹/₄" or more than ¹/₂".

**FROM EXPERIENCE**

Placing a rubber chair tip over the end of the trowel handle cushions the trowel handle when tapping block with the handle's end, preventing damaged handles and reducing hand fatigue.

A plumb alignment is confirmed for each block on the first course by observing the plumb vial. Both the end and the face of the block at the corner of the lead is aligned plumb (see Figure 11-7).

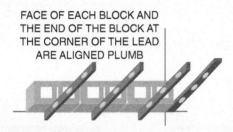

**Figure 11-7** Positioning the mason's level vertically against the face of each block on the first course permits creating a plumb alignment.

### FROM EXPERIENCE

The edge of the trowel blade may chip the face of a block if it is used to tap along a block's face shells to align it plumb. Using the trowel handle, cushioned with a rubber chair tip, prevents chipping the block.

The edge of the mason's level is then used to align straight the course of block (see Figure 11-8).

Checking the range again to affirm accurate alignment of this side of the corner with the lead at the opposite end of the wall is recommended (see Figure 11-9).

The opposite side of the corner is now built. A bed joint is spread, and block are bedded in the mortar and aligned level, plumb, and straight using the same procedures as for the first side (see Figure 11-10).

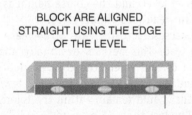

**Figure 11-8** Aligning the top edge of each block with the level's edge arranges the block in a straight line.

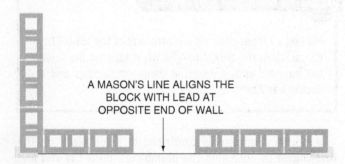

**Figure 11-9** The new lead (right) is aligned with the left end of the wall.

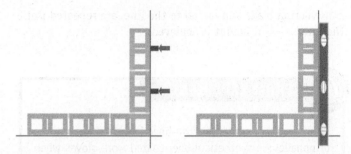

**Figure 11-10** After aligning level, the face of each block for the opposite side of the corner are aligned plumb (arrows) and then aligned straight (right).

The alignment for one side of a corner is completed before continuing with the opposite side. The side of the corner for which the length of the block at the corner of the lead is oriented is completed first (see Figure 11-11).

Building a well-aligned corner requires **facing the block,** which means aligning the bottom edge of each block as it is laid with the top edge of the block below it.

Facing each block with the top edges of those aligned below requires minimal repositioning of the block as they are aligned plumb, improving bond strength and production.

Above the first course, facing the block assures a plumb alignment of the lead once the top of the faces of all block is confirmed to be in straight alignment with the corner of the lead and the tail of the lead aligned plumb using the mason's level (see Figure 11-12).

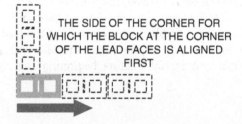

**Figure 11-11** The orientation of the length of the block at the corner of the lead determines the side of the corner first aligned.

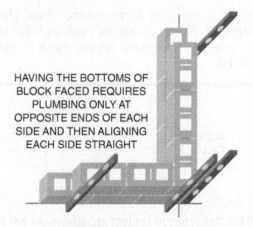

**Figure 11-12** Aligning the block with the two ends creates a plumb course.

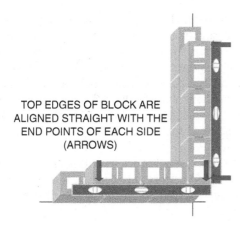

TOP EDGES OF BLOCK ARE
ALIGNED STRAIGHT WITH THE
END POINTS OF EACH SIDE
(ARROWS)

**Figure 11-13** The edge of the mason's level aligns each side of the corner straight with the plumb points established in the previous procedure.

Once aligned plumb, these end points are referenced for aligning all block straight with the edge of the mason's level. Tilting the top edge of each block also aligns them plumb (see Figure 11-13).

The three operations, aligning level, plumb, and straight, are repeated for each side of the remaining courses. Joints are tooled as they become thumb-print hard. Applying too much pressure on the joints can alter the wall alignment.

Upon completion, the accuracy of the work is carefully examined, giving special attention to the plumb alignments both at the corner and tail of the lead of each course and the straight alignments of the units between the corner of the lead and the tail of the lead on each side.

Observing the rack of the lead can indicate inaccuracies of lead construction. The **rack of the lead** is the block alignment at the tail end of the courses. The rack of the lead is observed by holding the edge of the level diagonally with the top edge of each block at the tail end of each course (see Figure 11-14).

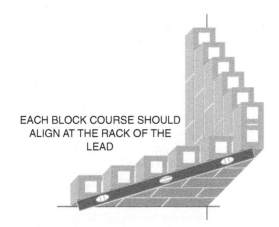

EACH BLOCK COURSE SHOULD
ALIGN AT THE RACK OF THE
LEAD

**Figure 11-14** The alignment of the top corner edge of the block at the tail of the lead of every course is examined for straight alignment. This is known as checking the rack of the lead.

If the rack of the lead is not uniform, careful examination of the corner should reveal inaccurate alignments. Realigning the corner level, plumb, and straight may require relaying the block to assure a strong mortar bond.

The materials and manufacturing processes produce square block with very accurate dimensions. Because of their close tolerances, aligning block on the outside of the corner results in accurate block alignment on the inside of the corner.

**FROM EXPERIENCE**

For interior walls, a "single bullnose" block, a block with one corner that has a 1" radius, rounded exposed corner, is recommended and may be required at the corner edge of CMU walls. The rounded corner is intended 1) for safety should one collide into the corner of walls, 2) to minimize chipping should something strike the corner, and 3) for aesthetics.

## Building a Block Jamb

A block jamb is built similar to a block corner. A block jamb is usually built next to a window or door opening, or at the end of a wall partitioning a larger area. With bond permitting, the block at the corner of the lead alternates between courses as being a full block or half-block (see Figure 11-15).

The same order of procedures is observed for building a block jamb as for either side of a block corner:

1. Align level and establish height.
2. Align plumb.
3. Align straight.

Where the end of the jamb is exposed, such as at window and door openings, it is as important to have the end of the jamb plumb as it is to have its face plumb. Where a jamb is butted at a right angle against another wall, plumbing the end of the block is not required. A joint separates the two perpendicular-built walls.

A block jamb is found at either side of an opening such as doors and windows. At locations such as these where block are below or above the opening, a uniform bond should be

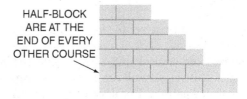

HALF-BLOCK
ARE AT THE
END OF EVERY
OTHER COURSE

**Figure 11-15** A half-block begins every other course of this block jamb.

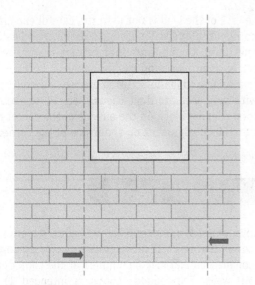

Figure 11-16 Each jamb on both sides of the window require cutting the block differently to maintain a plumb bond, joints of every other course aligned vertically.

continued at each jamb. This can necessitate cutting the end block for each course (see Figure 11-16).

Jambs such as these illustrated in Figure 11-16 are usually constructed as block are laid to the line on a longer wall during wall construction. The wall's face is aligned using a mason's line attached to leads at opposite ends of the wall. The jambs or block ends are aligned plumb using the mason's level.

Applying excessive pressure against the jamb when aligning with the level or when tooling the joints can misalign the jamb.

### FROM EXPERIENCE

For interior walls, a "double bullnose" block, a block for which both corners at one end have a 1" radius, are recommended and may be required at the end of jambs. The rounded corners are intended 1) for safety should one collide into the end of the jamb, 2) to minimize chipping should something strike the end of the jamb, and 3) for aesthetics.

## Simplifying Designs

The size of a block refers to the nominal size, or approximate block width. Actual sizes for blocks are ³⁄₈" less than nominal sizes. A block and ³⁄₈" head joint combined is 16" in length. Most block walls are constructed using a half-lap, running bond pattern. This design requires block to have an 8" overlap with adjoining courses. Designing walls whose lengths are multiples of 8" simplifies wall construction.

Establishing the half-lap running bond pattern at the corners is desirable. Even for walls whose lengths are multiples of 8", constructing block corners with block other than the 8" block require special provisions. Only the 8" block allows constructing corners with a half-lap pattern without making provisions for the block's dimensions. All other block sizes require either a special offset corner block or cutting a block at the corner to permit a half-lap, running bond pattern.

For walls whose lengths are multiples of 8" and wherever else the proposed patterns are desired, the following procedures can be observed for building the different sizes of block corners.

## 4" Block Corners

Unless a special offset 4" corner block is available, two options are considered for building 4" block corners at the end of walls requiring a half-lap pattern. Cutting 4" from the length of the block at the corner creates a half-lap pattern (see Figure 11-17).

Another method permitting a half-lap pattern with 4" block requires the addition of a piece 3⁵⁄₈" long and a ³⁄₈" head joint adjoining a full block's width (see Figure 11-18).

Either the addition of 4" adjoining the block's end or the removal of 4" from the end block's length creates a half-lap for 4" block.

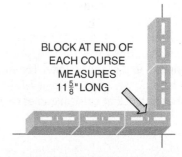

Figure 11-17 Cutting 4" from the length of the end block for every course creates a half-lap pattern for 4" block.

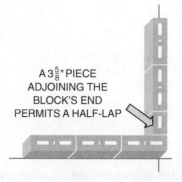

Figure 11-18 Having a 3⁵⁄₈" piece adjoining the block's end on every course creates a half-lap pattern. A ³⁄₈"-wide head joint between the two is necessary.

**FROM EXPERIENCE**

Use a masonry saw to create accurate, straight cuts instead of the mason's hammer, which can create fractures that weaken the block. A masonry saw should only be used by those complying with governing regulations, properly trained, and having personal protection, including respiratory, hearing, face, and eye protection.

# 6" Block Corners

Three options are available for building a 6" block corner and having a half-lap pattern. An offset, 6" corner block is designed to create a half-lap between courses (see Figure 11-19).

Cutting 2" from the length of the block at the corner creates a half-lap pattern (see Figure 11-20).

Another way to create a half-lap pattern with 6" block requires the addition of a piece $1\frac{5}{8}$" long and a $\frac{3}{8}$" head joint adjoining a full block's width (see Figure 11-21).

Either the addition of 2" adjoining the block's end or the removal of 2" from the end block's length creates a half-lap for a 6" block.

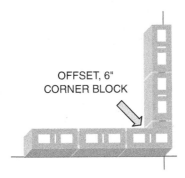

OFFSET, 6" CORNER BLOCK

**Figure 11-19** Using a specially designed offset corner block at the corner of the lead for each course establishes a half-lap pattern.

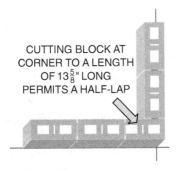

CUTTING BLOCK AT CORNER TO A LENGTH OF $13\frac{5}{8}$" LONG PERMITS A HALF-LAP

**Figure 11-20** Cutting 2" from the length of the end block for every course creates a half-lap pattern for 6" block.

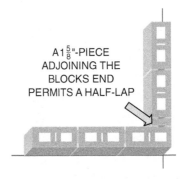

A $1\frac{5}{8}$"-PIECE ADJOINING THE BLOCKS END PERMITS A HALF-LAP

**Figure 11-21** Having a 1⅝" piece adjoining the block's end on every course creates a half-lap pattern. A ⅜"-wide head joint between the two is necessary.

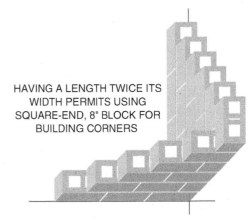

HAVING A LENGTH TWICE ITS WIDTH PERMITS USING SQUARE-END, 8" BLOCK FOR BUILDING CORNERS

**Figure 11-22** The dimensions of 8" block permit building corners with half-lap patterns.

# 8" Block Corners

The 8" block permits building corners with a half-lap, running bond pattern without any special offset corner blocks or cutting of the block. Simply orientating the length of a square-end block 90 degrees to the length of the corner block below it permits an 8" block corner with a half-lap pattern (see Figure 11-22).

# 10" Block Corners

A specially designed offset 10" corner block permits constructing 10" block corners with half-lap patterns. The end of the block intended to be oriented to the outside corner measures $7\frac{5}{8}$" wide, the same width as a standard 8" block. Laid at the very corner of the lead, one offset block is needed for each course of a corner (see Figure 11-23).

# 12" Block Corners

A specially designed offset 12" corner block permits constructing 12" block corners with half-lap patterns. As with the offset 10" corner block, the end of the 12" corner block

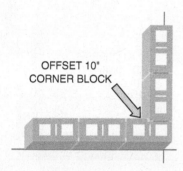

**Figure 11-23** The outside width of an offset 10" corner block is reduced to 7⅝" to create a half-lap pattern.

**Figure 11-24** The outside width of an offset 12" corner block is reduced to 7⅝" to create a half-lap pattern.

measures 7⅝" wide, the same width as a standard 8" block. Laid at the very corner of the lead, one offset block is needed for each course of a corner (see Figure 11-24).

## Aligning Cut Block

Frequently, the lengths of block walls are not multiples of 8". Such walls require cutting the blocks at the corners of the leads. A mason's hammer and chisel are often used to cut block manually. However, this practice can fracture the block, weakening its structural integrity. Using a power saw cuts block without sacrificing its structural integrity.

**CAUTION**

Wear protective equipment when using power saws, including eye, face, hearing, and respiratory protections. Only those trained by competent persons and complying with governing regulations should operate power saws. Electric saws must be provided with inline ground fault protection to protect the operator from electric shock and electrocution.

Placing the piece at the end of the wall rather than randomly along the length of the wall creates a uniform half-lap running bond. The piece is aligned vertically on alternating courses (see Figure 11-25).

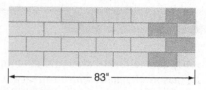

**Figure 11-25** Aligned vertically, the shaded block are cut to a length of 11", which creates a wall that is 83" long.

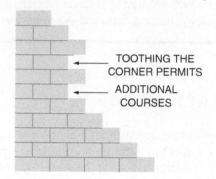

**Figure 11-26** A toothed corner is used to raise a corner above the courses limited by the racking of the tail end of the corner.

## Toothing Block Leads

The racking back of block at the tail of a lead limits the courses to which it can be built. Toothing permits completing a lead to any height before the adjoining walls are built. Toothing is accompanied by temporarily laying a block piece on the block below so that the lead can continue for another course (see Figure 11-26).

There are concerns for toothing a corner. First, the temporary block supports must be carefully removed once the lead can support its own weight. Removing the pieces before the mortar hardens is not recommended because the weight of the block necessitates their temporary support. In addition, the mortar joints between the toothed lead and wall must be fully compacted. Using a tuckpointer fills and compacts these joints.

## Using Corner Poles

Used less for block masonry than brickwork, placing metal corner poles or masonry guides at ends of walls eliminates having to construct leads. Commercially available, 2" × 2" aluminum poles with adjustable top and bottom fittings and diagonal bracings permit quick and accurate setups. Working near the corner poles and diagonal bracings requires care, because any movement of the poles or bracings can result in misaligned walls.

**CAUTION**

Metal corner poles are conductors of electricity. Maintain a minimum 10' distance between metal corner poles and overhead power lines.

# Installing Wire Reinforcement in Joints

Depending upon the project and governing codes, wire joint reinforcement may be required. See Chapter 10, "Laying Block to the Line," for more detailed information about wire joint reinforcement. The joint reinforcement should extend to the ends of the block at the tail end of leads so that it can be lapped by the wire later placed as the walls are constructed.

## ⚠ CAUTION

**The exposed ends of wire reinforcement can impale workers, causing serious injuries. Wear safety glasses and gloves when working with it. Metal wire reinforcement is a conductor of electricity. Make certain that wire reinforcement can come no closer than 10' to overhead high voltage electrical lines as it is handled.**

# Intersecting CMU Walls

A CMU wall adjoining another CMU wall at a right angle is typically anchored to it by one of two methods: intersecting it or butted against it and secured with steel strapping or wire fabric. A chase or recess is formed in a wall by cutting block where another wall is intended to intersect it (see Figure 11-27).

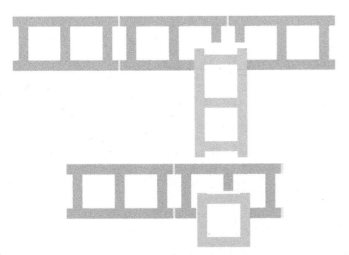

**Figure 11-27** Alternating block courses of a typical recess called a chase are formed by cutting the block. This procedure permits another CMU wall to intersect it at a right angle.

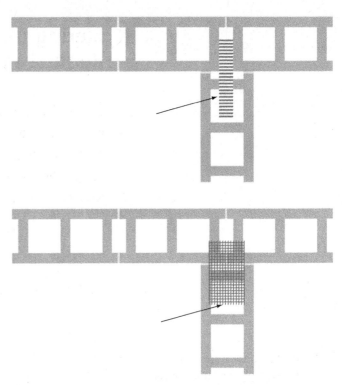

**Figure 11-28** Either steel straps or steel hardware cloth (arrows) secure the wall butted against another wall.

Zinc-coated steel straps or zinc-coated steel hardware cloth are used for lateral support of a wall that is built against the face of another wall (see Figure 11-28).

Depending upon the design requirements, the vertical joints where the two walls adjoin are either filled with mortar or a backer rod and caulk. The backer rod and caulked joint permit this vertical joint to remain sealed if there is differential movement between the two walls.

# Summary

- Constructing block leads requires the knowledge of layout and aligning a wall level, plumb, and straight using the 4' mason's level.
- As the term *lead* implies, accurate wall construction is lead, directed, or guided by accurate leads.
- Seek the supervision of an experienced mason when first attempting lead construction. Required proficiency should be demonstrated before attempting to construct leads without supervision.
- Consult a licensed engineer or local building codes, because conditions and building code regulations dictate appropriate block size for any given project.

## Procedures

# Laying Out a Block Wall and Corner Requiring a Block to Be Cut

*The wall is laid out to be 84" long. This requires cutting a block on each course to permit the desired length. For this illustration, each of the cut block (darker block) measures 12". The cut block is usually put at the end of the wall, and it is vertically aligned on every other course.*

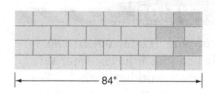

|← ———————— 84" ———————— →|

**A** This illustration shows proper placement of the cut block for courses one and three of the wall illustrated. The length of the block closest to the corner is cut shorter. Aligning the cut block vertically on every other course provides a uniform pattern bond.

**A**

**B** Laid in the stretcher position, the cut block on courses two and four are at the end of the corner. Aligning the cut block vertically on every other course provides a uniform pattern bond.

**B**

# Review Questions

*Select the most appropriate answer.*

**1** A type of block lead consisting of two walls connected at a right angle is called a

a. corner
b. jamb
c. pilaster
d. pier

**2** Using a mason's line to align leads at opposite ends of a wall with each other is called

a. bonding the wall
b. checking the range
c. centering the wall
d. facing the wall

**3** The standard course spacing for block walls is

a. 6"
b. 7⅜"
c. 7⅝"
d. 8"

**4** When it is necessary to increase or decrease course spacing, bed joint widths should be between

a. ⅛" and ½"
b. ¼" and ½"
c. ¼" and 1"
d. ⅜" and 1"

**5** The mason's level is used to align leads

a. level
b. plumb
c. straight
d. all of the above

**6** Comparing the alignment of all courses at the tail end of the lead by placing the level diagonally along the stepped end of the lead is called

a. checking the edge of the lead
b. checking the face of the lead
c. checking the rack of the lead
d. checking the range

**7** A 4" block corner with a half-lap pattern requires the length of the block at the corner to be

a. 6"
b. 7⅝"
c. 11⅝"
d. 12"

**8** A 6" block corner with a half-lap pattern requires the length of the block at the corner to be

a. 6"
b. 7⅝"
c. 11⅝"
d. 13⅝"

**9** The width of the ends of 10" and 12" offset corner blocks is

a. 7⅝"
b. 8"
c. 10"
d. 12"

**10** A procedure permitting building a lead to any number of courses is called

a. racking
b. stacking
c. spotting
d. toothing

**11** Intersecting block walls are adjoined using

a. chases
b. steel strap anchors
c. steel-welded fabric wire cloth
d. all of the above

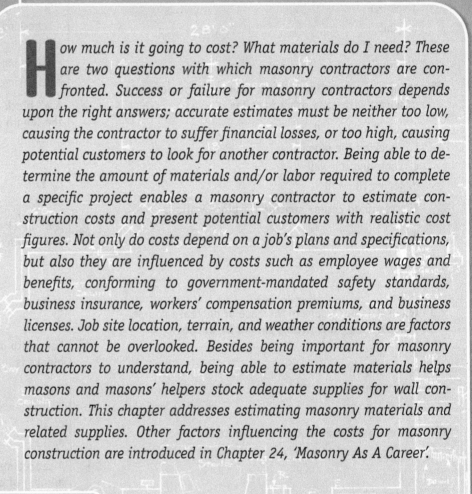

# Chapter 12 | Estimating Masonry Costs

**H**ow much is it going to cost? What materials do I need? These are two questions with which masonry contractors are confronted. Success or failure for masonry contractors depends upon the right answers; accurate estimates must be neither too low, causing the contractor to suffer financial losses, or too high, causing potential customers to look for another contractor. Being able to determine the amount of materials and/or labor required to complete a specific project enables a masonry contractor to estimate construction costs and present potential customers with realistic cost figures. Not only do costs depend on a job's plans and specifications, but also they are influenced by costs such as employee wages and benefits, conforming to government-mandated safety standards, business insurance, workers' compensation premiums, and business licenses. Job site location, terrain, and weather conditions are factors that cannot be overlooked. Besides being important for masonry contractors to understand, being able to estimate materials helps masons and masons' helpers stock adequate supplies for wall construction. This chapter addresses estimating masonry materials and related supplies. Other factors influencing the costs for masonry construction are introduced in Chapter 24, 'Masonry As A Career'.

## OBJECTIVES

After completing this chapter, the student should be able to:

- ⊗ Estimate quantities of brick, block, cementitious material, sand, steel reinforcement, and grout.
- ⊗ Estimate the amount of concrete needed for a concrete footing.
- ⊗ Estimate the amount of materials needed for a concrete slab.
- ⊗ Estimate the labor costs for given masonry projects.

## Glossary of Terms

**bidder** one who submits a bid price proposal for work on a project.

**bid price** the dollar amount of the proposal for supplying the materials and/or labor.

**cost estimate** a calculated cost of construction materials and/or labor, not a legally binding proposal.

**gable** the triangular area below a roofline.

**labor constant** the amount of labor required to perform a specific amount of work.

# The Importance of Accurate Estimates

A cost estimate is a calculated cost of construction materials and/or labor, not a legally binding proposal. Estimates are often used only as construction cost forecasts. But estimates are often calculated to prepare contracts that, once accepted by all parties involved, become legal documents to which all parties are bound.

Determining the amount of materials required to complete a specific project enables the mason to estimate construction costs. Higher estimations of materials than required result in overstated costs. Such errors result in unrealistic high pricing of projects. Projects are awarded to contractors with more accurate, lower estimates. Calculations of fewer materials than required result in estimations below that which is needed for project completion. Contractors experience financial losses on occasions where such estimations become a binding contract, requiring the contractor to fund costs of additional materials and labor.

Even when the material costs are not an issue, poor estimations reduce productivity because of the time required to reorder additional materials or remove surplus materials upon the project's completion. Also, schedule delays and material matching can become issues when too few materials are initially ordered.

Larger contractors rely on the expertise of an employee whose responsibility is to determine material and labor costs. Computer software is often used to calculate the costs. The bidder submits a bid price proposal for work on a project. The dollar amount of the proposal for supplying the materials and/or labor is called a bid price.

# Estimating Brick

Brick are typically delivered to the job site in cubes, bands of brick strapped together after production to enable trailer transportation, mechanical lifting, and delivery drop-off (see Figure 12-1). Typically, five straps of standard-size brick, each containing 105 brick, are contained in one cube of 525 brick. Similarly, five straps of oversize brick, each containing

**Figure 12-1 The brick cubes are delivered by truck and set about the jobsite with a fork lift truck.**

86 brick, are contained in one cube of 430 oversize brick. Elevating brick cubes above the ground onto wood pallets prevents soil stains and surface water migration.

The quantity of brick required depends on the size of the wall and the size of the brick. Calculating the number of square feet is the first step for determining brick quantity. To calculate the wall area in square feet, the length of a wall is multiplied by its height. For example, a wall measuring 20' long and 8' high equals 160 square feet.

Openings such as doors and windows must be considered or material estimates can be too high. For each wall, subtract the number of square feet represented by all openings from the square feet derived from the dimensions of the wall. Since brick must be cut at openings, and waste occurs, estimations can be simplified if opening dimensions exclude all measurements except foot measurements. For example, an opening measuring 3'4" wide and 6'8" high can be considered 3' × 6' for estimating purposes. The following is a sample problem for finding area.

## Sample Problem: Finding Area

Determine the number of square feet requiring brick for the wall illustrated in Figure 12-2. The wall measures 24' long

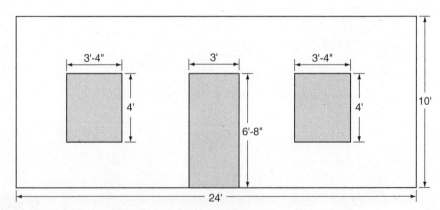

**Figure 12-2 The area to be calculated includes all but the gray areas representing the windows and door opening.**

and 10' high, has one door opening measuring 3' × 6'8", and has two window openings, each measuring 3'4" × 4'.

**Step 1**: Find total wall area.

24' × 10' = 240 square feet

**Step 2**: Find total area of openings.

3' × 6' = 18 square feet × 1 = 18 square feet
(total area of 1 door opening)
3' × 4' = 12 square feet × 2 = 24 square feet
(total area of 2 window openings)
18 + 24 = 42 square feet

**Step 3**: Subtract total area of openings from total wall area.

240 sq. ft. − 42 sq. ft. = 198 sq. ft.

The estimation of the quantity of brick is derived from the number of square feet multiplied by the number of brick required per square foot. The number of brick per square foot depends on the size of the brick. Brick sizes are explained in Chapter 1, "Basic Brick Positions and Brick Sizes." Including mortar joints, there are approximately

- 6.75 standard-size brick per square foot.
- 5.8 oversize brick or queen size brick per square foot.
- 3.0 economy brick per square foot.

The following is a sample problem for estimating brick.

## Sample Problem: Estimating Brick

How many *standard-size* brick are needed for an area of 100 square feet?

100 × 6.75 = 675 standard-size brick

How many *oversize* brick are needed for an area of 100 square feet?

100 × 5.8 = 580 oversize brick

How many *economy* brick are needed for an area of 100 square feet?

100 × 3 = 300 economy brick

Estimating the required number of brick for a project is simplified by adding the sum of materials required for each of the project's walls. A **gable** is the triangular area below a roofline. The formula for calculating the area of a triangle, ½bh, is used to calculate the area of a gable. As illustrated in Figure 12-3, *b* equals the length of the gable, and *h* equals the height of the triangle formed by the gable.

The following is a sample problem for estimating brick in a gable.

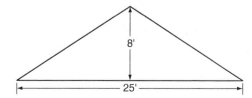

**Figure 12-3 The formula for calculating the area of a triangle is used to determine the area of a roof gable.**

## Sample Problem: Estimating Brick in a Gable

How many standard-size brick are needed for the gable illustrated in Figure 12-3?

In Figure 12-3, the overall width of the roof gable is 25', and the height at its center peak is 8'.

Area = ½ b'h'
Area = ½(25 × 8)
Area = ½(200)
Area = 100 sq. ft.
Brick = Area × 6.75
Brick = 100 × 6.75
Brick = 675

 **FROM EXPERIENCE**

Since brick are delivered to the jobsite in cubes of pre-bundled brick, more brick than estimated are typically ordered, since brick producers would rather not ship parts of full cubes of brick.

The surface of typical brick pavers, brick used on walkways and patios, measure 3⅝" × 7⅝". It takes approximately 4½ brick pavers having these dimensions for each square foot of surface area. Brick pavers are a topic of Chapter 17, "Brick Paving."

## Estimating Block

Block are typically delivered to the jobsite and stored on wood pallets to prevent soil stains and surface water migration (see Figure 12-4).

**Figure 12-4 These block are on a wood pallet, keeping the block clean and enabling them to be moved with a fork lift truck.**

Estimating concrete masonry units (CMUs or blocks) can be accomplished using the area method also. Since all standard block sizes have the same length and height, the quantity of block is the same regardless of the size selected. Block sizes are a topic of Chapter 9, Concrete Masonry Units. A block's face dimensions, including mortar joints, measures 16" long × 8" high. This is equivalent to 128 square inches. Since there are 144 square inches per square foot, one block represents 128/144th of a square foot, or 0.8888 square foot. 1.125 block are required for each square foot of wall area.

The product resulting from multiplying the square feet of the area by 1.125 equals the number of block needed. The following is a sample problem for estimating block.

## Sample Problem: Estimating Block

How many block are needed to construct a wall 25' long and 8' high?

**Step 1:** Determine the area.
$$25' \times 8' = 200 \text{ square feet}$$

**Step 2:** Multiply the area by 1.125.
$$200 \times 1.125 = 225 \text{ block}$$

# Estimating 1 Cubic-Foot Bags of Cementitious Material (Masonry Cement, Mortar Cement, or Portland Cement-Lime Mixes)

Masonry cements, mortar cements, and manufacturers' Portland cement-lime mixtures are called cementitious materials. These cementitious materials account for the durability and strength of the resulting mixed mortar that is used to bond masonry units to one another. For a better understanding of their properties, refer to Chapter 8, "Masonry Mortars." A 1 cubic-foot bag of cementitious material (masonry cement, mortar cement, or Portland cement-lime mix) is sufficient for making mortar to lay 125 brick or 40 block. Bagged cementitious material delivered to the jobsite should be covered with waterproof tarps or plastic to protect it from the elements of weather and raised on wood pallets to protect it from surface water (see Figure 12-5).

The estimated quantity of brick divided by 125 gives an estimate of the number of 1 cubic-foot bags of cementitious material required for mixing mortar to lay brick. Actual mortar usage may vary depending on the mortar joint widths, brick core sizes, and mortar management. The estimated quantity of block divided by 40 gives an estimate

**Figure 12-5** Bags of masonry cement are elevated off of the ground onto a wood pallet and covered with plastic tarp to keep moisture and rain from hardening the bagged cement.

of the number of 1 cubic-foot bags of cementitious material required for making mortar to lay block. Factors determining actual mortar requirements for block work are mortar joint widths, solid bedding versus face shell spreading below the first course, and mortar management. The following is a sample problem for estimating 1 cubic-foot size bags of cementitious material.

## Sample Problem: Estimating 1 Cubic-Foot Bags of Cementitious Material

How many 1 cubic-foot bags of masonry cement are needed to lay 1,000 brick?
$$1000/125 = 8 \text{ bags}$$

How many 1 cubic-foot bags of masonry cement are needed to lay 160 block?
$$160/40 = 4 \text{ bags}$$

**FROM EXPERIENCE**

Poorly placed, unlevel concrete footings reduce the quantity of block capable of being bedded with mortar resulting from the mixing of one bag of cementitious material. Inspect the concrete footings before estimating. Add additional bags to the estimate if footings are not level.

# Estimating Masonry Sand

Sand is measured and sold by volume or weight. It is typically delivered to a jobsite on a dump truck and placed

**Figure 12-6** Sand is proportioned and mixed with 1-cubic foot bags of cementitious material in the mortar mixer.

**Figure 12-7** Concrete, 8" deep, will be placed in this trench to support CMU load-bearing walls.

alongside the cementitious material and the mortar mixer (see Figure 12-6). Sand should be covered to protect it from contaminants, excessive wetting from rain, excessive drying, and wind displacement. Masonry sand is a topic of Chapter 8, "Masonry Mortars."

The ratio of cementitious material to sand, 1 part cementitious material and 2¼ to 3 parts sand, is the same recommendation for mixing mortar for either laying brick or block. One cubic-yard of sand, weighing approximately 2,080 pounds, is sufficient for mixing 8, 1 cubic-foot bags of cementitious material. Dividing the required number of 1 cubic-foot bags of cementitious material by 8 provides an estimate of the number of cubic yards of sand needed. The following is a sample problem for estimating the needed amount of sand.

## Sample Problem: Estimating Sand

How many cubic yards of sand are needed to mix 72, 1 cubic-foot bags of masonry cement?

    Cubic yards = # bags/8
    Cubic yards = 72/8
    Cubic yards = 9

### FROM EXPERIENCE

The true measured volume of sand varies depending upon the moisture content of the sand. Sand should be damp, clumping in the hand when squeezed, if volume measurements are to be accurate. Keeping sand piles covered with tarps prevents excessive drying, winds blowing away the sand, and debris contaminating it.

# Estimating Concrete for Concrete Footings

Placing the concrete footings that support masonry walls is not necessarily the responsibility of the mason. However, there are times when the mason may need to place the concrete footings. Refer to Chapter 13, "Residential Foundations," for more detailed information on concrete footings. One specification for residential concrete trench footings requires them to be a minimum of 8" thick and 24" wide. A typical concrete trench footing is illustrated in Figure 12-7.

The cubic yard is the unit of measure for calculating concrete. Cubic yards of concrete, frequently referred to simply as yards, are determined using the following formula:

Cubic Yards of Concrete = length' × width' × 0.66/27

Length equals the length of the footers, and width equals footing width. Both length and width must be in feet measurement units. For example, to construct a concrete footing 24' 6" long and 24" wide, length is considered as being 24.5' and width is considered as being 2'. The constant, 0.66, represents a concrete thickness of 8", equivalent to 8/12', or 0.66'. Since 27 cubic feet are equivalent to 1 cubic yard, the volume divided by 27 converts cubic feet to cubic yards. The following is a sample problem for estimating the amount of concrete required for a footing.

## Sample Problem: Estimating Concrete for Footings

The left and right sides of the footings intended to support foundation walls each measure 36' long, and the back and front sides of the footings each measure 24' long. Assuming the width of the concrete footings to be 2' and the depth of the concrete footings to be uniform and 8" thick, determine the number of cubic yards of concrete needed for the footings.

**Step 1:** Determine total length.

Total length = 36' + 36' + 24' + 24'

Total length = 120'

**Step 2:** For this problem, width is considered to be 2'.

**Step 3:** Use the formula:

Cubic Yards = length' × width' × 0.66/27

Cubic Yards = 120' × 2' × 0.66'/27

Cubic Yards = 158.4/27

Cubic Yards = 5.866

For the purpose of estimating, 6 yards of concrete are needed.

**The location of underground utilities, including but not limited to electric, fuel, sewer, telephone, and water lines, must be located, identified, and in some cases moved by the utilities companies before beginning ground excavations for footings and concrete slabs.**

# Estimating Concrete for Concrete Slabs

Concrete walkways, patios, and other surfaces are examples of concrete slabs, also called slabwork. In Figure 12-8, a 4"-thick concrete slab is intended as the garage floor.

Although it is not the scope of this chapter to address concrete finishing, masons often place concrete slabs for supporting mortared, brick paving. Residential concrete walkways and patios are a minimum of 4" thick. 4" is equivalent to ⅓ of a foot, often expressed as 0.33 feet when

**Figure 12-8 A concrete slab becomes the floor for this garage.**

performing mathematic calculations. Cubic yards of concrete for 4" thick concrete walkways and patios can be estimated using the following formula. The following is a sample problem for calculating the concrete required for a concrete slab.

## Formula for 4"-Thick Concrete Slabs

Cubic Yards = length' × width' × 0.33'/27

Values for length and width must be represented in feet. The constant, 0.33, represents 4". The constant, 27, represents the number of cubic feet equivalent to 1 cubic yard.

## Sample Problem: Estimating Concrete for Slabs

Determine the cubic yards of concrete needed for a 4" thick concrete walkway 32' long and 4' wide.

Cubic Yards = 32' × 4' × 0.33'/27

Cubic Yards = 42.24/27

Cubic Yards = 1.564

For this project, the estimated amount of concrete is 2 yards.

 **FROM EXPERIENCE**

Ready-mixed concrete suppliers delivering concrete to a job site may have minimum charges. Small orders are subject to additional delivery charges. Check with the supplier for additional charges before having them deliver ready-mixed concrete for smaller orders.

Concrete slab footings support such objects as masonry columns and chimneys. The thickness of the concrete is a factor of the weight load that it is to support. Slab footings supporting nonload-bearing columns, such as brick columns at driveway entrances, are often no more than 8" thick. Slab footings supporting fireplaces and chimneys are a minimum of 12" thick. Slab footings supporting masonry columns or posts and fireplaces and chimneys should extend at least 6" beyond the exterior face of each wall or each side.

## Formula for Slab Footings

Cubic Yards = length' × width' × thickness'/27

Values for length, width, and thickness are represented in feet. The length of the slab footing should be 12" more than the length of the structure it is to support. Likewise, the width of the slab footing should be 12" more than the width of the structure it is to support. Depending on the project, governing codes specify how thick the concrete slab is to be. The following is a sample problem for calculating concrete for a slab footing.

## Sample Problem: Estimating Slab Footings

Determine the cubic yards of concrete needed for a slab footing supporting a brick entrance column measuring 36" × 42" if the concrete slab is to be 8" thick.

**Step 1:** Add 12" to both the length and the width of the column.

$$36" + 12" = 48" = 4'$$
$$42" + 12" = 54" = 4' - 6" = 4.5'$$

**Step 2:** Apply the formula.

Cubic Yards = length' × width' × thickness'/27
Cubic Yards = 4' × 4.5' × 0.66'/27
Cubic Yards = 11.88/27
Cubic Yards = 0.44

For the previous example, an estimate of ½ cubic yard of concrete is required.

**FROM EXPERIENCE**

When ordering ready-mix concrete, give the supplier the dimensions for the pour. Compare their quantity estimates with those others have made. If discrepancies exist, have someone experienced make another estimate at the job site.

# Estimating Steel Reinforcements

Several types of steel reinforcements are used for residential masonry construction. Bar reinforcement is used to strengthen concrete trench footings, slab footings, and foundation walls. Welded wire fabric is used to prevent separation of concrete floor and walkway slabs as cracks develop. These reinforcement components are explained in Chapter 13, "Residential Foundations." Prefabricated wire joint reinforcement bedded in the mortar joints between the courses of block walls increases the wall's compressive strength and improves the wall's resistance to lateral forces. Corrugated veneer ties and veneer wire anchoring systems anchor 4" brick veneer walls to structural walls. Anchoring masonry veneer walls to structural back-up walls is a topic of Chapter 15, "Anchored Brick Veneer Walls." Compliance with governing regulations and engineered designs for specific projects are necessary when estimating types and quantities of reinforcement.

## Steel Rod Reinforcement

Steel rods or bars, typically no less than ½" in diameter, are bedded horizontally in wet concrete to strengthen footings. Bars or rods may also be required in CMU walls, placed

**Figure 12-9** Steel rods extend from the concrete footing and will be connected to additional rods continuing vertically through the open cells of block. More steel rods are embedded horizontally in the concrete footing.

in wet concrete as footings are placed and continuing vertically up through the open cells of CMUs filled with cement grout (see Figure 12-9). The spacing and size of bars in footings and walls depends upon the design of the structure, job specifications, and code regulations. Steel rods are available in 20' sections.

## Concrete Block Grout Fill Quantities

Grout is a mixture of cement and aggregates to which sufficient water is added to produce a pouring consistency without separation of the ingredients. A requirement for many walls, grout is used to fill hollow concrete masonry units for the purpose of increasing their strength. Grout is a topic of Chapter 8, "Masonry Mortars." The quantity of grout required to fill the open cells of concrete masonry units is based upon the cubic measure of void volume per 100 square feet of wall surface area. Refer to Table 9, Concrete Block Grout Fill Quantities in the Appendix for required amounts of grout.

## Welded Wire Fabric Reinforcement

Welded wire fabric reinforcement, also called wire mesh reinforcement, is placed in wet concrete to reinforce floor, walkway, and patio slabs (see Figure 12-10). The wire prevents the concrete from separating once cracks occur, making the cracks less noticeable and minimizing water migration.

A number 10 gauge wire welded at right angles and 6" on center is typically used. This type of wire is called 6-6-10-10 welded wire. Two common sizes available are 5' × 150' rolls and 5' × 10' sheets. The amount of wire required is equivalent to the square footage of the slab.

**Figure 12-10** Rolls of welded wire fabric reinforcement await placement in a concrete slab floor.

**Figure 12-11** Wire joint reinforcement is placed between every other course of this block wall.

## Prefabricated Wire Joint Reinforcement

When specified for strengthening masonry walls, prefabricated wire joint reinforcement is bedded in mortar joints between courses (see Figure 12-11). The standard length for sections of prefabricated wire joint reinforcement is 10'.

The total linear feet of wire joint reinforcement depends upon the design requirements. For walls requiring reinforcement above the first course of block and every other course thereafter, allow 80 linear feet of reinforcement per 100 square feet of wall area. To determine the total linear feet of wire reinforcement, divide the total wall area (minus the area of all openings) by 100, and multiply the quotient by 80.

**CAUTION**

The exposed ends of wire reinforcement can impale workers, causing serious injuries. Wear safety glasses and gloves when working with them. Metal rod and wire reinforcements are conductors of electricity. Uncoiled rolls of welded wire reinforcement must be secured to the ground at their ends to prevent them from recoiling and impaling workers in the process. Make certain that rod and wire reinforcements can come no closer than 10' to overhead high voltage electrical lines as they are handled.

## Corrugated Steel Veneer Ties

Corrugated steel ties anchoring brick veneer to wood wall framing are required (see Figure 12-12). The building code requires a minimum of one tie per 2.67 square feet of wall area. However, bedding ties above the first 8" of brickwork and horizontally every 16" thereafter and vertically 16" apart fastened to wall studs, a tie per 1.66 square feet is needed. Approximately 55 veneer wall ties are required for every 1,000 standard size brick. Approximately 65 veneer wall ties are required for every 1,000 oversize or queen size brick. One common nail capable of penetrating the wall stud a minimum of 1½" is required for each tie.

## Veneer Wire Anchoring System

A veneer wire anchoring system is a method for securing brick veneer to metal-stud framed structures. Allow 80 linear feet of veneer wire anchoring system per 100 square feet of wall area.

**Figure 12-12** Corrugated veneer wall ties anchor the brick veneer to the wood framework.

Corrugated steel veneer ties and veneer anchoring systems are topics in Chapter 15, "Anchored Brick Veneer Walls."

# Estimating Labor Costs

The amount of labor required to perform a specific amount of work is known as the **labor constant.** Many factors influence the labor constant:

- Site accessibility
- Project complexity
- Equipment expenses
- Employee wages and benefits
- Workers' compensation rates
- Insurance costs
- Weather conditions
- Business operating expenses

These factors are explained in Chapter 24, "Masonry as a Career."

The labor constant influences the labor costs. Four methods for quoting labor costs are the following:

1. Unit method
2. Square foot method
3. Per job method
4. Time and materials method

The "unit" method relies on calculating labor costs based on the quantity of brick or block to be laid. This amount is sometimes expressed as the cost per thousand for laying brick or the cost each for laying block. The "square foot" method relies on calculating labor costs based on a price per square foot of wall area for the project. The "per job" method is a definite labor charge quoted for a specific job. "Time and materials" is a method for determining material as well as labor costs. The masonry contractor agrees to furnish the materials in accordance with the contract plans and specifications. The contractor also agrees to provide the labor for completing the work as specified in the contract. The owner is obligated to compensate the contractor for furnishing the materials and labor at the mutually established and accepted rate or price.

# Summary

- Estimating construction costs is a complex process. Estimations are based on judgments from previous experiences as well as accurate calculations of materials, equipment, and labor for a given job.
- This chapter is intended to reveal the complexities involved in masonry construction estimations rather than to fully prepare you as an estimator.
- Poorly calculated estimates can result in unrealistically higher costs and can cause potential customers to seek better pricing.
- Calculations of fewer materials or less labor than is required result in estimations below that which is needed for project completion. Contractors can experience financial losses when estimations that become binding agreements are too low.
- As with developing masonry construction skills, it takes much experience to become qualified as an estimator.

# Review Questions

*Select the most appropriate answer.*

**1** The dollar amount of the proposal for supplying materials or labor for a specific job is called

  a. a bid price
  b. a contract agreement
  c. an estimated price
  d. a proposal

**2** The estimated number of brick per square foot is

  a. 5 standard-size brick, 4½ oversize brick, or 4 economy brick
  b. 6 standard-size brick, 5½ oversize brick, or 4½ economy brick
  c. 6.75 standard-size brick, 5.8 oversize brick, or 3 economy brick
  d. 8 standard-size brick, 6 oversize brick, or 5 economy brick

**3** The surface area of one standard-size CMU, including a head joint and bed joint, is

  a. 0.5 square foot
  b. 0.75 square foot
  c. 0.88 square foot
  d. 1.25 square feet

**4** A 1-cubic foot bag of masonry cement can be mixed for mortar to lay approximately

  a. 75 brick or 25 block
  b. 100 brick or 75 block
  c. 125 brick or 40 block
  d. 40 brick or 125 block

**5** The number of 1-cubic foot bags of masonry cement that can be mixed with 1 cubic yard of sand for making mortar is estimated to be

  a. 6
  b. 8
  c. 12
  d. 16

**6** The number of cubic feet in 1 cubic yard is

  a. 9
  b. 18
  c. 27
  d. 36

**7** The amount of labor required to perform a specific amount of work is called the

  a. employee productivity
  b. labor constant
  c. time factor
  d. wages contract

**8** A factor influencing the cost of the labor required to complete a job is

  a. employee wages and benefits
  b. site accessibility
  c. workers' compensation and insurance premiums
  d. all of the above

**9** The method for charging costs in which the owner is obligated to pay the contractor for furnishing the materials and labor at a mutually established and accepted rate or price is called

  a. materials and labor
  b. the per job method
  c. time and materials
  d. the unit method

**10** The area of a triangular gable-end of a house is calculated using the formula

  a. (length × height)
  b. ½ (base × height)
  c. 2( base × height)
  d. (base × height)/3

# Chapter 13 | Residential Foundations

*S*tart on the right foot. Build on a solid foundation. One often hears these and similar words of encouragement when beginning important tasks. The design and workmanship for the foundation are critical to the overall construction of any project. A foundation must be designed to support the weight of an entire structure, its contents, and its occupants. A foundation below ground level must be capable of resisting the lateral forces of the earth pushing against its walls. Its exterior walls must deter moisture and ground water from migrating through the walls. Foundation walls must be level and their corners square to permit accurate construction of the floors, walls, and roofs above them. The skills developed for building block leads and laying block to the line are necessary for constructing CMU foundation walls.

## OBJECTIVES

After completing this chapter, the student should be able to:

- Identify factors considered for the design of concrete footings.
- Describe methods for forming concrete footings.
- List design elements for foundation walls built with concrete masonry units (CMUs).
- Lay out and build a foundation wall.

# Glossary of Terms

**6-6-10-10 wire reinforcement** 10-gauge wire welded at right angles on 6" centers.

**anchor bolt** a threaded bolt secured in and projecting from the top of a foundation wall to which the wood sill plate is fastened.

**anchor strap** a sheet-metal strap secured in and projecting from the top of a foundation wall to which a wood sill plate is fastened.

**areaway** a walled area below grade that permits ventilation, light, or egress.

**backfilling** replacing the earth removed during construction against the outside of the foundation.

**bar reinforcement** steel rods placed in concrete for strengthening the concrete.

**batter boards** pairs of horizontal boards nailed to wood stakes beyond each corner to which layout string lines are fastened.

**brick shelf** a masonry ledge supporting the anchored brick veneer walls.

**compressive strength** expressed in pounds per square inch (psi), a measure of a material's resistance to compression as a result of the weight it supports.

**crawl space** the area between the ground surface and floor joists of houses without basements.

**damp proofing** the process of hindering the absorption and passage of water through the foundation walls below grade level.

**egress** a path of exit or rescue.

**footings** that part of the foundation supporting the foundation walls and transmitting the structure's weight to the soil.

**formed footings** footings resulting from the placement of concrete within temporary or permanent forms.

**foundation** that part of a building below the first floor framing.

**frost line** the greatest depth to which ground may be expected to freeze.

**grout** a mixture of cement, fine aggregates, and water with a consistency permitting pouring without separation of the ingredients.

**grouting** the process of manually pouring or machine-pumping grout in walls such as the hollow cells of CMU walls.

**parging** a cement-sand mixture coating applied to the exterior of foundation walls below finished grade.

**potential expansive soils** those soils that increase in volume significantly when wet.

**slump** a measure of the collapse of fresh concrete below the 12" formed by the slump cone.

**slump cone** a 12"-tall formed cone with a base diameter of 12" and a top diameter of 4" used to form a sample of fresh concrete for conducting a slump test.

**slump test** an assessment of the consistency of freshly placed concrete.

**trench footings** footings resulting from the placement of concrete directly into excavated trenches.

**waterproofing** an approved membrane extending from the top of the footing to ground level and preventing water penetration of the foundation wall; required for foundation walls below finished grade wherever a high water table or other severe soil-water conditions exist.

# Elements of a Foundation

A foundation is that part of a building below the first floor framing. It includes foundation walls and concrete footings. The concrete footings support the foundation walls and transmit the structure's weight to the soil. Three materials are recognized for constructing foundation walls: 1) concrete masonry units, 2) formed concrete, and 3) treated wood. All materials must meet the requirements of the governing building code regulations. This chapter addresses foundation walls constructed with CMUs only.

## CAUTION

**The details and data in this chapter are for general guidance only. Prior to using this information, a registered professional engineer should be consulted to ensure that the details are appropriate for a particular project or geographical area to correct any errors or inaccuracies in the details and to ensure that the details comply with governing building regulations and all laws.**

# Footings

This chapter is not intended to teach you how to become competent at designing and placing concrete footings. An engineer is responsible for the design of site-specific concrete footings, and the general contractor rather than the mason is usually responsible for placing the footings. However, since the mason's walls are supported by concrete footings, it is desirable for the mason to have an understanding of their design and construction.

Continuous concrete footings are placed on soils capable of supporting the structure. Conducting soil tests to determine the load-bearing capacity of the soil permits a properly engineered site-specific footing design. There are three considerations for designing continuous concrete footings: 1) size, 2) specified compressive strength, and 3) reinforcement.

## Footing Size

The footing size refers to the concrete's width and thickness. Building code regulations govern the width and thickness for concrete footings. Table R403.1 of the International Residential Code gives minimum width requirements for concrete footings. Factors determining a footing's width are 1) the load-bearing capacity of the soil, 2) the wall design, and 3) the number of building stories.

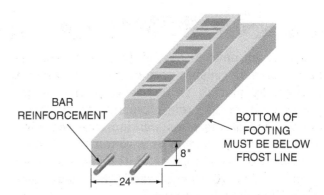

**Figure 13-1** This steel reinforced concrete footing is 24" wide and 8" thick.

Residential concrete footings are usually designed as being no less than 8" thick. Regions identified as having potential expansive soils, those soils that increase in volume significantly when wet, may require thicker footings. Figure 13-1 illustrates one example for a residential footing design.

## Compressive Strength

Compressive strength of concrete, expressed in pounds per square inch (psi), is a measure of concrete's resistance to compression, a force tending to shorten a member, as a result of the weight the footing supports. Building code regulations govern the compressive strength requirements for concrete used in footings. Minimum specified compressive strengths are given in Table R402.2 of the International Residential Code. Factors influencing the requirements are 1) type of construction, 2) location of the concrete, and 3) weathering potential. Strengths generally range between 2500 psi and 3500 psi.

## Reinforcement

Footings in certain seismic design categories prone to earthquakes or footings placed on engineered fill, soil that is added to build up low spots and adequately compacted to support the weight of the intended structure, may require steel bar reinforcement embedded the length of the footings. Bar reinforcement are steel rods placed in concrete for strengthening the concrete. Bar reinforcement is also called rebar or reinforcement rods. A registered professional engineer can specify the size and number of bar reinforcements required. Reinforcement diameters and number designations are included in this book's Appendix, Table 10, Steel Reinforcement Sizes. Depending on seismic design categories and soil classification, vertical reinforcement bars embedded in the wet concrete for the footings and extending into the open cells of the CMU foundation walls may be required.

# Methods for Constructing Footings

Two methods of constructing concrete footings are perimeter footing construction and monolithic construction. Perimeter footings are cast in place, either in trenched excavations or forms provided, and function only to support the foundation walls. Monolithic construction permits footings supporting foundation walls and floors to be cast as one element known as a monolithic slab.

## Perimeter Footings

Although usually required to be much deeper, the bottom of concrete footings should be no less than 12" below finished grade. In regions experiencing freezing temperatures, the bottom of the footing must be below the frost line (see Figure 13-2). Always obey local building code regulations governing footer placement.

The **frost line** is the greatest depth to which ground may be expected to freeze. The top of footings should be level. Footings may be stepped alongside sloping elevations. It is recommended to have no one step higher than 24" or shorter than 24" (see Figure 13-3).

Steps whose heights are multiples of 8" permit full block coursing (see Figure 13-4).

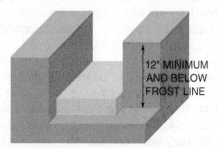

**Figure 13-2** The bottom of the footing should be at least 12" below finished grade but also below the regional designated *frost line.*

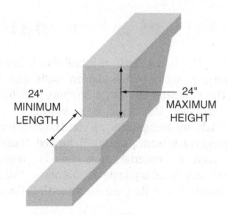

**Figure 13-3** *Steps,* recommended to be no shorter than or higher than 24", enable the elevation of footings to change.

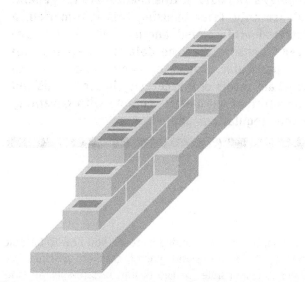

**Figure 13-4** The footing steps align level with each course of CMUs.

## Monolithic Slab Foundations

Monolithic slab foundations are designed having deeper concrete perimeter beams and less concrete depth serving as the floor (see Figure 13-5). Together, the floor slab and perimeter beams support the loads imposed by the structure.

Unlike perimeter footings, monolithic slabs may not be below the frost line. Monolithic slabs are designed with steel

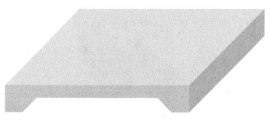

**Figure 13-5** A cross-sectional view of a monolithic slab footing shows additional concrete around the perimeter, providing additional support for walls.

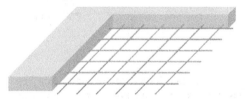

**Figure 13-6** Welded wire reinforcement helps prevent separation of the concrete slab.

bar reinforcement and steel wire reinforcement. Wire reinforcement known as **6-6-10-10 wire reinforcement**, meaning 10-gauge wire welded at right angles on 6" centers, is commonly used (see Figure 13-6).

**CAUTION**

The exposed ends of wire reinforcement can impale workers, causing serious injuries. Wear safety glasses and gloves when working with it. Uncoiled rolls of welded wire reinforcement must be secured to the ground at their ends to prevent them from recoiling and impaling workers in the process. Metal wire reinforcement is a conductor of electricity. Make certain that wire reinforcement can come no closer than 10' to overhead high voltage electrical lines as it is handled.

# Concrete Masonry Foundation Walls

Concrete masonry foundation walls are constructed of hollow CMUs or blocks. Factors considered for properly designed CMU foundation walls include 1) CMU size, 2) mortar type, 3) vertical bar reinforcement and grouting, 4) provisions for anchoring the wood sill plates, 5) damp-proofing or waterproofing the exterior walls, 6) exterior water drainage, 7) ventilation, and 8) egress routes.

## CMU Size

Building code regulations govern the specified size or width of the CMUs. Factors determining the required block size are 1) wall height, 2) unbalanced backfill height, 3) soil classification, and 4) wall design. Greater wall heights require larger blocks. Increasing unbalanced backfill height, or the difference in height of the exterior and interior finish ground levels, necessitates larger block sizes. The volume change potential for different soil classifications is another determining factor of block size. Tables R404.1.1(1), R404.1.1(2), and R404.1.1(3) in the International Residential Code address foundation wall thickness. When anchored brick veneer walls are intended to be supported by the block foundation, the block size must allow for a brick shelf, a ledge to support the brick.

## Mortar Type

Building code regulations govern the mortar strength required for foundation walls. Type M or Type S masonry mortar is generally specified for mortars used to build CMU foundation walls.

**CAUTION**

Cementitious materials can be hazardous to the respiratory system, eyes, and skin. Inhalation can cause throat and upper respiratory system irritations as well as certain lung diseases. Use MSHA/NIOSH-approved respirators if general ventilation is not adequate to control dust. Wear eye protection with side shields that meet the intent of OSHA standards. Wear tight-fitting, unvented, or indirectly vented eye protection where there is danger of excessive amounts of dust coming into contact with the eyes. Wear alkali-resistant protective clothing, gloves, and boots. Wash skin areas contaminated by masonry cements with ph neutral soap and water.

# Vertical Bar Reinforcement and Grouting

Depending on soil classification, unbalanced backfill, and adjacent slopes, hollow masonry units may require bar reinforced grouting. **Grouting** is the process of mechanically pouring or machine-pumping grout in the hollow block cells of CMU walls. **Grout** is a mixture of cement, fine aggregates, and water with a consistency that permits pouring without separation of the ingredients. Steel bar reinforcement put in wet concrete when footings are placed strengthens the walls (see Figure 13-7).

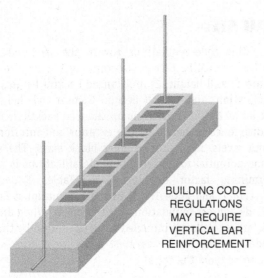

**Figure 13-7** Steel bar reinforcements are required for foundations in severe shrink-swell soils and some other circumstances.

BUILDING CODE REGULATIONS MAY REQUIRE VERTICAL BAR REINFORCEMENT

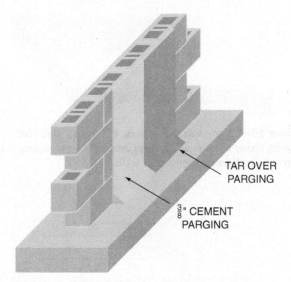

TAR OVER PARGING

$\frac{3}{8}$" CEMENT PARGING

**Figure 13-9** Coating the outside of the foundation with cement and then tar *damp-proofs* the foundation. An impermeable membrane is required for a *waterproof* foundation.

## Anchorage

A wood sill plate is anchored to the top of the foundation with foundation **anchor bolts** or foundation **anchor straps.** Threaded anchor bolts are placed a maximum of 6' on center and not more than 12" or less than 7 bolt diameters from each end of a plate section. Bolts must be a minimum of ½" in diameter and extend a minimum of 7" into the CMUs. Foundation anchor straps must provide equivalent anchorage as ½"-diameter foundation anchor bolts. Anchor bolts and anchor straps are secured in the block cells using mortar (see Figure 13-8). The different seismic design categories may have additional requirements. Refer to The International Residential Code, Section R403.1.6 for category-specific requirements.

*For step-by-step instructions on installing anchor bolts, see the Procedures section on page 188.*

## Damp-Proofing and Waterproofing

**Damp-proofing** is the process of retarding the absorption and passage of water through the foundation walls below grade level. Damp-proofing is required for foundation walls enclosing habitable or usable spaces below grade. Section R406.1 of the International Residential Code requires masonry foundation walls to be damp-proofed from the top of the footing to the finished grade with a minimum ³/₈"-thick Portland cement **parging** "dampproofed with a bituminous coating, 3 pounds per square yard of acrylic modified cement, ¹/₈" coat of surface-bonding mortar complying with ASTM C 887 or any material permitted for **waterproofing** in Section R406.2," (see Figure 13-9). Waterproofing for foundation walls requires an approved membrane that not only retards or slows down water intrusion, but one that totally prevents any water intrusion.

Section R406.2 of The International Residential Code requires waterproofing for foundation walls below finished grade wherever a high water table or other severe soil-water conditions exist. An approved membrane extending from the top of the footing to ground level is required for waterproofing foundation walls.

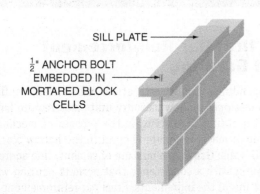

SILL PLATE

½" ANCHOR BOLT EMBEDDED IN MORTARED BLOCK CELLS

**Figure 13-8** Embedded in fully mortared block cells, *anchor bolts* secure the sill plate to the top of the foundation.

### ◄◄◄ CAUTION ►►►

Wear personal protection equipment as recommended by the product manufacturer, and comply with all safety and environmental regulations when exposed to foundation coatings and their vapors or fumes.

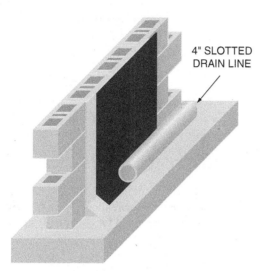

**Figure 13-10** A plastic, 4" slotted drain line collects water along the bottom of the foundation's exterior walls.

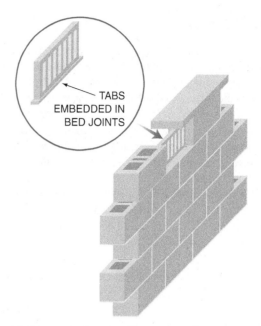

**Figure 13-11** Permitting the sill plate to bridge, the foundation vent is aligned level with the tops of the CMUs.

## Water Drainage

The accumulation of surface water and ground water may enter foundations below grade unless the water is collected and diverted. Except for locations with well-drained ground or sand-gravel mixture soil classifications, foundations below finished grade enclosing habitable or usable spaces require drainage systems along their entire exterior perimeters. Two materials designed to collect and discharge water are 4" drain tiles or 4" perforated plastic pipe. Installed at or just below the basement floor level, the tile or pipe are placed on a 2" minimum crushed stone bed and covered with a 6" minimum crushed stone layer one sieve size larger than the tile joints or perforated pipe openings (see Figure 13-10). More specific requirements for foundation drainage can be found in Section R405 of the International Residential Code.

An approved filter membrane material covering the stone prevents soil from blocking spaces between the stones. The open joints between tiles require approved coverings preventing stone from blocking the pipe. The water is discharged from the tiles or pipe by gravity or mechanical pumping.

## Ventilation

Houses with no basements require ventilation between the floor joists and the ground surface. For this area, known as a crawl space, Section R408 of the International Residential Code requires a minimum of 1 square foot of ventilation opening for every 150 square feet of floor area, with one such opening within 3' of each corner of the building. Foundation vents measuring 8" tall and 16" wide, dimensions equivalent to the face of a CMU and its joints, are placed along the top course of block (see Figure 13-11).

To permit the tops of the CMUs to support the foundation sill plate, the tops of the foundation vents should be no higher than the top of the last course of block. The author beds the tabs of the foundation vents in the mortar bed joint below the top block course for better anchorage. Installing the vents with the tabs at the top results in having the vent above the tops of the CMUs and interferes with the sill plate. A 4" cap block placed below the vent provides a proper fit where 4" cap block are specified for the top course of a foundation (see Figure 13-12).

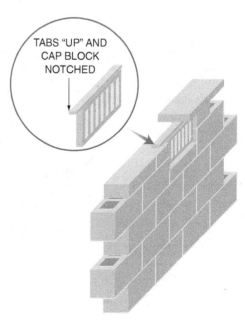

**Figure 13-12** The vent tabs rest in notches cut in the top of the cap block to permit the sill plate to lie flat.

## Egress Route

An egress is defined as a path of exit or rescue. Basements are required to have an egress. Egress must comply with building code regulations. Section R310.1 of the International Residential Code states that "basements with habitable space and every sleeping room shall have at least one openable emergency escape and rescue opening." An areaway is a means for egress below grade. An areaway is a walled area below grade permitting ventilation, light, or egress. Carrying no loads of the structure as do the foundation walls, areaways should be constructed from properly designed grout and rebar reinforced CMU or concrete walls to prevent collapsing from pressure behind the walls.

## Laying Out the Footings

The foundation dimensions and the required footing specifications must be known before laying out the footings. The outside dimensions of the footings are greater than the outside dimensions of the foundation walls since the footing surface is to be wider than the CMUs. The author uses mason's lines for identifying the outer edges of the footings. Batter boards, pairs of horizontal boards nailed to wood stakes beyond each corner, secure the layout lines (see Figure 13-13).

*For step-by-step instructions on setting up batter boards, see the Procedures section on pages 189–191.*

Having lines representing opposite sides parallel with one another and having the diagonal measurements equal assures a square layout. Lines are removed before placing the concrete.

Two methods are used for constructing footings. These methods include trench footings and formed footings. A ground excavation forms trench footings. The concrete is placed directly into the open trenches. The trenches are excavated to the widths and depths required for the footings. Trenches are dug with mechanical backhoes. The outer edges of the intended trenches can be recognized with a trail of lime or masonry cement before digging (see Figure 13-14).

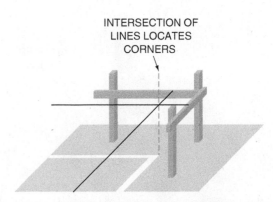

Figure 13-14 Dry masonry cement aligned with the lines on the batter boards marks the ground for accurate digging.

### CAUTION

**Only those trained by competent persons and complying with OSHA regulations should operate backhoes or other mechanical construction equipment. Stay clear of construction equipment as the trenches are dug. Underground utilities, including but not limited to electric, fuel, sewer, telephone, and water lines, must be located, identified, and in some cases moved by the utilities companies before excavation begins.**

Aligning stakes level in the center of the trenches helps to place level concrete footings (see Figure 13-15).

Formed footings require concrete forms before placing the concrete. Two-inch framing lumber securely fastened to stakes driven into the ground can be substituted for actual concrete forms (see Figure 13-16).

Having the tops of the forms level and placing the concrete level with the top of the forms permits level footings. Temporary forms are removed after the concrete hardens. Permanently installed combination *footing forms and foundation drainage systems* are available. After the concrete is placed, these permanently installed polyvinyl chloride (PVC) systems perform the function of a foundation drainage system. The outer walls of the drainage forms are slotted, permitting water to enter into the form's open channels and

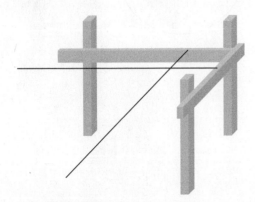

Figure 13-13 Mason's lines attached to *batter boards* identify the outside of corners and walls.

Figure 13-15 Stakes aligned level at the tops permit an accurate depth of concrete for *trench* footings.

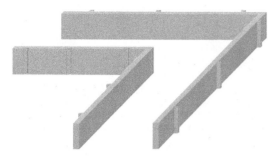

**Figure 13-16 Aligned level at their tops, 2" × 8" framing lumber is used to confine concrete for *formed* footings.**

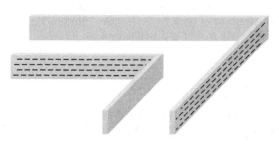

**Figure 13-17 This permanently installed footing form serves as a foundation drainage system.**

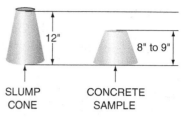

**Figure 13-18 Using a slump cone to perform a slump test reveals the concrete's consistency.**

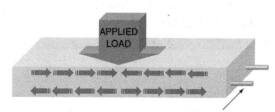

BAR REINFORCEMENT PLACED NEARER BOTTOM HELPS CONCRETE RESIST THE FORCES OF SEPARATION

**Figure 13-19 Applied loads result in forces of compression nearer the top and forces of separation nearer the bottom of a concrete footing.**

then exit through installed drain outlets (see Figure 13-17). Being below the foundation wall and above the bottom of the footing, the system claims to provide optimum drainage.

# Placing the Concrete

Placing concrete in the forms is often referred to as pouring concrete. However, concrete mixed with a volume of water permitting pouring results in ingredient separation. The mixture of heavier crushed stones separates from the lighter Portland cement paste, resulting in a weaker product. Higher strength, stiffer concrete is placed rather than poured in the forms. The proper, stiff consistency should necessitate using shovels, hoes, or rakes to position the concrete in the forms. Performing a slump test determines the freshly mixed concrete's consistency. A slump cone, a 12" tall formed cone with a base diameter of 12" and a top diameter of 4", is used to form a sample of the freshly mixed concrete. Removing the form reveals the concrete's consistency. Concrete for footings should have a slump not less than 3" or more than 4". The slump is a measure of how much the concrete displaces below the 12" formed by the slump cone (see Figure 13-18).

Where design specifications require, bar reinforcement is bedded horizontally in the concrete. Loads or weights supported by the footings create compression forces within the concrete nearer its top and tension forces tending to pull it apart nearer the bottom (see Figure 13-19). The reinforcement is placed nearer the bottom of the concrete because

the imposed forces and loads result in stress cracks first occurring at the bottom of the concrete.

Depending on building code regulations and soil classifications, bar reinforcement extending vertically through the top of the footings may be required to strengthen grout placed in the open cells of the CMU walls (see Figure 13-20).

The tops of the footings should be level. The author prefers a rough surface rather than a smooth surface. Using the tines of a garden rake to create grooves in the concrete's surface increases the mortar's bond with the footing. Covering the concrete with plastic tarps for the first twenty-four hours prevents excessive water evaporation from the concrete.

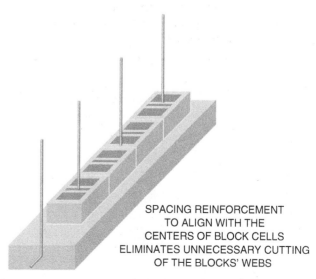

SPACING REINFORCEMENT TO ALIGN WITH THE CENTERS OF BLOCK CELLS ELIMINATES UNNECESSARY CUTTING OF THE BLOCKS' WEBS

**Figure 13-20 When building code regulations require, vertical bar reinforcement is spaced to align with the block's open cells.**

Excessive loss of water results in shrinkage cracking and inferior concrete hydration, the chemical reaction between the water and the cement necessary for developing strong concrete.

**CAUTION**

**Wet concrete can irritate and damage both skin and eyes. Wear eye protection with side shields. Wear alkali-resistant protective clothing, gloves, and boots to prevent skin contact.**

# Laying Out the Block Walls

Reattaching mason's lines to the batter boards helps to locate the corners and walls of the foundation (see Figure 13-21).

**CAUTION**

**Wear eye protection with side shields when laying out and constructing foundations.**

The corners of the foundation are established and visually marked. Masonry nails or masonry screws can be secured in the concrete to permanently mark each outside corner.

**FROM EXPERIENCE**

Drilling a hole into the concrete and screwing a self-threading masonry wall screw into it permanently identifies each corner of the foundation.

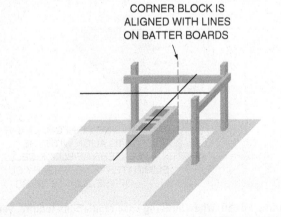

CORNER BLOCK IS ALIGNED WITH LINES ON BATTER BOARDS

**Figure 13-21** Lines attached to the batter boards locate the foundation corners.

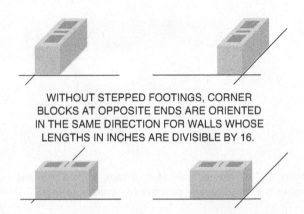

WITHOUT STEPPED FOOTINGS, CORNER BLOCKS AT OPPOSITE ENDS ARE ORIENTED IN THE SAME DIRECTION FOR WALLS WHOSE LENGTHS IN INCHES ARE DIVISIBLE BY 16.

**Figure 13-22** Two options for proper orientation of CMU faces at opposite ends of walls whose lengths in inches are divisible by 16 or length in feet is divisible by 4 are illustrated.

The dimensions are confirmed and equal diagonal measurements of rectangular areas are verified to assure square corners. The block bond is determined and marked along the footing. Chapter 11, "Constructing Block Leads," addresses corner constructions and the placement of cut blocks to accommodate walls having lengths requiring cutting block to length.

Constructing crawl spaces using 8" blocks or basements using 10" or 12" blocks result in walls whose lengths are multiples of 8". Designing foundations wall lengths in 8" increments eliminates cutting blocks to accommodate wall lengths. The following guidelines are observed for walls constructed on footings with no steps.

- The face of CMUs at opposite ends of a wall whose length in inches is divisible by sixteen or length in feet is divisible by four without a remainder in the quotient should be oriented in the same direction to permit a running bond pattern (see Figure 13-22).
- The faces of CMUs at opposite ends of a wall having a length in inches allowing dividing by eight and resulting in a quotient having a whole number without a remainder but is not divisible by sixteen without a remainder in the quotient should be oriented in opposite directions to permit a running bond pattern (see Figure 13-23).
- If working with dimensions expressed as feet rather than inches, the faces of the CMUs at opposite ends of a wall whose length in feet is divisible by two but not by four without a remainder in the quotient should be oriented in opposite directions to permit a running bond pattern (see Figure 13-23).

The previous guidelines are not necessarily true for walls built on stepped footings. The guidelines apply for walls with an even number of 8" steps (2, 4, 6). Just the opposite orientation is observed for walls whose footing has 8" steps of an odd number (1, 3, 5). A 16" step is considered as two 8" steps. A 24" step is considered as three 8" steps.

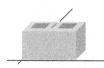

WITHOUT STEPPED FOOTINGS, CORNER
BLOCKS AT OPPOSITE ENDS ARE
ORIENTED IN OPPOSITE DIRECTIONS FOR
WALLS WHOSE LENGTHS IN INCHES ARE DIVISIBLE
BY ONLY 8 OR LENGTHS IN FEET ARE DIVISIBLE BY ONLY 2.

**Figure 13-23 Two options for proper orientation of CMU faces at opposite ends of walls whose *lengths in inches are divisible without a quotient remainder by 8 but not 16* or *length in feet are divisible without a quotient remainder by 2 but not 4* are illustrated.**

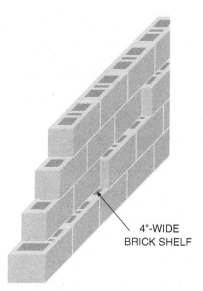

4"-WIDE
BRICK SHELF

**Figure 13-24 Both 12" and 8" CMUs have equal lengths and heights.**

# Building the Block Walls

The leads are constructed first, permitting the walls to be raised using the procedures of laying block to the line. Having the final course of all leads at the same level ensures level framework above the foundation. Chapter 10, Laying Block to the Line, and Chapter 11, Constructing Block Leads, can be reviewed for these procedures.

*For step-by-step instructions for checking the level alignment of foundations using a rotary laser level, see the Procedures section on pages 192–195.*

 **FROM EXPERIENCE**

Often, and almost always when footings have no steps, it is best to lay the entire first course before building the corners. This is preferred for the following reasons:

1. Having the top of the first course of CMUs level permits laying all remaining courses of the leads on equal course spacing.
2. An accurately laid out running pattern bond is confirmed before leads are built.
3. The first course elevates the walls above the surface of the footings, eliminating the potential for soil and rain water on top of footings and hindering wall construction.

Foundation walls supporting anchored brick veneer walls require a 4" **brick shelf.** A foundation wall designed for 12" block below finished grade and reduced to 8" block walls at finished grade level is a standard practice for foundations supporting anchored brick veneer walls.

**CAUTION**

**Comply with all federal and state safety regulations before entering excavations and trenches. The area between foundation walls below grade and the adjacent soil embankment resulting from excavations are susceptible to cave-ins that can result in serious injuries or death. Refer to OSHA Construction Industry Regulations 1926.651 and 1926.652 for specific standards.**

Because the 12" and 8" blocks have equal lengths and heights, the running bond pattern and course spacing are not affected when block size, meaning block width, is reduced (see Figure 13-24).

However, the blocks at the corners require a 4" shortening since the overall outside lengths of the walls are reduced by 4" on each side. The first 8" block for each side of each corner should measure 4" less than a standard block's length of 15⅝", or 11⅝" long (see Figure 13-25).

 **FROM EXPERIENCE**

Rather than cutting an 8" block to a length of 11⅝", you can use 12" half blocks at the corners. Turned sideways, the 12" half blocks have the needed dimensions. This eliminates the time required to cut the 8" blocks and assures accurate lengths for the blocks at the ends of the corners (see Figure 13-26).

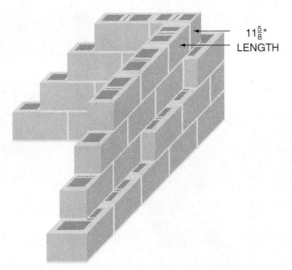

**Figure 13-25** A 4" reduction of the block width requires shortening the first block on each side of the corner by 4".

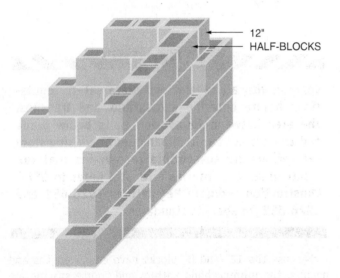

**Figure 13-26** In this instance, 12" *half-block* are used rather than shortening the two 8" blocks at the corner.

Veneer wall ties are spaced according to building code regulations for supporting the brick veneer. Section R703.7.4.1 of the International Residential Code requires that ties be not less than No. 22 U.S. gage by ⅞" corrugated and be spaced not more than 24" on center horizontally and vertically. Typical installation procedures are to bed wall ties in the bed joints of every other course of block and 16" apart along each course.

To permit a uniform bond, blocks may require cutting beside openings for doors or windows. A concrete masonry lintel is placed across the top of an opening to support the wall above it (see Figure 13-27).

Concrete masonry lintels should have a minimum bearing of 8" on each side of the opening. CMU lintels have a definite top side and bottom side. Rebar is placed in the lintel nearer the bottom side when it is cast. The rebar is most needed to

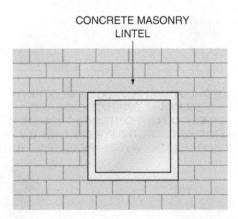

**Figure 13-27** A lintel supports the wall above it.

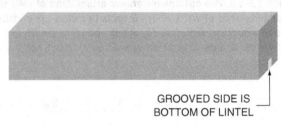

**Figure 13-28** The positioning of steel reinforcement bars within the lintel requires positioning the *grooved side* as the bottom.

help the lintel resist separation or pulling apart forces occurring along the bottom of the lintel, a factor influenced by the loads it supports. The top side of some lintels is imprinted with the word "top." The bottom side of other lintels has a ¾" × ¾" center slot or groove that runs the length of the lintels. The groove or slot of these lintels should be positioned toward the bottom (see Figure 13-28). This assures having the bar reinforcement in a lintel positioned nearer its bottom where it is most needed for allowing the lintel to resist tension forces, forces tending to pull it apart. Lintels are not required above foundation vents 16" in length.

Determined by the crawl space area, the required numbers of 8" by 16" foundation vents are placed along the last course of blocks. Two methods for installing foundation vents are illustrated in Figures 13.11 and 13.12.

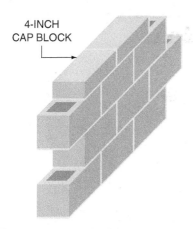

**Figure 13-29** At a height of 4", including the bed joint, cap blocks provide a solid bearing for a wooden sill plate.

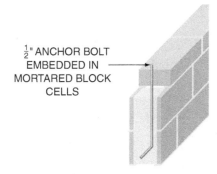

**Figure 13-30** Anchor bolts or anchor straps are embedded in fully mortared block cells.

**Figure 13-31** Bracings placed against the insides of foundation walls help prevent walls from collapsing when backfilling.

Solid, cap blocks are often specified for the top course of block on a foundation. Cap blocks measure $3\frac{5}{8}$" tall. With $\frac{3}{8}$" bed joints, the cap block course measures 4" in height (see Figure 13-29).

Placed between the cap blocks and extending above them, anchor bolts or anchor straps are embedded in fully mortared block cells of the course below the cap blocks (see Figure 13-30).

The installation of anchor bolts or anchor straps completes the block foundation.

# Preventing Walls from Collapsing

Before **backfilling,** replacing the earth removed during construction, a drainage system and damp-proofing or waterproofing are to be completed. Having the insides of the foundation walls below grade laterally braced and having the weight of the structural framing in place helps prevent the backfill from collapsing the walls (see Figure 13-31). Foundation walls above grade should be laterally braced on both sides to prevent high winds collapsing the walls before the structural framing is completed.

Extreme care should be taken when backfilling to prevent large stones or large dirt clods from hitting the walls. Their forces can crack or collapse the walls.

# Summary

- Although it is not customary for the mason to place the concrete footings, it is beneficial for the mason to recognize properly placed footings.
- Having knowledge of factors influencing design and building code regulations may prevent the mason from constructing foundation walls supported by poorly designed and improperly placed footings.
- The three primary considerations for designing continuous concrete footings are 1) size, 2) specified compressive strength, and 3) reinforcement.
- The mason is given the responsibility to build block or CMU foundation walls in accordance with acceptable engineered design and governing building code regulations.
- Factors considered for properly designed CMU foundation walls include 1) CMU size, 2) mortar type, 3) vertical bar reinforcement and grouting, 4) provisions for anchoring the wood sill plates, 5) damp-proofing or waterproofing the exterior walls, 6) exterior water drainage, 7) ventilation, and 8) egress routes.
- Because foundation walls have numerous details, a competent mason should supervise the construction of block foundation walls.

## Procedures    Installing Anchor Bolts

**A** Place block pieces in the block cell below the level for the bottom of the anchor bolt.

**B** Place the angled end of the bolt down into block cells.

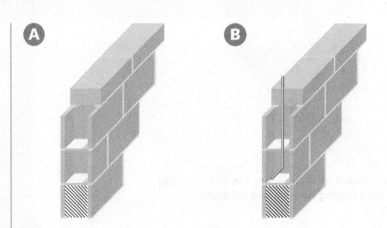

**C** Fill the block cells with mortar. Install the anchor bolt and align between head joints of cap block when used, allowing enough of the threaded end of the bolt exposed to extend above the top of the cap block and sill plate so that a washer and nut can secure the plate to the foundation wall.

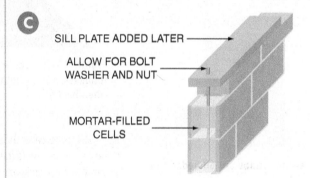

SILL PLATE ADDED LATER

ALLOW FOR BOLT WASHER AND NUT

MORTAR-FILLED CELLS

**D** Continue laying cap block, installing anchor bolts as required.

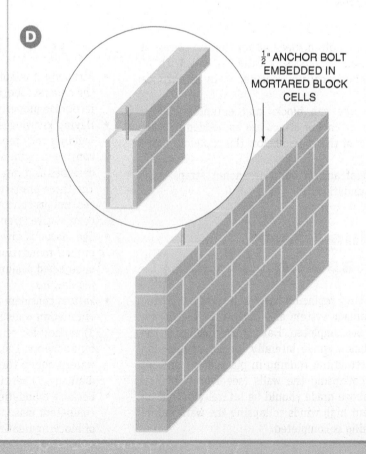

$\frac{1}{2}$" ANCHOR BOLT EMBEDDED IN MORTARED BLOCK CELLS

## Procedures

# Setting Up Batter Boards

**A** Drive three stakes securely into the ground beyond the outsides of the proposed corners. The stakes should be located far enough from the proposed walls to permit working without disturbing them, and 2 × 4 framing lumber is often used for this purpose.

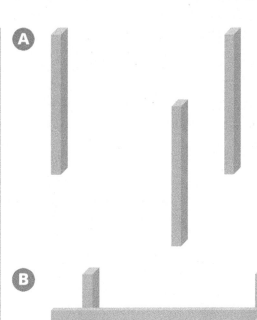

**B** Fasten a horizontally placed 2 × 4 between adjacent stakes at each corner. The resulting configurations are called *batter boards*.

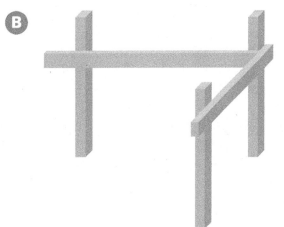

**C** Attach a mason's line from one set of batter boards to the set opposite it so that the attached line represents the desired orientation for a given wall of the proposed building. The line can be secured to nails or screws driven into the boards.

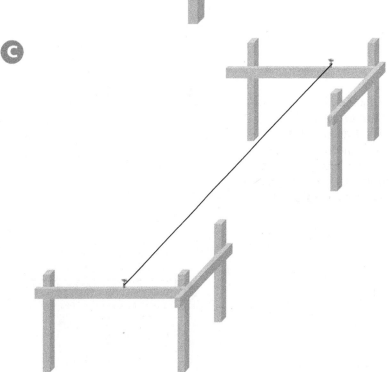

## Procedures   Setting Up Batter Boards (continued)

**D** Attach a line between two sets of batter boards perpendicular to the line established in Procedure C, representing the first of four corners. A line perpendicular to the first line is confirmed with the principle of a 3-4-5 right triangle. The two lines form a right angle if, when measuring 3' from the intersection of the two lines in one direction and 4' from their intersection in the opposite direction, the diagonal line is equivalent to 5'.

**D**
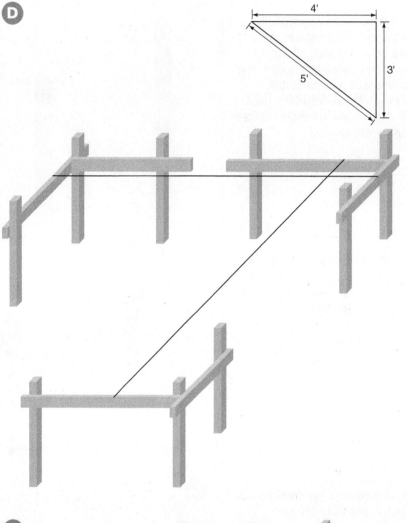

**E** Attach another line perpendicular to the line established in Procedure C and parallel to the line established in Procedure D so that the two lines intersect at a point equivalent to the specified length of the proposed wall.

**E**
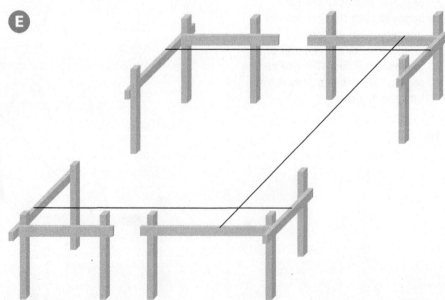

**F** Attach a line between two sets of batter boards parallel to the line attached in Procedure C and intersecting lines established in Procedures D and E at a point equivalent to the specified width of the proposed end walls.

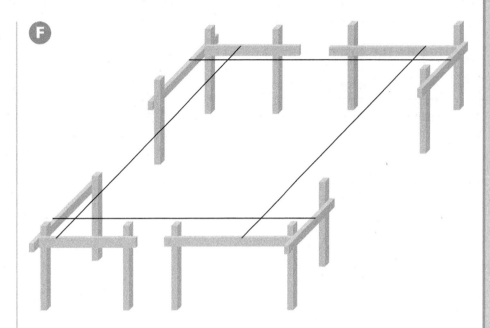

**G** Confirm a square layout. The four lines form square corners only if the two diagonal measurements are equal. If the diagonal measurements are not equal, and the lengths between the intersections of the lines are not accurate lengths, repeat Procedures C, D, E, and F. If the diagonal measurements are not equal, but the lengths between the lines are accurate lengths, realign the corners square using the Pythagorean theorem, and check the diagonal lengths again.

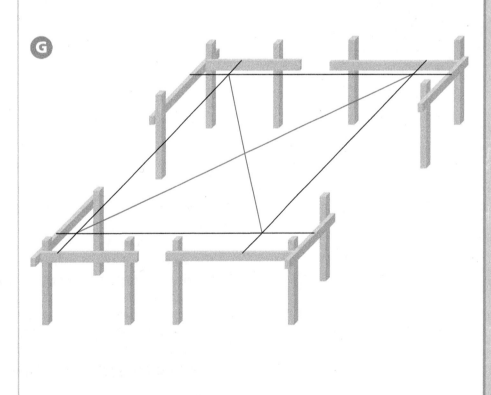

## Procedures

# Checking the Level Alignment of Foundations Using a Rotary Laser Level

### Setting Up the Instrument

**A** Adjust the three legs of the tripod so that the flat-head instrument mount appears level. Its height should not put the laser at eye level. Stepping firmly on the footpads at the base of each leg anchors them into the ground.

- Hold the instrument carefully with both hands and set the instrument's base onto the tripod's instrument mount. Tighten the threaded connector below the instrument mount into the instrument's base.

- Turn the battery-powered instrument on to activate it. Automatic self-leveling instruments complete an automatic leveling cycle before becoming operational. Once the leveling cycle is complete, the level's head begins to rotate. Adjust manually leveled instruments according to manufacturer's instructions.

**A**

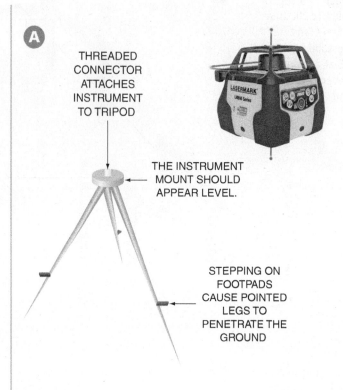

THREADED CONNECTOR ATTACHES INSTRUMENT TO TRIPOD

THE INSTRUMENT MOUNT SHOULD APPEAR LEVEL.

STEPPING ON FOOTPADS CAUSE POINTED LEGS TO PENETRATE THE GROUND

### CAUTION

Adjust the tripod height so that the instrument's laser beam is not at eye level. Do not stare at the light beam. Post warnings for others in the area. Avoid eye exposure to the laser light beam, and wear eye protection, "laser safety goggles which will protect for the specific wavelength of the laser and be of optical density (O.D.) adequate for the energy involved" meeting the intent of OSHA standards as defined in the OSHA Construction Industry Regulations, Section 1926.102(b) (2)(i) and Table E-3.

## Calibrating the Instrument

**B** Rotary laser levels are precision measuring instruments requiring careful handling. It is important to check the instrument's calibration before it is used. The accuracy as stated by the manufacturer should be confirmed. For instruments with a stated accuracy of $\pm\frac{1}{8}''$ at 100', the following operations test the accuracy of the level's calibration.

- Set up the instrument as previously described at a distance of 100' from a wall or the electronic detector. Turn the laser on, and allow it to self-adjust. Once the head begins to rotate, mark the location of the beam on the wall, or record the measurement taken using the leveling rod and attached detector. Carefully loosen the instrument on the instrument mount without altering the tripod's position, and rotate the instrument 180°. Retighten the instrument on the mount, and mark the location of the beam on the wall near the first mark, or record the measurement taken using the leveling rod and attached detector held in the exact location as before. There should be no more than a $\frac{1}{4}''$ difference between the two marks on the wall. Likewise, there should be no more than a $\frac{1}{4}''$ difference in the measurements recorded with the detector and leveling rod. Differences more than $\frac{1}{4}''$ require recalibrating the instrument as instructed in the owner's manual.

**B**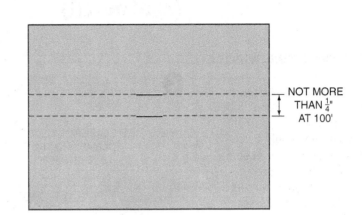

NOT MORE THAN $\frac{1}{4}''$ AT 100'

**Procedures**

# Checking the Level Alignment of Foundations Using a Rotary Laser Level (continued)

## Using the Laser Beam Detector

**C** Laser detectors are necessary in bright light conditions and especially outdoors. Bright light prevents seeing the light beam projected by the laser level onto surrounding surfaces. *Laser detectors* are electronic sensors capable of recognizing a laser beam. The laser detector clamps to a leveling rod. Sliding the detector up or down the leveling rod permits the detection zone of the detector to locate the laser beam. Once the laser beam is within its detection zone, the dual-sided detector is moved slightly up or down the leveling rod as indicated by the arrows of the LCD. A horizontal line appears once the detector and the laser beam are at the same level. In addition to the LCD, the detector emits an audible tone confirming it and the laser beam are at the same level. The measurement aligned with the detector is read on the leveling rod.

## Reading the Leveling Rod

**D** A leveling rod is a telescoping rod graduated in feet, inches, and fractions. The fractions may be either eighths or tenths of inches. Sectional telescoping rod lengths up to 25' are common. The rodman, or person holding the rod, records the measurement targeted by the attached laser detector. Rod readings are explained in the following illustration.

**C**

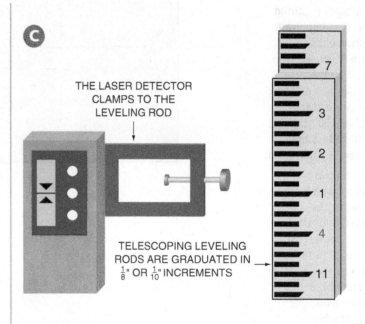

THE LASER DETECTOR CLAMPS TO THE LEVELING ROD

TELESCOPING LEVELING RODS ARE GRADUATED IN $\frac{1}{8}$" OR $\frac{1}{10}$" INCREMENTS

**D**

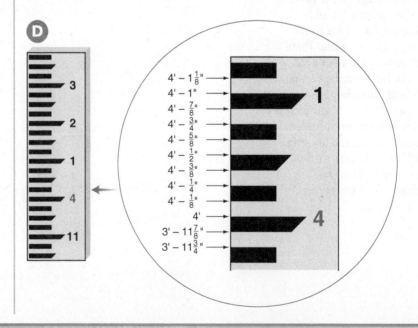

$4' - 1\frac{1}{8}"$
$4' - 1"$
$4' - \frac{7}{8}"$
$4' - \frac{3}{4}"$
$4' - \frac{5}{8}"$
$4' - \frac{1}{2}"$
$4' - \frac{3}{8}"$
$4' - \frac{1}{4}"$
$4' - \frac{1}{8}"$
$4'$
$3' - 11\frac{7}{8}"$
$3' - 11\frac{3}{4}"$

## Interpreting Readings Along the Top of a Concrete Footing

**E** With the instrument accurately calibrated, record measurements along the top of the footings. Measurements taken along footings that have no steps should be alike in reference to the laser level. In the left side of this illustration, the laser beam is detected at the reading of 63½". In the right side of this illustration, the laser beam is detected on the leveling rod at 63 ¾". The difference between the two readings is ¼", equivalent to the difference in elevation of the two points. The *larger* reading of 63¾" indicates a surface ¼" *lower* than the reading of 63½".

## Confirming Level Block Coursing

**F** Place the leveling rod with detector attached on top of the block coursing. Compare the rod reading with those of other corners. Note: The instrument must be accurately calibrated and not moved while taking these readings. Identical readings on the leveling rod confirm level block coursing at the leads.

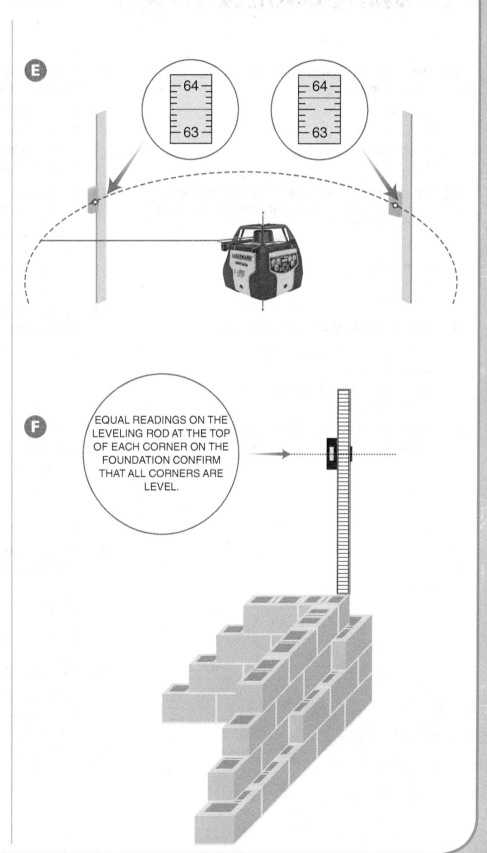

EQUAL READINGS ON THE LEVELING ROD AT THE TOP OF EACH CORNER ON THE FOUNDATION CONFIRM THAT ALL CORNERS ARE LEVEL.

# Review Questions

*Select the most appropriate answer.*

**1** That part of a building below the first floor framing is known as the

a. excavation
b. egress
c. foundation
d. areaway

**2** The concrete structure supporting the foundation walls and transmitting the structure's weight to the soil is called the

a. lintel
b. bridging
c. shelf
d. footing

**3** The size of a footing is determined by

a. the load-bearing capacity of the soil
b. the wall design
c. the number of building stories
d. all of the above

**4** The minimum thickness of concrete for residential footings is

a. 4"
b. 8"
c. 12"
d. 16"

**5** The compressive strength of concrete used for footings is determined by

a. the type of construction
b. the location of the concrete
c. the weathering potential
d. all of the above

**6** Steel rods embedded in concrete to strengthen the concrete are called

a. bar reinforcement
b. rebar
c. reinforcement rods
d. all of the above

**7** The minimum depth for the bottom of a concrete footing below finished grade is

a. 8"
b. 12"
c. 16"
d. 24"

**8** The greatest depth to which ground may be expected to freeze for a given region is called the

a. frost line
b. freezing point
c. frost cap
d. frost zone

**9** The maximum height or minimum length for a single step along a concrete footing is

a. 8"
b. 16"
c. 24"
d. 32"

**10** The size of block used for a foundation depends on

a. the wall height
b. the height of unbalanced backfill
c. the soil classification
d. all of the above

**11** Masonry cement generally specified for building foundation walls is

a. Type M
b. Type S
c. Type N
d. either a or b

**12** A mixture of cement and aggregates mixed with water to produce a pouring consistency that is placed into blocks' open cells to strengthen the wall is known as

a. flowable fill
b. parging
c. cement
d. grout

**13** The structure supported by the foundation walls is secured to the top of the foundation using

a. nails
b. anchor bolts
c. anchor straps
d. either b or c

**14** A material used to help prevent water migrating from the outside and through foundation walls is

a. cement parging
b. perforated plastic pipe
c. an approved membrane
d. all of the above

**15** The area between the ground surface and floor joists of houses without basements is called

a. the areaway
b. the egress
c. the crawl space
d. none of the above

**16** A walled area below grade permitting ventilation, light, or exit route is called

a. an areaway
b. a retaining wall
c. a stoop
d. none of the above

**17** An assessment of the consistency of freshly placed concrete is called the

a. cone test
b. slump test
c. compression test
d. tensile test

**18** Steel bar reinforcement embedded horizontally in concrete footings should be placed nearer the

a. top
b. middle
c. bottom
d. sides

**19** It is recommended to cover new concrete for

a. at least the first 12 hours
b. at least the first 24 hours
c. at least the first week
d. none of the above

**20** Foundation walls supporting anchored brick veneer walls require a brick shelf having a width of

a. 2"
b. 4"
c. 6"
d. 8"

**21** Anchored brick veneer walls are anchored to the block foundation walls by means of

a. wick ropes
b. veneer wall ties
c. anchor bolts
d. rebar

**22** Placed across the top of an opening in a masonry wall to support the wall above it is a

a. rowlock
b. lintel
c. truss
d. joist

# Chapter 14

# Constructing Water-Resistant Brick Veneer Walls

ater penetration is a leading cause of masonry wall failures. The sources of this water include rain, snow, ice, and air moisture. Water corrodes and weakens steel wall reinforcement and structural steel, rots wood framing, damages interior wall finishes, decreases the effectiveness of insulation, and permits the unsightly crystallization of water-soluble salts on the faces of walls. Water penetration can also be unhealthy for a building's occupants. Moisture promotes the growth of molds, microscopic organisms present almost everywhere in the environment. Molds can present respiratory problems as well as nose, throat, skin, and eye irritations.

## OBJECTIVES

**After completing this chapter, the student should be able to:**

- Identify the sources of water behind exterior masonry walls.
- Identify means of minimizing water migration through masonry walls.
- Define the terms *flashing* and *weeps*.
- Describe installation procedures for flashing and weeps.
- List different materials used as flashing.
- Explain how mortar collection systems can improve the effectiveness of weeps.
- Discuss the types of water repellants and their recommended applications.

## Glossary of Terms

**air infiltration barrier** a sheathing membrane installed to the exterior of wall framing for preventing air infiltration and resisting moisture migration from outside while permitting the escape of interior moisture.

**air space** the space between the backside of the brick veneer wall and the exterior side of the backer wall, permitting air circulation.

**backer wall** the wall to which the brick veneer is anchored.

**base flashing** flashing at the base of the wall just above the finished grade level.

**drainage type wall system** a wall design that has a means of diverting water from the air space to the outside of the exterior, single-wythe masonry wall.

**efflorescence** a deposit of water-soluble salts appearing on the surface of masonry walls.

**film-formers** water repellants adhering to the surface of masonry walls.

**flashing** a water impermeable material installed in the wall system, diverting collected water to the exterior of the wall.

**head flashing** flashing above openings.

**initial rate of absorption (IRA)** a test determining the moisture content of brick.

**lime run** a white or gray, crusty calcium carbonate deposit on the faces of masonry walls as a result of any of several calcium compounds in solution with water brought to the surface of the masonry through an opening, where the solution reacts with carbon dioxide in the air.

**mortar bridging** the result of excessive mortar protruding from joints and contacting the backer wall, blocking the air space between the brick veneer wall and the backer wall.

**mortar collection systems** also called mortar deflection systems or mortar breaks, products designed to prevent mortar from obstructing the performance of weep holes, weep vents, or wick ropes.

**mortar protrusions** that mortar squeezed beyond the face and backside of the brick as the brick is bedded in mortar.

**new building bloom** efflorescence appearing soon after construction because of the increased presence of water during construction.

**penetrants** water repellant coatings that penetrate the surface of masonry walls.

**sill flashing** flashing below masonry sills at windows and doorframes.

**suction rate** the rate at which a brick absorbs water.

**water repellants** liquids applied to masonry walls for the purpose of reducing water absorption.

**weep holes** openings left in the head joints of brickwork to permit water drainage and allow air circulation in the air space.

**weep vents** vented covers, corrugated materials, or open mesh polyester installed in open head joints to permit water drainage and to allow air circulation.

**white scum** white or gray stains on the face of brick masonry, typically related to the cleaning of brickwork with unbuffered hydrochloric (muriatic) acid solutions or inadequate prewetting or rinsing of the brickwork during cleaning.

**wick ropes** pieces of $1/4$" to $3/8$" diameter cotton rope routed from the air cavity through head joints beyond the face of the wall to divert water from the air cavity above the flashing.

# Controlling Water Migration

It is a fact that water migrates through single-wythe brick walls. According to the Brick Industry Association, "Under normal conditions, it is nearly impossible to keep a heavy wind-driven rain from penetrating a single wythe of brickwork, regardless of the quality of the materials or the degree of workmanship used." Following recommended masonry construction practices minimizes moisture-related problems. Water penetration may not be completely eliminated, but it must be minimized, and any water intruding into a masonry wall should be diverted to the outside. There are several recommendations for minimizing water penetration of 4" anchored brick veneer walls:

- Ensure proper moisture content of the brick at the time it is bedded in mortar.
- Ensure fully mortared joints.
- Maintain an unobstructed air cavity.
- Install flashing and weeps.

## Moisture Content of Brick

Studies show that most water penetrating a brick wall occurs at separations or cracks between the mortar and the brick or at junctures with other materials. Very little water enters directly through the brick or mortar. During the manufacturing process, high kiln temperatures fuse the raw materials into a highly water-resistant mass. It is therefore important to ensure a good bond between the brick and the mortar. A good bond requires brick to have proper moisture content. Brick laid when too dry, especially during warmer weather, causes the mortar to set too quickly. This results in shrinkage of the mortar, forming small cracks that allow water to enter the wall. It is recommended to dampen very dry brick before bedding in mortar. However, laying brick that have too high water content or whose surfaces are saturated can reduce the absorption of the mortar into the brick. This also results in a weaker bond between the brick and the mortar.

The mason can conduct an initial rate of absorption test at the jobsite to determine the moisture content of the brick before they are to be bedded in mortar. An initial rate of absorption test, called an IRA test, is conducted in the field using water and a medicine dropper to determine the moisture content of brick.

According to the Brick Industry Association Technical Notes 7B, Specification for Masonry Structures ACI 530.1/ASCE 6-05/TMS 602-05 require that the initial rate of absorption (IRA) of brick at the time of laying not exceed 1 gram per minute.

**Figure 14-1** Absorption rates exceeding 20 drops in less than 1½ minutes require one of the alternative procedural adjustments described.

The Brick Industry Association recommends the following procedures to perform a jobsite IRA test. Using a wax crayon, a circle the size of a quarter coin is drawn on the brick's surface, the side to be bedded in mortar. The time required for the circled area to absorb 20 drops from a medicine dropper is recorded (see Figure 14-1). The suction rate, the rate at which the brick absorbs the water, is considered too high if the 20 drops are absorbed within the circle in less than 1½ minutes.

The following are three recommendations for brick that have too high an IRA:

1. Wet units several hours before laying.
2. Mix mortar with the maximum amount of water.
3. Use approved mortar retentivity admixtures.

## Mortar Joints

Most building codes and engineering specifications require full mortar joints. Applying full head joints to the brick before they are laid and avoiding deeply furrowed bed joints are both recommended. Unfilled joints are frequently found in walls experiencing water intrusion. Joint width variations do not contribute to water penetration as long as they are full and within the tolerances of The Masonry Standards Joint Committee's Specification for Masonry Structures. The Masonry Standards Joint Committee's Specification for Masonry Structures (ACI 530.1-99/ASCE 6-99/TMS 602-99) in paragraph 3.3G.1.b, provides variations of head joint thickness from $-\frac{1}{4}$" to $+\frac{3}{8}$".

Joints that are not tooled to compact the outer surface of the joint are likely to experience more water penetration. Mortar joint finishes recommended for exterior walls include the concave, "V", grapevine, and the troweled weather joint. Depending on geographic location, other joint finishes may not be considered weather resistant. Consult the brick manufacturer, industry standards, or a professional engineer before selecting joint finishes other than those identified here.

Remember that when it comes to mortar, "wetter is better." In most cases, mortar having a high water content results in a stronger bond strength than mortar that is dryer. Keep mortar well tempered, and spread no more mortar than is necessary for brick to be bedded within "1 minute or so" of the time of spreading, the time recommended by the Brick Industry Association. Leave mortar in tarp-covered wheelbarrows or mortar boxes, putting no more mortar on mortar boards at a time than can be used without tempering.

Minimizing mortar protrusions and eliminating mortar bridgings reduces moisture-related problems in the air space between brick veneer walls and backer walls, the walls to which brick veneer is anchored. A mortar protrusion is that mortar squeezed beyond the face and backside of the brick as the brick is bedded in mortar. The trowel removes the protruding mortar at the face. But mortar protrusions within the air space between the brick and backer walls often remain, reducing optimum airflow within the air space. Mortar bridging is the result of excessive mortar protruding from joints and contacting the backer walls. Mortar bridging blocks airflow and provides a direct path to the backer wall for water that has penetrated the exterior masonry wythe.

## Air Space

There should be an air space, a space permitting air circulation between the brick veneer and the backer wall. The air space also provides a drainage path to the outside of brick veneer walls where water has migrated through the wall system. The air space also increases air circulation, permitting removal of air moisture that can result in water condensation behind the brick veneer. Section R703.7.4.2 of the International Residential Code requires the air space to be a minimum of 1" wide. The Brick Industry Association Technical Notes 7 recommends a 2" air space between brick walls and backup walls that have steel studs. Care must be taken to prevent mortar droppings, mortar protrusions, and mortar bridgings from obstructing the air space, providing a direct route for water to reach the framing or backup wall (see Figure 14-2). Although the building code has provisions for mortar or grout to fill the air space, the air space is recommended.

Mortar protrusions can be minimized by repositioning furrowed bed joints (see Figure 14-3).

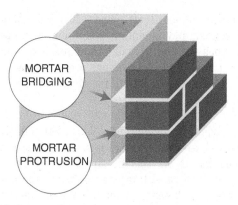

**Figure 14-2 Excessive mortar protrusions and bridging are evident in this wall system.**

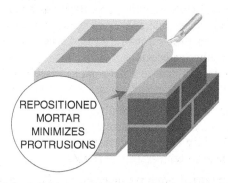

**Figure 14-3 Positioning and passing the trowel blade in such a manner as to return furrowed mortar from behind the brick and onto the brick minimizes mortar protrusions once brick are bedded.**

## Flashing and Weeps

Flashing is a water impermeable material installed in the wall system that diverts collected water to the exterior of the wall. Section R703.7.5 of the International Residential Code requires flashing beneath the first course of masonry above finished ground level above the foundation wall or slab and at other points of support, including structural floors, shelf angles, and lintels.

Following are the three categories of flashings:

1. Sheet metals
2. Composite materials
3. Plastic or rubber compounds

The Brick Industry Association Technical Notes 7A addresses flashing materials mentioned here. Copper is recognized as a superior flashing material because it withstands all harmful acid and alkali action that is present in mortar. Copper is easily formed to shape at the jobsite. Aluminum, galvanized steel, and sheet lead are not recommended as flashing materials because their reaction with mortar leads to corrosion and disintegration of these metals.

Composite materials can be a combination of two or more materials. Some of these materials are glass or cotton fabric, waterproofed creped kraft paper, blended asphalt compounds, copper, and aluminum. Aluminum or copper coated on both sides with an asphalt compound is one such product. The asphalt coating provides protection against electrolysis, a chemical reaction in which the metal causes flashing failure.

Rubber compounds include polyvinyl chloride (PVC) materials, Ethylene Propylene Diene Monomer (EDPM), rubber, and assorted rubberized materials. Selection of materials should be based on specific applications. For example, EDPMs provide excellent resistance to weathering, chemicals, ozone, and UV exposure but are not considered to be oil resistant. Flashings are available in thicknesses from 20 mil to 40 mil (0.5 mm to 1 mm). Semi-rigid, extruded polypropylene flashings are available for windows and doorjambs in cavity wall construction. Some flashings are attached to walls using mechanical anchoring systems or adhesives approved by the flashing manufacturer, whereas other flashings have adhesive backings.

Although cost can be a determining factor in the selection of flashing material, there are other factors to consider:

- Lifetime and resistance to ultraviolet light.
- Resistance to alkaline masonry mortars.
- Compatibility with joint sealants.
- Tolerances to temperature extremes.

Only approved products identified as flashing should be used. Polyethylene sheeting or asphalt impregnated building felt should not be used as flashing. Sections of flashing should be lapped 6" and the joints properly sealed using only those sealants recommended for the specific flashing.

Three locations for installing flashing in brick veneer walls are at the base of walls, below brick sills, and at the head of or above openings.

### CAUTION

**Wear industrial gloves that have a high tear rating when handling metal flashings to protect your hands from cuts. Wear eye and skin protection when exposed to sealants. Avoid breathing vapors and wear respiratory protection where adequate ventilation is not available.**

### Base Flashing

Penetrating water and condensing moisture in the air cavity drain to the bottom of the wall. Without flashing, this moisture flows into the foundation's hollow concrete masonry units (CMUs). As a result, foundation walls and basements become damp. Also, the water damages interior wall materials, including insulation, wallboard, and paint (see Figure 14-4).

Base flashing stops the downward travel of water and diverts it to the outside of the wall, preventing its downward

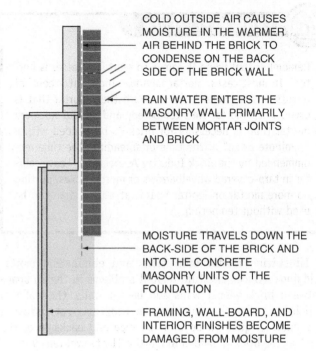

COLD OUTSIDE AIR CAUSES MOISTURE IN THE WARMER AIR BEHIND THE BRICK TO CONDENSE ON THE BACK SIDE OF THE BRICK WALL

RAIN WATER ENTERS THE MASONRY WALL PRIMARILY BETWEEN MORTAR JOINTS AND BRICK

MOISTURE TRAVELS DOWN THE BACK-SIDE OF THE BRICK AND INTO THE CONCRETE MASONRY UNITS OF THE FOUNDATION

FRAMING, WALL-BOARD, AND INTERIOR FINISHES BECOME DAMAGED FROM MOISTURE

**Figure 14-4 Water causes moisture-related problems behind brick veneer walls.**

travel into the open cells of CMUs used for the foundation walls. Section R703.7.5 of the International Residential Code requires base flashing "beneath the first course of masonry above finished ground level above the foundation wall or slab".

### FROM EXPERIENCE

Install base flashing above ground level to be higher than potentially future mulched plant beds later to become "ground level". Mulch can block weeps at ground level, preventing the discharge of water and restricting air circulation in the air space.

To form a drip edge, the flashing should start ½" from the outside face of the wall, continue through the wall below the bed joint, turn upward against the framing or adjacent wall a minimum of 8", and be securely fastened. It must be continuous for the entire perimeter of the structure. The flashing extending beyond the exterior face of the wall can be cut flush with the face of the wall, specifications permitting. All joints should be lapped at least 6" and sealed with approved mastic or adhesives that are compatible with the flashing material. These and more recommendations can be found in the Brick Industry Association Technical Notes 7B. Without proper drainage, the base flashing collects the water and holds it like a dam. Weep holes or weep vents are used for a drainage system. A wall design with means of

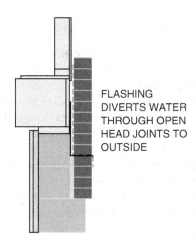

FLASHING
DIVERTS WATER
THROUGH OPEN
HEAD JOINTS TO
OUTSIDE

**Figure 14-5** Base flashing stops the downward travel of water and weep holes permit the water to return to the outside.

diverting water from the air space to the outside of the exterior wythe is called a **drainage type wall system** (see Figure 14-5).

### Weeps

Weeps are included in masonry walls to permit water drainage from behind the exterior wall to its face. Three methods of weeps are used: weep holes, wick ropes, and weep vents.

*Weep Holes.* Weep holes are openings left in the head joints of brickwork to permit drainage. They are typically created by omitting mortar in the third head joint on the course just above the base flashing. The Brick Industry Association Technical Notes 7B recommends open head joint weeps to be at least 2" high and no more than 24" apart. The openings permit water to return outside while allowing air circulation in the cavity for quicker drying. Section R703.7.6 of the International Residential Code requires weep holes at a maximum spacing of 33" on center and not less than $^3/_{16}$" in diameter, and to be located directly above the flashing.

Sometimes, oiled rods or short pieces of rope are mortared into the head joints and then removed after the mortar has hardened. This creates smaller, less noticeable openings in the wall. Such weep holes should be at least $^1/_4$" in diameter. Weep holes can create problems.

- Open head joints can become blocked by debris.
- The unfilled joints create a dark shadow and give an unfinished appearance to the wall.
- Insects enter into the cavity through such unfilled joints.
- Sometimes, they are mistakenly thought of as a source for water penetration and therefore filled.

Weep holes can be filled with materials that prevent insects and debris from entering the joint yet still allow them to work properly. These materials include cotton rope and assorted vents.

As the course of brick above the flashing are laid, place short pieces of $^3/_8$" diameter ropes between the brick at the bottom of mortared head joints just above the flashing, extending from the backup wall and beyond the face of the wall 2" to 3". After the mortar becomes thumbprint, carefully pull the ropes from the joints so as not to misalign the brick. The resulting weep holes are uniformly shaped and adequately sized.

*Wick Ropes.* Pieces of $^1/_4$" to $^3/_8$" diameter cotton rope routed from the air cavity through head joints beyond the face of the wall to divert water are called wick ropes. The Brick Industry Association Technical Notes 7B recommends wick and tube weeps to be spaced no more than 16" apart. This is equivalent to every other head joint. Water travels through the rope to the face of the wall. Like kerosene passing through the wick of a kerosene lamp, the water flows through the rope. This process is referred to as wicking. The ropes are often referred to as **wick ropes.** It is important that cotton rope be used because other roping materials may not wick the water.

*Weep Vents.* Weep vents can be used instead of weep holes. Weep vents are vented covers, corrugated materials, or open mesh polyester installed in the open joints. Weep vents perform the same functions as weep holes but prevent the problems associated with the open holes. Manufacturers can duplicate the color and the texture of the mortar and therefore disguise the vents.

Weep vents allow air to enter at the base of the wall. Air flows by convection currents through the wall cavity and out the top through vents at the soffit. The airflow reduces moisture-related problems, including mold growth within the wall. Weep vents having vented covers are typically installed every third head joint. But the recommended spacing for weep vents made of corrugated or open mesh materials may be the same as that for wick ropes, placed every other head joint, since airflow is somewhat restricted.

### Sill Flashing

**Sill flashing** is installed below masonry sills under windows and doorframes (see Figure 14-6). Horizontally laid brick sills can be more likely to experience water penetration because water does not run off them as quickly as it runs down a vertical, brick veneer wall. The Brick Industry Association Technical Notes 7 recommends that the ends of the flashing should extend beyond the jamb lines on both sides and turn up into the head joint not less than 1" to form a dam. As with base flashing, the flashing should start $^1/_2$" from the outside face of the wall. Cotton ropes placed horizontally on the flashing behind the sill and routed below the sill to the face of the wall wick water to the outside.

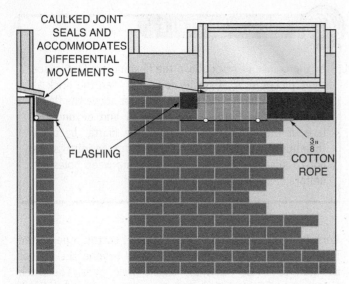

**Figure 14-6** Sill flashing stops water and diverts it to the exterior through weeps.

### Head Flashing

Head flashing is installed above openings (see Figure 14-7).

Head flashing protects steel lintels supporting brick masonry from corrosion and rust. Head flashing should be continued 6" beyond both ends of the steel lintel and turned up into the head joint not less than 1" to form a dam. The flashing should extend vertically a minimum of 8" above the lintel, where it is to be anchored to the framing or backup masonry wall. As with base and sill flashings, head flashing should extend through the wall to the outside face of the wall. Open weeps are to be installed every third head joint, but if using wick ropes, mesh vents, or corrugated vents, then the spacing should be every other head joint.

## Mortar Collection Systems

Some mortar always falls into the air cavity, ending up on the base flashing or atop the head flashing. This mortar blocks the weeps, inhibiting their intended performance. Mortar collection systems, also called mortar deflection systems or mortar breaks, are products designed to prevent mortar from obstructing the performance of weep holes, weep vents, or wick ropes. Placed on top of the base flashing or head flashing inside the wall cavity, an open-mesh polyester material suspends the mortar droppings above the level of the flashing (see Figure 14-8). This permits mortar from blocking the weeps. A dovetail-design product claims better water and air flow because it results in mortar droppings suspended at two different levels. Regardless of design preference, a mortar-collection device should be considered for preventing mortar-clogged weep holes or weep vents at base and head flashing.

## Efflorescence

Efflorescence is a deposit of water-soluble salts appearing on the surface of masonry walls. Although efflorescence may have no effect on a building's structural integrity, this chalky white coating detracts from the appearance (see Figure 14-9).

Two conditions must be present before efflorescence can occur.

1. Soluble salt must be in the masonry.
2. A significant source of water must be present to dissolve the salt.

In Technical Notes on Brick Construction 23A, the Brick Industry Association reports that "when chlorides occur as

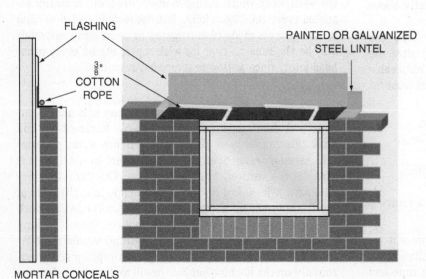

MORTAR CONCEALS STEEL LINTEL

**Figure 14-7 Head flashing is placed above the steel lintel.**

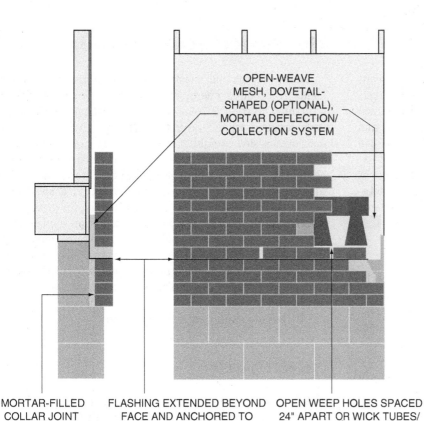

OPEN-WEAVE MESH, DOVETAIL-SHAPED (OPTIONAL), MORTAR DEFLECTION/ COLLECTION SYSTEM

**Figure 14-8** The open weave mesh mortar collection material prevents falling mortar from blocking the weeps.

MORTAR-FILLED COLLAR JOINT

FLASHING EXTENDED BEYOND FACE AND ANCHORED TO BACKING A MINIMUM OF 8" VERTICALLY ABOVE WEEPS

OPEN WEEP HOLES SPACED 24" APART OR WICK TUBES/ MATERIALS 16" APART

**Figure 14-9** The face of this brick wall exhibits efflorescence.

- Using water free of salts for mixing mortar
- Installing flashing and weeps
- Properly filling and tooling mortar joints
- Using recommended cleaning agents and procedures

Efflorescence should not be mistaken for lime run, calcium carbonate deposits, or white scum, silicate deposits. Both appear as white or gray discolorations. According to the Brick Industry Association, lime run results as "the water takes any of several calcium compounds into solution and brings them to the surface of the masonry through an opening. At the surface, the solution reacts with carbon dioxide in the air, thus forming the crusty calcium carbonate deposit." The Brick Industry Association reports that "white scum is typically related to the cleaning of brickwork with unbuffered hydrochloric (muriatic) acid solutions or inadequate prewetting or rinsing of the brickwork during cleaning."

efflorescence, it is usually the result of the use of calcium chloride as a mortar accelerator, contamination of masonry units or mortar sand by salt water, or the improper use of hydrochloric acids in cleaning solutions."

Since masonry materials are made from materials mined from the earth, the presence of salts is always a possibility. Efflorescence can be prevented by doing the following:

- Protecting materials and walls from rain during construction

## New Building Bloom

Efflorescence may appear soon after construction because of the increased presence of water during construction. This occurrence is known as new building bloom. The bloom disappears as the materials and wall system dry. Removing bloom with water often prolongs the problem. If efflorescence persists, then it can be an indication of other problems.

# Water Repellants

Water repellants are liquids applied to masonry walls for the purpose of reducing water absorption. Water repellants should not be used to remedy water penetration until an inspection finds the wall to be properly designed and constructed and all repairs have been made. Chimneys can be an exception. Because of their exposed surfaces and exposure to severe weather, water repellants are sometimes recommended to prevent water penetration leading to deterioration of chimneys. The two categories of water repellants are film-formers and penetrants.

## Film-Formers

Film-formers adhere to the surface of masonry walls. Their large molecular size prevents them from penetrating the surface. Film-formers can prevent water from entering hairline cracks, but surface preparation is important for proper performance. Film-formers may prevent evaporation of moisture within the masonry through the exterior of the wall. For this reason, film-formers should not be used in climates experiencing freezing and thawing cycles.

## Penetrants

Penetrants are coatings that penetrate the surface of the wall and perform by repelling water rather than forming a film. Penetrants are more resistant to ultraviolet degradation. Because penetrants coat the pores rather than bridge them, moisture within the wall can migrate to the outside and evaporate. Penetrants cannot stop water migration through hairline cracks because they do not bridge the cracks.

The thickness of the film or the depth of penetration and the severity of the weather influences the life span of the coatings. Reapplication of coatings can be complicated. A penetrating coating may not be able to be applied over a film-forming coating. Reapplication of the same coating can cause clouding of the surface. It may be necessary to strip the existing surface of any prior coating, and this procedure can involve hazardous chemicals regulated or restricted from use. The brick manufacturer should be consulted before applying any repellant.

**CAUTION**

Read and comply with recommendations of the brick manufacturer before using any water repellant. Test first on a small inconspicuous area before applying to larger areas. Comply with all safety and health personal protection standards as well as environmental regulations when using chemicals.

# Air Infiltration Barriers

Preventing air and moisture infiltration are key requirements for today's energy-efficient, brick veneered, frame structures. Section R 703.2 of the International Residential Code requires asphalt-saturated felt paper free from holes and breaks and weighing not less than 14 pounds per 100 square feet or other approved weather-resistant material be applied over studs or sheathing of all exterior walls. An air infiltration barrier or housewrap is a sheathing membrane installed to the exterior of framing for preventing air infiltration. It also resists moisture migration from outside while permitting the escape of interior moisture. Housewrap is a specially coated, woven polyethylene fiber material. Installing the housewrap is a task not generally performed by the mason. However, the mason needs to apply the specially formulated tape used to seal the seams and edges to tears or rips before proceeding with the brickwork.

# Summary

- Proper design and quality construction are important factors for preventing water migration through masonry walls.
- Key design elements include 1) properly installed flashings and weeps diverting water to the outside, 2) unobstructed air cavities permitting air circulation, and 3) air infiltration barriers restricting moisture migration.
- Quality construction procedures include 1) applying full mortar joints, 2) insuring proper moisture content of brick at time of laying, and 3) observing recommendations of materials manufacturers, engineered specifications, industry standards, and building code requirements.
- Remediation of inferior wall design and construction are costly processes.

## Review Questions

*Select the most appropriate answer.*

**1** A possible result of water penetrating a brick veneer wall is

a. damaged structural framing materials
b. decreased performance of thermal insulations
c. respiratory problems developed by a building's occupants
d. all of the above

**2** Studies show that water is likely to penetrate a brick veneer wall mostly

a. through the brick
b. through the mortar
c. between the mortar joints and the brick
d. both a and b

**3** A test determining the moisture content of a brick at the time of laying is called the

a. average water content
b. initial rate of absorption
c. moisture percentage
d. suction rate

**4** To minimize water penetration, it is recommended to

a. apply fully mortared head joints
b. apply minimally furrowed mortared bed joints
c. use a rake jointer to tool mortar joints
d. both a and b

**5** The air space between an anchored brick veneer wall and wood framing should be at least

a. $\frac{1}{2}$"
b. 1"
c. $1\frac{1}{2}$"
d. 2"

**6** Excessive amounts of mortar protruding from the bed joints, becoming attached to the backup wall and blocking the air space between the backside of a brick veneer wall and the backup wall is called mortar

a. bridgings
b. protrusions
c. troughs
d. veins

**7** Water-impermeable materials installed in masonry walls to control water penetration are called

a. coats
b. flashings
c. liners
d. membranes

**8** Open head joints permitting water drainage are called

a. drain holes
b. spillways
c. weeps
d. wicks

**9** Open head joints minimize water within the air space by permitting

a. water drainage beyond the face of the wall
b. air circulation within the air space
c. greater air pressures within the air space than outside the wall
d. both a and b

**10** Flashing should be installed

a. at the base of a wall
b. below sills at openings in a wall
c. on lintels above openings in walls
d. all of the above

⑪ Products designed to prevent mortar droppings from blocking weep holes or vents are called

a. mortar breaks
b. mortar collection systems
c. mortar deflection systems
d. all of the above

⑫ The source for water-soluble salts found on the surface of masonry walls can be

a. contaminated building materials
b. water used for mixing mortar
c. cleaning agents
d. all of the above

⑬ A condition exhibited as water-soluble salts on the face of a newly constructed wall is called

a. chalking
b. new building bloom
c. salt corrosion
d. powder coating

⑭ Water-repellant coatings that penetrate the surface of masonry walls are called

a. film-formers
b. penetrants
c. surficants
d. sash coats

⑮ Sheathing membranes covering the exterior of framed walls intended to prevent air infiltration are called

a. air infiltration barriers
b. housewraps
c. raincoats
d. both a and b

# Chapter 15

# Anchored Brick Veneer Walls

Brick is considered by many as the premier exterior facade for quality homes. Brick does not rot, twist, warp, or dent. The damaging forces of hail do not penetrate it. Brick does not burn. Less noise is transmitted through the walls of brick homes. For these reasons, residential construction, including homes, apartments, town houses, and condominiums, often relies on anchored brick veneer walls to provide low-maintenance, attractive, and desirable exterior walls.

## OBJECTIVES

After completing this chapter, the student should be able to:

- Identify factors influencing the spacing of brick veneer from the backup wall.
- Identify and describe different methods for anchoring brick veneer to wall framing.
- Demonstrate the use of masonry spacing scales for proper brick course spacing.
- Demonstrate base, sill, and head flashing installations.
- Set steel lintels, and lay brick above openings.
- Identify areas where differential movements are likely to occur, and describe brick veneer construction details necessary for permitting differential movements of building materials.
- Construct brick veneer walls.
- Construct brick rowlock sills.

# Glossary of Terms

**adjustable anchor assemblies** assemblies used to anchor brick veneer to wood or steel framing.

**adjustable reinforcement assemblies** ladder and truss-type joint reinforcement assemblies used to anchor brick veneer to CMU walls.

**air infiltration barrier** a sheathing membrane installed to the exterior of framing for preventing air infiltration.

**anchored brick veneer** a single-wythe brick wall supporting no weight other than its own, requiring vertical support and anchored to either a load-bearing or non-load-bearing wall, having an air space between the brick wall and the backup wall.

**arching action** forces within a brick wall permitting the brick to partially support its own weight.

**brick binding** forces created by differential movements, exerting forces on window frames and doorframes, often leaving them inoperable.

**brick mold** an exposed trim of door and window jambs adjacent to the ends of brick on a brick veneer wall.

**brick shelf** that part of the top of a foundation wall supporting the anchored brick veneer wall.

**corrugated steel ties** sheet metal anchors for securing brick veneer to wood framing.

**dead loads** permanent loads, such as structural framing, wallboard, floor and roof systems, and permanent attachments, such as heating, ventilation, and air conditioning components.

**differential movements** unequal movements of building materials caused by changing temperatures or moisture contents of materials.

**drip** a cutout on the underside of the projection intended to prevent water from traveling beyond it and back to the face of the wall.

**eaves** those parts of the roof projecting beyond the exterior face of the wall.

**fascia** a board installed on the exposed ends of rafters or roof trusses.

**frieze board** a board extending from the soffit to a point below the top of the anchored brick veneer wall.

**gable** the triangular-shaped end of a wall.

**head** the top of an opening such as above doors and windows.

**jamb** the exposed sides of a door or window frame.

**live loads** nonpermanent loads, such as occupants, furnishings, rain, and snow.

**load-bearing wall** a wall that supports its own weight and the weight and forces placed on it by other parts of the structure as well as the weights and forces imposed by nature.

**masonry sills** masonry installed below window and door openings, usually brick rowlocks or precast stone.

**non-load-bearing wall** a wall supporting its own weight and no other weights or forces.

**rake board** trim enclosing the open space at the ends of the brick coursing and the sloping roof line in a gable wall.

**seismic loads** loads or forces acting on buildings due to the action of earthquakes.

**soffit board** a board attached to the underside of the rafters or roof trusses extending beyond the face of the exterior wall.

**steel angle lintel** a steel angle iron supporting the brickwork above an opening.

**veneer wire anchoring system** a system for securing brick veneer to both wood-framed and metal-stud framed structures.

**wind loads** calculated as wind pressures, increasing with wind velocity.

# Anchored Brick Veneer

**Anchored brick veneer** is a single-wythe brick wall, vertically supported and anchored to either a load-bearing or non-load-bearing wall, with an air space between the brick wall and the backup wall. Anchored brick veneer must not be confused with 1" or thinner brick used as exterior cladding and bonded directly to a backup wall. In this chapter, any reference to "brick veneer" is considered to mean anchored brick veneer. A brick veneer wall is designed to support no weight other than its own. Since brick veneer walls support no other weight than their own, they are classified as non-load-bearing walls. A **non-load-bearing wall** supports its own weight and no other weights or forces.

Anchored brick veneer walls are attached to the structural load-bearing walls of homes. A **load-bearing** or structural wall supports its own weight and the weight and forces placed on it by other parts of the structure as well as the weights and forces imposed by nature. These include live loads, dead loads, wind loads, and earthquakes. **Live loads** are nonpermanent loads, such as occupants, furnishings, rain, and snow. **Dead loads** are permanent loads, such as structural framing, wall-board, floor and roof systems, and permanent attachments such as heating, ventilation, and air conditioning components. **Wind loads** are calculated as wind pressures and increase with wind velocity. **Seismic loads** apply to buildings constructed in specific seismic categories subjected to the action of earthquakes and are addressed in Section R301.2.2 of the International Residential Code.

For residential construction, the structural wall to which the brick veneer is anchored may be wood frame, steel frame, CMUs, or cast concrete. Brick veneer is often anchored to both the CMU foundation walls and wood or steel framed walls of homes (see Figure 15-1).

**Figure 15-2** Oriented-strand board wood sheathing, the visible 4' × 8' stack in the foreground, is attached to wood studs and covered with an air infiltration barrier. Brick veneer walls are later anchored to the wall studs.

For years, brick veneer has been anchored to wood-framed structural walls (see Figure 15-2).

In recent years, there has been more construction where the brick veneer is anchored to steel studs. This trend is due partly to the variety of metal framing products available as well as the demands for more fire-resistant, noncombustible construction. To meet noncombustible framing requirements, weather-resistant exterior grade gypsum wall sheathing instead of plywood or other wood products is applied to the steel studs (see Figure 15-3).

**Figure 15-1** Supporting no weight other than its own, brick veneer is the choice of exterior material for this wood frame home.

**Figure 15-3** The design for the top portion of these exterior walls specifies brick veneer anchored to non-load-bearing, light gage steel framing covered with moisture-resistant, noncombustible gypsum sheathing.

# Supporting the Weight of Brick Veneer

Concrete or CMU foundation walls usually support the weight of anchored brick veneer walls. That part of the top of a foundation wall supporting the brick veneer is called the brick shelf. Typically, a foundation below *grade* or ground level begins with 12" CMUs. At grade level, 8" CMUs replace the 12" CMUs. This results in a 4"-wide brick shelf to support the brick veneer wall. Refer to Chapter 13, "Residential Foundations," for more detailed information about CMU foundation walls having a brick shelf.

Typically, the shelf aligns level with the finished grade outside the foundation walls. A stepped shelf is formed along sloping ground outside of foundation walls. A stepped shelf permits starting the brickwork at grade level while continuing the CMU foundation above grade (see Figure 15-4).

Beginning the brickwork at the lowest shelf on the foundation permits the mason to eventually lay the entire length of the wall (see Figure 15-5).

 **FROM EXPERIENCE**

Covering brick shelves with 2 × 4 framing lumber or other available materials as CMU foundation walls are constructed keeps them free of mortar droppings and other debris, eliminating the need to remove hardened mortar before beginning to build the supported brick walls.

For instructions on foundation construction, see Chapter 13 "Residential Foundations." In some instances, corrosion-resistant structural steel supports the weight of anchored

**Figure 15-5** Spacing three courses of standard size brick equivalent in height to each course of the CMU foundation simplifies alignment of the brick with the CMUs.

brick veneer walls. Section R703.7.2 of the International Residential Code addresses exterior veneer support.

# Provisions for an Air Space

The face of brick veneer supported by CMU foundations extends beyond the CMU wall, creating an air space between the brick and the CMU foundation (see Figure 15-6). The air space helps reduce water condensation by permitting air circulation. The air space also provides room for the mason's fingers while handling each brick as it is laid.

While the air cavity should be at least 1" wide for brick veneer walls anchored to wood framing studs or 2" wide for

**Figure 15-4** This CMU foundation with a stepped shelf is designed for beginning brick veneer above ground level.

**Figure 15-6** The brick extend beyond the CMU wall to create an air space between the two walls.

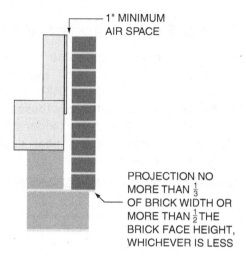

Figure 15-7 **These two details, air space and projection, must be considered when establishing a wall line.**

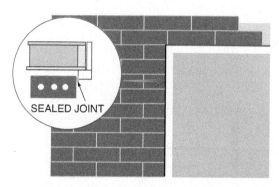

Figure 15-8 **Trim for this opening extends beyond the back edge of the brick. Properly sealed, water cannot penetrate the joint.**

brick anchored to steel studs, a brick should project no more than the lesser of $\frac{1}{3}$ its width or $\frac{1}{2}$ its height beyond the front edge of the brick shelf, (see Figure 15-7). Section R703.7.4 of the International Building Code requires a maximum distance separating the veneer from the sheathing material of 1" where corrugated sheet metal ties are used to anchor the brick veneer.

Although often specified on the building plans, it may be the mason who determines the width of the air cavity and the projection of the brick wall beyond the brick shelf. Factors influencing the air cavity width and brick projection include the following:

- Brick width
- Door and window trim thickness
- The accuracy of the plumb alignment of the exterior wall framing
- Design of roof eaves

 **FROM EXPERIENCE**

Avoid spreading excessive amounts of mortar for bedding brick. The mortar protruding from the back side of the brick blocks the intended air space.

## Brick Width

Standard size and oversize brick are approximately $3\frac{1}{2}$" wide. Because the brick shelf on a CMU wall is typically 4" wide, providing a 1"-wide air space requires projecting the brick beyond the face of the CMU wall (see Figure 15-7).

Approximately $2\frac{3}{4}$" inches wide, the face of queen-size brick may be spaced 4" from the exterior of the foundation

wall. Since the brick shelf is 4" wide, queen-size brick do not need to project beyond the face of CMU foundation walls.

## Trim Thickness

Brick veneer walls must prevent water from entering at the edges of door and window openings. This is accomplished by having window and door trim extended beyond the backside of the brick and sealed with a waterproof sealant (see Figure 15-8).

> **CAUTION**
>
> Avoid skin contact with sealants and caulking products, and avoid breathing their fumes. Read the manufacturer's printed instructions for safety precautions.

Narrower trim may require less spacing between the exterior structural walls and the backside of the brick. However, a minimum 1" air space should remain.

## Accuracy of the Plumb Alignment of the Exterior Wall Framing

Although desirable, exterior structural walls are not always plumb. Checking the exterior walls for plumb before beginning the brickwork enables the mason to determine the wall line for the first course. Establishing a plumb line from the top of the wall where the trim is intended to the bottom is possible using a plumb bob. In some cases, it may be necessary to begin the first course of brick further from or closer to the structural wall. This is necessary if the brick veneer is to be plumb while maintaining an appropriate width cavity. A structural wall leaning out at the top may require greater spacing between it and the face of the brick at the bottom (see Figure 15-9). However, at least $\frac{2}{3}$ of the brick width must be supported by the foundation wall.

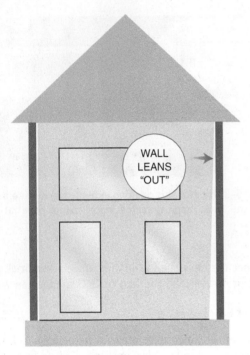

Figure 15-9 With the right side wall leaning out toward the top, brickwork is spaced further than normal at the bottom while complying with building code regulations.

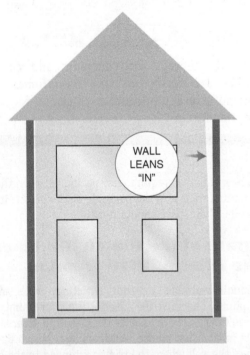

Figure 15-10 Brickwork may start closer to a structural backing wall leaning in towards the top, as illustrated in the right side wall, while complying with building code regulations.

A structural wall leaning in at the top may require starting the brick closer to this wall to permit a plumb wall while permitting window and door trim to extend beyond the backside of the end of brick (see Figure 15-10).

**CAUTION**

Ladders may be needed to climb to heights required for establishing a plumb line using a plumb bob. Comply with OSHA standards 1926.1053 when using ladders. Only those trained by a competent person should use ladders. Do not use metal or conductive ladders near overhead high voltage electric lines.

Consulting an engineer is advisable before proceeding with the brickwork if the conditions illustrated in Figure 15-9 and Figure 15-10 are evident. It could be that additional anchorage of the brick veneer is recommended or that other structural details require addressing.

## Design of Roof Eaves

The roof eaves are those parts of the roof projecting beyond the face of the wall. Roof eaves can include soffit boards, rake boards, and frieze boards. The truss or rafter design can determine the width of the soffit board and the attachment of the rake board at the roof eaves. These details control the face of the brickwork. A fascia board is a board installed on the exposed ends of rafters or roof trusses. A soffit board is a board attached to the underside of the rafters or roof trusses extending beyond the face of the exterior walls. A frieze board is a board extending from the soffit board to a point below the top of the brick veneer, closing the opening between the top of the brick wall and the soffit board. A rake board is trim enclosing the open space at the ends of the brick coursing and the sloping roof line in a gable wall. Establishing a plumb line from either the frieze board or rake board to the bottom of the wall helps determine the location for the face of the brick veneer. Using a plumb bob and string simplifies this procedure (see Figure 15-11).

## Laying Out the Brick Walls

Once the face of the brick wall lines are established, dry bonding eliminates needless brick cutting while assuring uniform, properly sized head joints for the bond pattern. Relying on the mason's scales not only assures appropriately sized, uniform bed joints, but also it allows constructing brick walls to desired heights. This consideration ensures appropriate allowances for brick rowlock sills below doors and windows and ensures full brick coursing above doors and windows (see Figure 15-12). Using the mason's scales for course spacing is explained in Chapter 7, "Masonry Spacing Scales."

Marking the desired scale onto the backup wall at their ends before the leads are built helps the mason to monitor the height of the leads as they are constructed without having to unfold the mason's rule to check course height for each course.

## Building the Walls

Attaching a mason's line to the brick leads permits laying brick to the line. Using masonry corner poles eliminates constructing brick leads at the corners (see Figure 15-13). Corner poles increase productivity by permitting the entire wall to be laid to the line.

Properly installing base flashing, sill flashing, and head flashing along with weeps at required locations eliminates moisture-related problems in the walls. A mortar dropping collection device should be considered for preventing blockage of weep holes or vents. Preventing air and moisture infiltration are key requirements for today's energy-efficient, brick-veneered frame structures. An **air infiltration barrier** or **housewrap** is a sheathing membrane installed to the

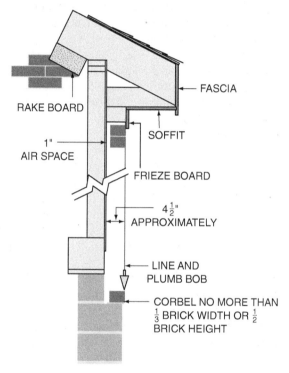

**Figure 15-11** The backside of the frieze board or rake board is the reference point for this wall's plumb line.

**Figure 15-13** The mason's line is attached to the corner pole and permits the entire wall to be laid to the line.

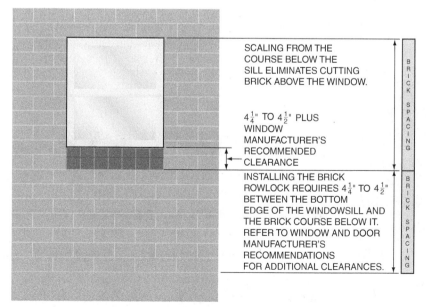

**Figure 15-12** Scaling permits desired spacing and prevents cutting brick below the sill and above the opening.

exterior of framing to prevent air infiltration. It also resists moisture migration from outside while allowing interior moisture to escape. Housewrap is a specially coated, woven polyethylene fiber material. Installing the housewrap is a task not generally performed by the mason. However, the mason needs to apply the specially formulated tape used to seal the seams and edges of tears or rips before proceeding with the brickwork. These details are explained in Chapter 14, "Constructing Water-Resistant Brick Veneer Walls."

**CAUTION**

**Avoid skin contact with adhesive products, and avoid breathing their fumes. Read manufacturer's printed instructions for safety precautions.**

## Anchoring the Brick Walls

The brick veneer must be anchored securely to the wall framing. The two types of framing are wood stud framing and light gage metal framing. Because of its resistance to fire, galvanized or zinc-coated steel framing is frequently specified for public buildings and commercial projects. There are four methods for anchoring brick veneer: 1) corrugated steel ties, 2) veneer wire anchoring systems, 3) adjustable anchor assemblies, and 4) adjustable ladder and truss-type joint reinforcement assemblies.

### Corrugated Steel Ties

Corrosion-resistant, zinc-coated corrugated steel ties are generally used to anchor brick veneer to wood framing (see Figure 15-14). Corrugated ties are not intended for anchoring brick veneer to steel stud framing, multiwythe masonry walls, or cavity walls. Corrugated stainless steel ties are an option for galvanized steel ties.

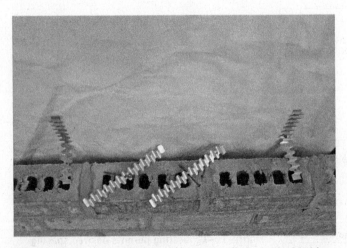

Figure 15-14 Corrugated metal veneer wall ties embedded in mortar anchor the brickwork to the wood framing.

Corrugated steel ties are made of sheet metal, approximately $7/8$" wide and 6" to 7" long. They are made of mill-galvanized steel, hot-dipped galvanized steel, or stainless steel. The thicknesses range from thinner 28-gage steel to heavier 16-gage steel. Section R703.7.4.1 of the International Residential code requires sheet metal ties to be "not less than No. 22 U.S. gage by $7/8$" corrugated". Remember that the thickness of sheet metal increases as the gage number decreases. The following are recommendations or building code requirements for the installation of corrugated wall ties.

- Each is secured to a wood stud with a corrosion-resistant nail penetrating the stud a minimum of $1^1/2$" and within $1/2$" of the bend.
- Section R703.7.4.1 of the International Residential Code requires ties to be spaced no more than 24" on center horizontally and vertically and support no more than 2.67 square feet of wall area. Wall ties are typically installed on every sixth course of standard size brick or every fifth course of oversize brick, and secured to each wall stud.
- Section R703.7.4 of the International Building Code requires a maximum distance separating the veneer from the sheathing material of 1" where corrugated sheet metal ties are used to anchor the brick veneer.
- It is recommended to place additional ties within 8" of openings, windows, shelf angles, and vertical expansion joints.
- Ties should penetrate at least half the veneer thickness but no closer than $5/8$" to the face of the wall.
- Mortar should completely surround the ties.
- Hot-dipped galvanized steel or stainless steel should be the choice of materials on exterior walls and other places where metal corrosion is likely.

 **FROM EXPERIENCE**

Keeping the heads of nails well above the brick below by driving them at an angle helps prevent accidentally striking the brick below it, causing it to become misaligned in the wall.

**CAUTION**

**Avoid having veneer wall ties exposed where they can cause injuries by cutting the skin.**

## Veneer Wire Anchoring System

A corrosion-resistant veneer wire anchoring system is a method for securing brick veneer to both wood-framed

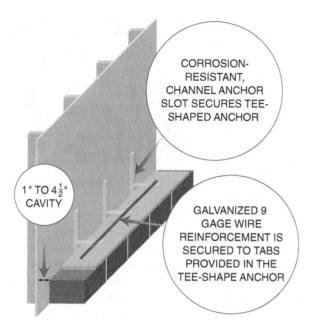

**Figure 15-15** This wire anchoring system secures brick veneer walls to both wood and metal wall studs while permitting a wider cavity.

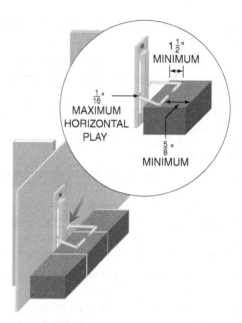

**Figure 15-16** Attention must be given for a properly positioned back-plate and "v" adjustable anchor assembly.

and metal stud-framed structures. This anchorage system permits a wider cavity between the back of the veneer and the framing. These assemblies can also be used for wood framed walls, permitting a maximum 4½" spacing separating the veneer from the sheathing material. Channel anchor slots are first attached to the studs. Tee-shaped anchors, available in lengths of 2", 3", 4", and 5", accommodate different cavity widths and are secured to the anchor slots. Tabs in the tee-shaped anchors secure a galvanized, No. 9 U.S.-gage wire reinforcement (see Figure 15-15). A 12" maximum vertical spacing is recommended.

Wire reinforcement conducts electricity. Handle wire so that it can come no closer than 10' to high voltage overhead lines. Be aware of the ends of wire reinforcement and prevent them from impaling others.

## Adjustable Anchor Assemblies

Made of hot-dipped galvanized steel or stainless steel, adjustable anchor assemblies are used to anchor brick veneer to steel framing (see Figure 15-16). These assemblies should provide veneer anchorage no more than 18" vertically or 32" horizontally. The slotted back-plate permits vertical adjustment for the wire anchor. It is secured to a steel stud using approved corrosion-resistant screws. A ³/₁₆" diameter wire anchor engaged through the open slot of the back-plate

is embedded at least 1½" into the mortar bed joint and having a minimum of ⅝" mortar coverage at the face of the wall. Adjustable anchor assemblies may also be used to anchor brick veneer to wood stud framing, and in some cases, CMU walls and concrete walls.

## Adjustable Ladder and Truss-Type Joint Reinforcement Assemblies

Adjustable ladder and truss-type joint reinforcement assemblies connected by eye and pintle to rectangular tabs are sometimes used in place of corrugated veneer ties for CMU-backed brick veneer walls. These assemblies are explained in Chapter 16, "Composite and Cavity Walls."

# Sill, Jamb, and Head Details

Brick veneer walls typically include door and window openings. Careful attention must be given to prevent water from entering the wall below such openings at the sills, at jambs beside these openings, and above them at the heads. Differential movement of the building materials must also be addressed.

## Masonry Sills

Masonry sills are installed below window and door openings. The masonry sill serves to divert water away from the structure. Sills used for veneer construction are usually brick, concrete, or stone. Frequently, brick are laid in the rowlock position to create sills. Some brick manufacturers stock special shaped brick for sills or are capable of making desired shapes. The following are recommendations for constructing brick sills:

- Install sill flashing and weeps below the sill as described in Chapter 14, "Constructing Water-Resistant Brick Veneer Walls."
- Slope the brick a minimum of 15 degrees for adequate water drainage.
- Extend the sill brick a minimum of 1" beyond its closest point to the face of the wall.
- Use solid brick at the ends of sills to eliminate exposed brick cores.
- Provide spacing as recommended by the window manufacturer between the bottom of the window or door frame and the brick sill to accommodate differential movements of the building materials.

Figure 15-17 illustrates a properly detailed brick sill.

Having a horizontal rather than sloping bottom, concrete or stone sills should have a drip on the bottom of the sill to prevent water from traveling to the face of the wall. A **drip** is a cutout on the underside of the projection intended to prevent water from traveling beyond it and back to the face of the wall. The inner edge of the drip should be a minimum of 1" from the face of the wall (see Figure 15-18).

The bottom, front-sloping edge of the brickwork serves as the drip for a brick sill. Flashing and weeps should be installed within the wall and beneath brick rowlock sills, concrete sills, and stone sills to collect any water penetrating the sill.

To prevent water intrusion while allowing for differential movement of building materials, a water-sealed space, having a width recommended by the window or door manufacturer between the masonry sill and the bottom edge of the window or door frame, is needed. This space is typically ¼" to ⅜" wide. **Differential movements** are unequal movements of building materials caused by changing temperatures or moisture contents of materials. Both shrinkage and expansion of materials are results of differential movements.

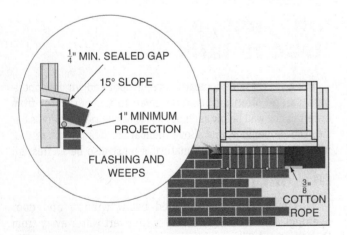

**Figure 15-17** The brick sill in this illustration includes recommended details, slope, projection, and spacing below the window, for preventing water penetration and brick binding. Comply with window manufacturer's recommendation and job specifications for brick spacing below the window.

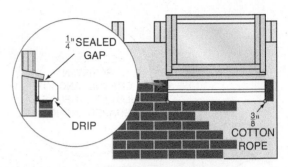

**Figure 15-18** A cutout drip below this concrete sill prevents water from returning to the wall.

Inoperable windows and doors can be the result of inadequate allowances for differential movements of building materials. Forces created by such movements can be exerted on window frames and doorframes, leaving them inoperable. This condition is known as **brick binding**.

**FROM EXPERIENCE**

A strip of plywood with the same thickness as the desired clearance between the bottom edge of the window frame and the brick sill can help maintain that desired clearance. Adhere the plywood to the bottom edge of the window frame temporarily with tape and remove it when the mortar for the sill brick cures.

## Masonry Jambs

The exposed sides of a door or window frame are called the **jambs**. The **jamb trim, called brick mold** is the exposed trim of door and window jambs adjacent to ends of brick on a brick veneer wall. The door or window frame is attached to the structure's wood or steel framing. Ends of brick form a brick jamb adjacent to the vertical brick mold. The Brick Industry Association Technical Notes 28 recommends a space no less than ¼" or more than ½" separating the brick mold from the brickwork (see Figure 15-19). This detail permits the proper application of a waterproof sealant, isolates the trim from the brick, and eliminates improper door and window operation resulting from differential movements by permitting horizontal movements of the building materials.

**FROM EXPERIENCE**

A piece of plywood with the same thickness as the desired gap placed temporarily at the ends of jambs permits butting brick against the plywood and leaving a uniform spacing between brickwork and jambs once the plywood is removed.

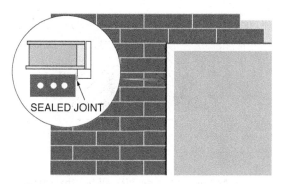

**Figure 15-19** A ¼" clearance between the ends of brick and the brick mold provides a sealed joint and accommodates differential movements.

**Figure 15-21** Brick to the left of the opening are cut to fit. To the right of the opening, a half-brick is substituted for the small piece, requiring shortening the adjacent brick.

Where an opening interrupts the brick course, it becomes necessary to cut the brick next to the vertical door and window jambs. However, the bond pattern should not be altered because of the opening (see Figure 15-20). This is known as holding the bond.

Very small brick pieces can be more difficult than larger pieces to install. As an optional method, the small pieces are eliminated by substituting them with half-brick and shortening the full brick beside the half-brick (see Figure 15-21).

For procedures on holding the bond next to openings, refer to Chapter 5, "Laying Brick to the Line," Procedure 5, "Holding Bond at Openings in a Wall."

## Heads

The top of an opening is referred to as the **head.** Brickwork above door and window heads requires a masonry support system. Two means of supporting the brickwork are structural steel and masonry arches. A **steel angle lintel**

is placed horizontally across an opening to support the brickwork above the opening (see Figure 15-22).

The following are recommendations for steel angle lintels:

- Galvanized steel lintels should be used to resist corrosion.
- The steel lintel should be a minimum of ¼" thick.
- The horizontal leg supporting the brick should be at least 3½" wide for 4" nominal thick brick or 3" wide for 3" thick brick.
- The vertical leg of the angle lintel should be 4" for spans no greater than 4' and 6" for spans not exceeding 10'.
- The ends of the angle lintel should rest on the supporting wall a minimum of 4" for spans no greater than 4' and 6" for spans exceeding 4'.

The steel lintel can rest directly on the brickwork, requiring no mortar bedding. Keeping the front edge of the

**Figure 15-20** Brick on each side of the opening are cut, and the vertical head joint alignment remains consistent.

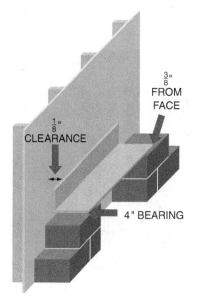

**Figure 15-22** This illustrates a properly installed steel lintel. Increase bearings to 6" for spans greater than 6'.

lintel ³/₈" behind the face of the wall allows mortar to conceal the edge. Also maintain ¹/₈" minimum clearance between the lintel and the structural backing (see Figure 15-22).

Table R703.7.3, "Allowable Spans For Lintels Supporting Masonry Veneer," in the International Residential Code give the size of steel angle lintel requirements based on the span and number of stories above the angle lintel.

Studies show that steel exposed to the atmosphere corrodes at the rate of 1 mil per year, and steel in marine environments corrodes at the rate of 5 mils per year. According to *Corrosion Engineering* (McGraw Hill, 1978), increasing lintel thickness ¹/₁₆" increases service 62¹/₂ years in most environments or 12¹/₂ years in marine environments.

**FROM EXPERIENCE**

Flush hot steel lintels with cold water before setting and bedding brick in mortar above them to prevent the high temperatures of the steel from drying the mortar too quickly.

Flashing and weeps should be installed above the lintel to collect and divert any moisture to the outside. The reveal trim above the window should not support the lintel. A waterproof, caulked joint between the reveal trim and the lintel allows for differential movements of building materials while blocking water intrusion.

Temporary shoring or support is necessary for longer spans to prevent the lintel from deflecting or sagging near its center. A centered prop or post positioned temporarily beneath the steel lintel serves as temporary shoring (see Figure 15-23). Shoring should remain in place for 1 to 3 days.

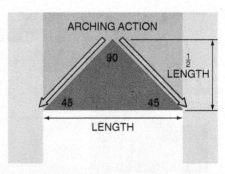

**Figure 15-24** Arching action enables all but the weight in the triangular area above the opening to be supported by the wall. Without arching action, a lintel must additionally support the shaded area above the triangle.

# Arching Action

A condition known as arching action permits part of the brickwork above the opening to become self-supporting once the mortar hardens. **Arching action** exists when the weight of the brickwork supported by the steel lintel is minimized to a triangular area above the opening (see Figure 15-24).

Arching action requires several courses of brickwork above this triangular area to resist arching thrust, or the counter forces tending to eliminate arching action. If arching action is not expected or if temporary shoring is not provided, a larger steel lintel capable of supporting the entire weight of the brick veneer above the opening is required. These considerations require only qualified individuals to specify the structural design and sizing for steel lintels. Always comply with applicable building code regulations for lintels supporting brick veneer.

Depending on the building's design, the brickwork either continues above lintels or terminates at the heads of doors and windows. For single-story structures, ending the brickwork at the heads of doors and windows reduces construction costs by eliminating steel lintels and additional brickwork (see Figure 15-25).

**Figure 15-23 Temporary shoring prevents this lintel from deflecting or sagging in the middle.**

**Figure 15-25 Eliminating brickwork above the windows and door reduces masonry materials and labor.**

**Figure 15-26** A brick arch supports the brickwork above this window.

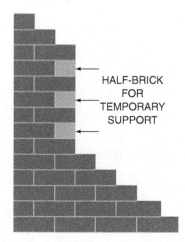

**Figure 15-28** Half-brick on alternating courses temporarily support the tail end of the lead.

Masonry arches are another means of supporting brickwork above openings (see Figure 15-26). Masonry arches are addressed in Chapter 22, "Brick Masonry Arches."

## Toothing a Brick Wall

Completing the brickwork on one side of a structure before beginning another side is a widespread practice. It eliminates having to raise all of the walls together. This requires less scaffolding and avoids the constant moving of scaffolding from one side to another. A process known as toothing a brick wall is often relied upon to limit construction to one wall at a time (see Figure 15-27).

The procedure of toothing a brick corner is accomplished using a half-brick as temporary support for the tail-end of the lead on every other course (see Figure 15-28).

**Figure 15-27** Toothing the front wall allows the wall to be completed before constructing the adjacent wall.

Toothing presents several concerns. Unintentional movement of the brick and pliable mortar when removing the half-brick weakens the mortar bond. Providing adequate bonding with adjoining walls necessitates compacting the mortar. The author uses a tuckpointer and pointing trowel for filling the joints of the adjoining wall. Issues dealing with toothing brick corners are explained in Chapter 6, "Constructing Brick Leads."

 **FROM EXPERIENCE**

Always remove brick temporarily supporting the toothed lead before the mortar hardens. Using a "sawing" motion with the edge of the trowel blade to loosen mortar on temporary supports helps prevent unintentional misalignments of the lead.

## Holding Bond in Gables

Brickwork may extend into a gable, or the triangular-shaped end of a wall. Although the length of each course becomes shorter in a gable, alternating courses of brick should align. This is known as holding bond. Head joints on every other course should remain vertically aligned, maintaining a plumb bond. Cutting the brick at each end of the course permits proper alignment of the brick for a uniform bond. The brick at each end of the course should extend near the gable line, permitting the rake board trim to cover the ends of the brick wall (see Figure 15-29). Brick must extend beyond the plane of the soffit board for walls that do not have a rake board.

**Figure 15-29** A half-lap running bond pattern continues with each successive brick course in the gable area.

## Summary

- A correctly detailed brick veneer wall requires minimal maintenance. Proper planning, layout, and implementation are necessary for successful results.
- Brick veneer walls must have adequate flashing and weeps to prevent moisture-related problems.
- A registered professional engineer should be consulted to ensure details appropriate for a particular project or geographical area and to ensure compliance with governing regulations and building codes.
- Not until required proficiency is demonstrated should you attempt to construct brick veneer walls without supervision.

# Review Questions

*Select the most appropriate answer.*

1. A brick veneer wall
   a. is anchored to some type of backup wall
   b. consists of a single wythe of brick
   c. supports no weight other than its own
   d. all of the above

2. A wall that supports its own weight and the weight and forces placed on it is called a
   a. load-bearing wall
   b. structural wall
   c. non-load-bearing wall
   d. both a and b

3. Loads imposed on a structure, including occupants, furnishings, rain, and snow, are examples of
   a. dead loads
   b. live loads
   c. permanent loads
   d. variable loads

4. Structural framing, wall-board, floor and roof systems, and permanent attachments are examples of
   a. dead loads
   b. live loads
   c. permanent loads
   d. variable loads

5. The minimum width for the air cavity between a brick veneer wall and the backup wall should be no less than
   a. ½" if anchored to either wood stud or steel stud backup walls
   b. 1" if anchored to wood studs or 2" if anchored to steel studs
   c. 1½" if anchored to either wood stud or steel stud backup walls
   d. 2" if anchored to either wood stud or steel stud backup walls

6. Brick veneer walls should project no further beyond the supporting foundation wall than the lesser of
   a. ⅕ its width or ½ its height
   b. ¼ its width or ½ its height
   c. ⅓ its width or ½ its height
   d. ½ its width or ½ its height

7. Moisture-related problems within wall systems can be minimized by installing
   a. flashings
   b. weeps
   c. mortar collection systems
   d. all of the above

8. Corrugated metal veneer ties should be anchored to wall framing with corrosion-resistant nails penetrating the studs a minimum of
   a. 1"
   b. 1½"
   c. 2"
   d. 2½"

9. Corrugated metal veneer ties should embed in the veneer thickness
   a. at least ¼ the wall thickness
   b. at least ⅓ the wall thickness
   c. at least ½ the wall thickness
   d. at least ¾ the wall thickness

10 **Corrugated metal veneer ties should remain from the face of the wall at least**

a. ¼"

b. ³⁄₈"

c. ½"

d. ⁵⁄₈"

11 **A masonry window sill should have a minimum slope of**

a. 15°

b. 25°

c. 30°

d. 45°

12 **Differential movements requiring clearances between masonry sills and the bottom edges of windows and doors anchored to wood framing are a result of**

a. settlement of brick veneer walls

b. shrinkage of wood building materials once installed

c. settlement of masonry foundation walls

d. none of the above

13 **The minimum width of the clearance between brick jambs and windows or doors is recommended to be**

a. ¼"

b. ³⁄₈"

c. ½"

d. ⁵⁄₈"

14 **The top of an opening is referred to as the**

a. eve

b. head

c. plate

d. sill

15 **The ends of steel angle lintels should rest on the supporting walls a minimum of**

a. 2" for spans up to 4' and 4" for spans greater than 4'

b. 4" for spans up to 4' and 6" for spans greater than 4'

c. 6" for spans up to 4' and 12" for spans greater than 4'

d. 8" for spans up to 4' and 16" for spans greater than 4'

16 **Steel angle lintels should have minimum clearances of**

a. ⅛" from the backup wall and ⅛" from the face of the brick wall

b. ⅛" from the backup wall and ³⁄₈" from the face of the brick wall

c. ³⁄₈" from the backup wall and ⅛" from the face of the brick wall

d. ³⁄₈" from the backup wall and ³⁄₈" from the face of the brick wall

# Chapter 16

# Composite and Cavity Walls

**N**ot only are masonry composite walls and cavity walls chosen for their strength, but their resistance to rain, fire, and sound transmission keep these two wall designs at the forefront of popular wall selections. The designs for composite walls and cavity walls are similar. Both are multiwythe masonry walls, which means there are two separate wythes of different masonry units, and each wythe has different properties and strengths. Tying the wythes together with joint reinforcement and insulating the air cavity between them creates excellent exterior structural bearing walls. Frequently, aesthetically appealing low-maintenance brickwork is used for the outside wythe, and cost-saving concrete masonry units are used for the inside wythe, also becoming the interior wall face. Although similar, there are differences between composite walls and cavity walls, affecting both their physical properties and selection in wall design.

## OBJECTIVES

**After completing this chapter, the student should be able to:**

- Compare similarities and differences between cavity walls and composite walls.
- Identify the types of thermal insulation used in cavity walls and composite walls.
- Explain how a rain screen wall is designed to reduce water migration through its exterior wythe.
- Identify the types of joint wire reinforcements and their applications.
- Explain how brick veneer walls with masonry backing are different from cavity walls or composite walls.
- Explain the differences between control joints and expansion joints and where each is used.
- List the procedures for constructing both cavity walls and composite walls.
- Explain how grout is used to reinforce cavity walls and composite walls.

# Glossary of Terms

**adjustable assemblies** masonry wall reinforcement consisting of ladder-type or truss-type reinforcement embedded in the inner wythe with extended tabs or eye hooks to which rectangular adjustable sections are connected and embedded in the joints of the outer wythe.

**bearing plates** steel plates with welded anchors on their bottom sides embedded in the tops of masonry bearing walls to which structural steel beams or joists are secured.

**bond beam** a horizontal reinforced beam comprised of specially formed bond beam concrete masonry units, grout, and horizontally placed steel reinforcement bars designed to strengthen a wall, support loads above openings, or distribute imposed loads uniformly.

**cap** an architectural concrete or stone member for part of or all of a wythe that is terminated below the top of an adjoining wall.

**capital** an architectural concrete or stone top for piers and columns, protecting them from the weather and enhancing their appearances.

**cavity wall** a masonry wall consisting of an inner and an outer wythe bonded with corrosion-resistant metal ties and separated by an air space not less than 2" nor more than 4½".

**cleanouts** openings in the face-side of CMU walls, enabling the removal of mortar droppings before grouting.

**composite masonry wall** a masonry wall consisting of two wythes of different masonry units having different strength characteristics connected with corrosion-resistant wire reinforcement or brick headers and acting as a single wall in resisting forces and loads.

**control joint** a vertical separation completely through a concrete masonry wall that is filled with an inelastic substance controlling the location of cracks caused by volume changes resulting in the shrinkage of block.

**coping** the projecting top cap of a wall.

**differential movement** the unequal movement of the building materials in a wall system.

**drainage wall systems** masonry walls designed to divert water from their air cavity to the exterior of the outer wythe.

**drip** a cutout in the underside of caps, copings, and sills preventing water from traveling back to and running down the face of the wall, causing water to drip beyond the face of the wall.

**expansion joint** a horizontal or vertical separation completely through a brick wythe that is filled with an elastic substance permitting the expansion of brick walls caused by thermal movements or increasing volume of brick.

**fire wall** a fire-resistant rated wall usually built within a structure from the foundation and extending above the roof to restrict the spread of fire from one part of a structure to another.

**granular fill insulations** lightweight, inorganic, perlite or vermiculite granules treated for water repellency used as wall insulation.

**high-lift grouting** grouting done once a wall is built to its story height or final height, whichever is less, in a single pour.

**ladder-type wire reinforcement** masonry wall reinforcement consisting of two or more longitudinal rods welded to perpendicular cross rods, forming a ladder design.

**load-bearing wall** a wall supporting its own weight and the structural loads, weights, and forces to which the structure is subjected.

**low-lift grouting** grouting done in multiple pours as the wall is built.

**multiwythe grouted masonry wall** a multiwythe wall whose air cavity is filled with grout.

**parapet wall** that part of a wall extending above the roofline.

**rain screen wall** a masonry cavity wall containing protected openings permitting the passage of air but not water into the cavity, permitting equal air pressures of the outside air and that within the air cavity.

**rigid insulation** polystyrene materials produced from extruded foam or molded bead processes placed between the wythes of cavity walls as wall insulation.

**shelf angles** steel structural members on which brick are bedded for wall support.

**soft joints** horizontal expansion joints minimizing wall cracks by permitting both the expansion of brick masonry walls and the deflection of the shelf angles.

**through-wall flashing** flashing extending completely across the air cavity separating masonry wythes and through the outer wall beyond the exterior face of the wall.

**truss-type wire reinforcement** masonry wall reinforcement consisting of two or more longitudinal rods welded to diagonally oriented cross rods, forming a truss design.

**wythe** a masonry wall.

## Multiwythe Walls

Figure 16-1 illustrates a double-wythe wall with concrete masonry units (CMUs) used on the interior wythe and brick for the outer wythe. Similar wall designs are used frequently in masonry construction.

There are major differences in the design and construction of composite walls and cavity walls.

## Comparing Composite and Cavity Walls

A composite masonry wall is a masonry wall consisting of two wythes of different masonry units. Each wythe has different strength characteristics connected with corrosion-resistant wire reinforcement or masonry headers, but they act as a single wall in resisting forces and loads (see Figure 16-2).

A cavity wall is a masonry wall consisting of an inner and an outer wythe bonded with corrosion-resistant metal ties and separated by an air space not less than 2" nor more than 4½". The space can be insulated or left open (see Figure 16-3).

Cavity walls are classified as noncomposite walls because the outer wythe is designed to resist forces developed within that wall rather than transfer the forces to the inner wythe,

**Figure 16-4** The masonry fire wall extends above the structure's roof line.

as does a composite wall. Similar to brick veneer walls, brick facing tied to another wall, the outer wythe of a cavity wall is mostly a non-load-bearing wall. Unlike brick veneer walls, the outer wythe of a cavity wall is designed to resist loads and forces rather than transfer them to the backing wall. The inner wythe supports structural loads.

Both composite walls and cavity walls are used as load-bearing walls. A load-bearing wall supports its own weight and the structural loads, weights, and forces to which the structure is subjected. A load-bearing wall supports weights, including floor and roof materials, furnishings, and inhabitants. A load-bearing wall is designed to resist such forces as wind and seismic earth movements. Brick are often specified for the outer wythe, while CMUs are popular for the inner wythe.

Both composite walls and cavity walls provide excellent resistance to fire. Since masonry walls have excellent fire resistance, these walls are a popular choice among architects and engineers for the construction of fire-resistant, load-bearing walls and fire walls. A fire wall is a fire-resistant rated wall usually built within a structure from the foundation and extending above the roof to restrict the spread of fire from one part of a structure to another (see Figure 16-4).

Cavity wall construction exhibits important advantages over composite wall construction. Cavity walls provide superior resistance to moisture and sound transmission. Properly insulated cavity walls exhibit superior thermal resistance.

**Figure 16-1** This illustrates a multiwythe masonry wall consisting of concrete masonry units and brick.

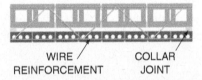

WIRE REINFORCEMENT　　COLLAR JOINT

**Figure 16-2** The two wythes are tied together with joint reinforcement. A mortar-filled collar joint closes each end of the wall.

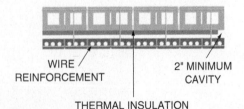

WIRE REINFORCEMENT　　2" MINIMUM CAVITY

THERMAL INSULATION

**Figure 16-3** Having wythes separated by at least 2" permits thermal insulation in the cavity.

## Controlling Moisture

Cavity walls are classified as drainage wall systems because wind-driven rain penetrating the outer wythe drains down its backside to the bottom and is diverted to the outside by base flashing and weep vents installed approximately

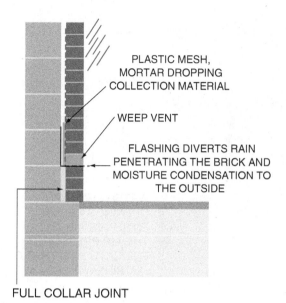

PLASTIC MESH,
MORTAR DROPPING
COLLECTION MATERIAL

WEEP VENT

FLASHING DIVERTS RAIN
PENETRATING THE BRICK AND
MOISTURE CONDENSATION TO
THE OUTSIDE

FULL COLLAR JOINT

**Figure 16-5** **About 8" above finished grade, base flashing and weep vents divert water from the cavity.**

8" above finished grade. The air space prevents water from migrating to the inner wythe (see Figure 16-5).

Flashing should also be installed below and above all wall openings and at the tops of walls beneath copings, the cap or top course of the wall. The flashing should extend vertically a minimum of 8" above its base. Flashing is either embedded in or adhered to the inner wythe. Adhesive-backed flashing additionally secured to the inner wythe with corrosion-resistant metal straps and fasteners is one practice. Overlapping and properly sealing the ends of flashings ensures watertight seals. It is recommended to extend the flashing through the exterior wythe slightly beyond the face of the wall. The ends of flashings above and below openings should be turned up at least 1", forming an end dam and preventing water seepage (see Figure 16-6).

> ◤◤◤ **CAUTION** ◢◢◢
>
> **Avoid skin contact with adhesive products and avoid breathing their fumes. Read manufacturer's printed instructions for safety precautions.**

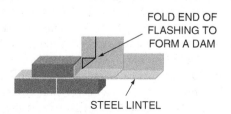

FOLD END OF
FLASHING TO
FORM A DAM

STEEL LINTEL

**Figure 16-6** **Folding the end of the flashing prevents water leakage. This produces tray flashing.**

Open weeps, spaced no more than 24" apart, are placed above and level with the flashing. Wick ropes or round tubes should be spaced no more than 16" apart. Plastic mesh above the flashing prevents mortar droppings from blocking the weeps. A 2" to 3" layer of pea gravel is sometimes used in place of the plastic mesh. Care must be taken to ensure that the gravel does not interfere with air circulation in the cavity.

# Insulating Composite and Cavity Walls

Composite walls and cavity walls can be insulated to permit thermal requirements for today's energy-efficient designs. Two types of insulation are granular fill and rigid.

## Granular Fill Insulations

Granular fill insulations include lightweight, inorganic, perlite, or vermiculite granules treated for water repellency. Poured into the cavity, they form a nonsettling thermal barrier while blocking the transmission of moisture and sound. Once installed, care must be taken not to penetrate either of the two wythes because the small insulating granules will discharge from the wall cavity.

> ◤◤◤ **CAUTION** ◢◢◢
>
> **Wear a respirator to avoid breathing airborne granular fill insulations. Wear eye protection to keep airborne dust from injuring the eyes.**

## Rigid Insulations

Rigid insulations are polystyrene materials produced from extruded foam or molded bead processes (see Figure 16-7).

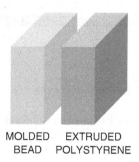

MOLDED   EXTRUDED
BEAD   POLYSTYRENE

**Figure 16-7** **Round beads are visible in molded bead insulation; whereas extruded polystyrene appears as a dense foam.**

**Figure 16-8** This rain screen design for a cavity wall minimizes moisture migration.

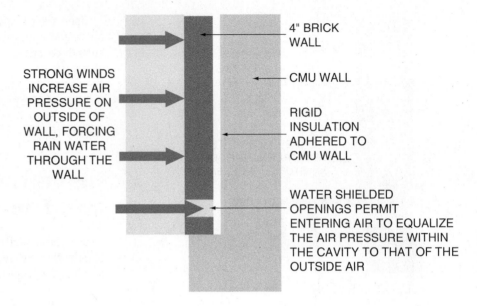

STRONG WINDS INCREASE AIR PRESSURE ON OUTSIDE OF WALL, FORCING RAIN WATER THROUGH THE WALL

4" BRICK WALL

CMU WALL

RIGID INSULATION ADHERED TO CMU WALL

WATER SHIELDED OPENINGS PERMIT ENTERING AIR TO EQUALIZE THE AIR PRESSURE WITHIN THE CAVITY TO THAT OF THE OUTSIDE AIR

Extruded polystyrene insulations are reportedly more moisture-resistant because they have a closed cell structure with no voids between the cells. And because moisture reduces thermal efficiency, extruded insulation is considered superior over molded-bead rigid insulation. Rigid insulating boards are placed horizontally in the cavity between wall ties. These insulating boards are secured to the inner wall with mechanical fasteners or bonded to the inner wythe with an appropriate adhesive. A 1" space between the insulation and the outer wythe is recommended for proper moisture drainage and ventilation. Polyisocyanurate foam core sheathings can be used where a narrower insulation is needed. For comparable thickness, this material has a significantly higher insulating value than expanded polystyrene or molded-bead foam board. For maximum performance, the insulation must be securely anchored to the exterior side of the interior wythe.

**CAUTION**

**Use extreme care when holding on to large pieces of foam board when winds prevail or are expected. High winds cause the foam board to act as a wind sail and may cause you to lose your balance and fall.**

## Rain Screen Walls

A cavity wall can be constructed as a rain screen wall, which is so called because the wall design enables the outer wythe to prevent water migration into the wall system better than conventional cavity walls. Like other cavity wall systems, rain screen walls rely upon flashing and weeps to return water and moisture within the air cavity to the exterior of the wall. But in addition to flashing and weeps, rain screen walls contain protected openings permitting the passage of air but not water into the cavity. These openings help equalize the air pressure within the cavity to that of greater outside air pressure during windy weather. Pressure equalization is needed to minimize water infiltration through the exterior wythe. Otherwise, an increase in air pressure on the outside of the wall caused by strong winds and a lower air pressure within the air cavity between the inner and outer wythes promotes water migration into the air cavity and eventually into the inner wythe.

Vents are sometimes placed near the top of rain screen walls to permit air circulation in the cavity. To maintain the required air space, rigid board insulations rather than granular fill insulations should be adhered to the outer side of the inner wythe whenever insulating any cavity wall, including those designed as rain screen walls (see Figure 16-8).

## Wall Reinforcement

The inner and outer wythes of composite and cavity walls are connected with horizontal joint reinforcement or unit ties. For exterior walls and wherever corrosion is a concern, the metal reinforcement should be either corrosive-resistant, hot-dipped galvanized, or stainless steel. Uncoated or mill-galvanized wire, which has less coatings than hot-dipped galvanized wire, should only be used where corrosion is not a factor and not on exterior walls. The likely corrosion of mill-galvanized wire or uncoated joint reinforcement wire leads to bed joint failures between the wire and the face of the wall. The wire enlarges in diameter as it corrodes, loosening the joint and allowing excessive water penetration. Embedded in the bed joints at specific intervals, the reinforcement strengthens the wall by connecting the two wythes and permits the loads of one wythe to be shared by the other. Horizontal joint reinforcement is available in 10' lengths. The ends of sections should be lapped 6". The following are the three types of joint reinforcement and ties:

1. Ladder type
2. Truss type
3. Adjustable assemblies

## Ladder-Type Wall Reinforcement

Ladder-type wire reinforcement consists of two or more longitudinal rods welded to perpendicular cross rods, forming a ladder design. As both wythes are raised together, 10' sections of ladder-type wire reinforcement, approximately 2" narrower than the width of the wall, are installed (see Figure 16-9).

Ladder-type reinforcement is recommended for composite walls and cavity walls. Ladder-type reinforcement is available with a bend or drip in the cross rods for preventing water from traveling across the wire to the interior (see Figure 16-10). According to Technical Notes 21 of the Brick Industry Association, "drips in ties reduce the strength of the tie, therefore, the wall area per tie must be reduced by 50%". The bend may decrease the transfer of lateral loads. A reduced spacing is recommended for cavity walls reinforced with wall reinforcement having a drip.

## Truss-Type Wall Reinforcement

Truss-type wire reinforcement consists of two or more longitudinal rods welded to diagonally oriented cross rods, forming a truss design (see Figure 16-11).

Truss-type reinforcement is used for composite walls, but the Brick Industry Association does not recommend it for cavity walls. According to Technical Note 21A of the

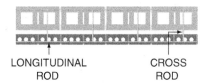

LONGITUDINAL      CROSS
ROD               ROD

**Figure 16-9** Ladder-type reinforcement has cross rods welded perpendicular to the longitudinal rods.

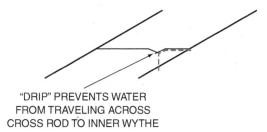

"DRIP" PREVENTS WATER
FROM TRAVELING ACROSS
CROSS ROD TO INNER WYTHE

**Figure 16-10** The bend, placed between the wythes, stops water migration toward the inner wythe. Eliminating mortar droppings on the wire helps assure performance of the drip.

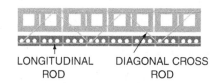

LONGITUDINAL    DIAGONAL CROSS
ROD             ROD

**Figure 16-11** Cross rods welded diagonally to the longitudinal rods form triangular or truss shapes.

Brick Industry Association, "...tests also indicate that truss-type joint reinforcement used in brick cavity wall construction helped to develop a degree of composite action in the horizontal span but did not contribute to any composite action in the vertical span. This restraint in the horizontal direction will reduce the amount of in-plane movement and possibly result in bowing of the masonry walls. Thus, truss-type joint reinforcement is not recommended for brick and block cavity walls." The truss diagonals can restrain differential movement between the wythes. The design of a cavity wall should enable both wythes to share part of any lateral load, such as the force created by high winds on the outer wall.

## Adjustable Assemblies

Adjustable assemblies consist of ladder-type or truss-type reinforcement embedded in the inner wythe with extended tabs or eye hooks to which rectangular adjustable sections are connected and embedded in the joints of the outer wythe (see Figure 16-12).

There are several advantages to using the adjustable assemblies:

- The inner wall can be erected ahead of the outer wall, reducing the time required for enclosing the building.
- Rigid insulation can be held in place with the proper mechanical attachments.
- The two wythes can be tied together where bed joints do not align.
- Larger differential movements between the wythes can be accommodated.

**FROM EXPERIENCE**

Use bolt cutters to cut wire joint reinforcement. Avoid striking the wire with the edge of the trowel blade because the blade's edge may become chipped.

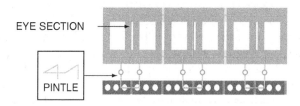

EYE SECTION

PINTLE

**Figure 16-12** Pintle tabs connected to the inner wythe reinforcement are embedded in the outer wythe.

# Masonry-Backed Brick Veneer Walls

Masonry-backed, anchored brick veneer walls should not be confused with cavity walls or composite walls. In residential construction, brick veneer walls are frequently anchored to block foundations using veneer wall ties. Such walls are not considered to be cavity walls or composite walls because of their design properties. Unlike cavity walls and composite walls, in which wire joint reinforcements and adjustable assemblies enable both wythes to resist loads, Brick Industry Association Technical Notes 21 reports that "the exterior wythe of a brick veneer wall transfers out-of-plane loads to the backing...."

# Accommodating Differential Movements

**Differential movement** is the unequal movement of the building materials in a wall system. Both temperature and moisture cause differential movements. All building materials expand at higher temperatures and contract at lower temperatures. Clay brick are less affected by temperature changes than CMUs. Steel experiences greater dimensional changes due to temperature changes than either brick or CMUs.

Brick and CMUs expand as their moisture contents increase and contract as moisture content decreases. Metal is unaffected. However, studies show that brick and CMUs are affected differently by moisture. Brick are never smaller than when taken from the kiln. They expand over time due to air moisture and rain. Normal air temperatures do not permit brick to contract as the moisture content decreases. In contrast, CMUs tend to become smaller over time due to moisture loss and carbonation, a reaction between carbon dioxide present in the air and calcium compounds in the masonry unit. There is a tendency for composite and cavity walls constructed of brick and CMUs to exhibit opposing forces. The brick wall is increasing in length as the concrete masonry wall is shortening. The effects of these opposing movements require both expansion joints in brick walls and control joints in CMU walls.

## Expansion Joints

An **expansion joint** is a horizontal or vertical separation completely through a brick wythe that is filled with an elastic substance permitting the expansion of brick walls caused by thermal movements or increasing volume of brick. Expansion joints divide longer brick masonry walls into shorter sections. Vertical separations entirely through brick walls, such as that in Figure 16-13, are required in longer brick masonry walls. Horizontal expansion joints are seen below shelf angles, the steel support for some masonry walls.

**Figure 16-13** The expansion joint in this brick wall is filled with an approved sealant whose color blends with the mortar color.

The separations, usually ⅜"- to ½"-wide, are filled with a polyurethane foam material known as a *backer rod* and sealed with an approved sealant. The expansion joint is compressed as the brick masonry wall expands, minimizing cracking in the brick wall. Joint reinforcement should not cross over expansion joints.

## Control Joints

A **control joint** is a vertical separation completely through a concrete masonry wall that is filled with an inelastic substance controlling the location of cracks caused by volume changes resulting in the shrinkage of block. Control joints are formed like expansion joints. However, control joints have the tendency to open rather than close (see Figure 16-14).

Technical Notes 21 of the Brick Industry Association reports that "these vertical movement joints do not have to be

**Figure 16-14** Separation of this control joint is evidence of the block wall shortening.

**Figure 16-15** The horizontal steel framework sections of this structure serve as masonry shelf angles. *(Courtesy of Old Virginia Brick Company)*

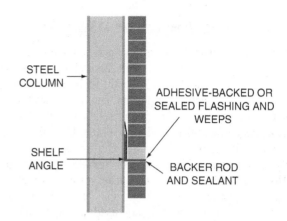

**Figure 16-16** This illustrates a soft joint below a properly sized shelf angle.

placed at the same location in a wall." For instance, expansion joints near the corners of brick walls may be specified without having control joints aligned in the inner wythe of the concrete masonry walls behind them.

## Shelf Angles

Masonry walls supported on **shelf angles,** steel structural members on which brick are bedded for wall support, require horizontal expansion joints below the shelf angles. Welded or riveted to steel I-beams, shelf angles transfer the weight of masonry walls to the structure's steel framework (see Figure 16-15).

Referred to as **soft joints,** these horizontal expansion joints minimize wall cracks by permitting both the expansion of the brick masonry walls and the deflection of shelf angles. Soft joints are filled with a backer rod and waterproof sealant. Shelf angles must bear at least two thirds of the supported wythe's thickness (see Figure 16-16).

## Sound Resistance

Cavity walls provide excellent resistance to sound transmission, the passage of sound from one side of a wall to the opposite side. The separation between the outer and inner

walls greatly reduces sound-causing vibrations due to the dampening effect of the air space and heavy mass of the masonry.

## Constructing Composite and Cavity Walls

Several concerns must be addressed when building composite or cavity walls. These concerns include course spacing, raising the two wythes, installing horizontal joint reinforcement, insulating the cavity, installing flashings, and installing weeps.

### Course Spacing

Modular spacing scales ensure aligning the bed joints of two dissimilar materials at 16" intervals (see Figure 16-17).

This enables the two wythes to be connected with wire joint reinforcement.

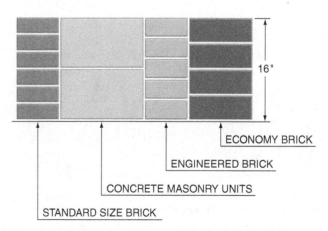

**Figure 16-17** Modular scaling permits aligning dissimilar materials every 16".

## Raising the Adjacent Wythes

The inner wythe is usually constructed ahead of the outer wythe. Using an adjustable assembly type of reinforcement allows the inner CMU wall to be raised ahead of the outer brick wall. Completing the inner wall wythes ahead of the outer wythes permits other trades to begin their activities sooner (see Figure 16-18). In addition, completing the inner wythe ahead of the outer wythe enables applying sprayed-on bituminous or flexible, acrylic coatings on the cavity side of the inner wythe, a typical practice for stopping water migration. Adjustable assembly horizontal wire reinforcement systems are used so that the outer wythe can be anchored later.

When the two wythes are raised together, a two-course CMU inner wythe is usually laid ahead of the brick outer wythe. The face of the CMU wall exposed to the air cavity is parged with mortar or sealed with an applied coating to inhibit water migration. Then, the outer wythe is constructed level with the inner wythe.

*For step-by-step instructions on constructing a 12" composite jamb, see the Procedures section on pages 239–240.*

Regardless of the method for raising the two wythes, the vertical spacing between joint reinforcement may be any of the following

- Six courses of standard-size brick
- Five courses of engineered/oversize brick or queen-size brick
- Four courses of economy brick
- Three courses of structural facing tile
- Two courses of CMUs

**Figure 16-18 The CMU walls are completed before beginning the brick exterior walls, enabling other aspects of interior construction to begin. The eye hooks of the adjustable assembly horizontal joint reinforcement are evident. The lower portion of the CMU wall is already coated with a black-color, sprayed-on moisture barrier.**

## Installing Horizontal Joint Reinforcement

Modular scaling permits 16" intervals for each of these five options. Spacing joint reinforcements at 16" height intervals permits joining together the two wythes.

Where the two wythes are raised together, the joint reinforcement should lay flat across the top of the two wythes, embedded in the mortar bed joints. Sections of horizontal joint reinforcement are lapped 6" at their ends. The reinforcement is positioned to provide equal spacing between the reinforcement's longitudinal rods and the width of the wall.

### CAUTION

**Horizontal joint reinforcement is a conductor of electricity. Keep it a minimum of 10' from overhead high voltage electrical lines. Carry the wire in such a manner as to prevent impaling other workers with the ends of the wire.**

## Insulating the Cavity

When rigid board insulation between the two wythes is specified, the insulation is placed against the outer side of the inner wythe for optimum insulating performance. Using recommended adhesives or mechanical fasteners keeps the rigid board insulation in place.

## Installing Flashings

Installing base flashing above grade level is required in most cases. The cavity or collar joint below the base flashing should be filled with mortar or grout. To prevent the pressure of the mortar or grout from misaligning the wall, it may be necessary to permit the mortar joints to harden before filling the cavity. From the face of the outer wythe, the flashing extends to the cavity side of the inner wythe and then vertical a minimum of 8". This flashing method is called **through-wall flashing.** Technical Notes 21 of the Brick Industry Association recommends that flashing extends 8" down the back side of the inner wythe, across the air cavity separating masonry wythes and through the outer wythe, continuing beyond the exterior face of the wall (see Figure 16-19). Technical Notes 21 also recommends that "where the flashing is not continuous, such as over and under openings in the wall, the ends of the flashing should be turned up approximately 1" (25 mm) into a head joint in the exterior wythe to form an end dam."

Overlapping the ends of the flashing a minimum of 6" and sealing the laps with a sealant recommended by the manufacturer are necessary procedures. The flashing is securely attached to the exterior side of the inner wythe using an approved adhesive or adhesive-backed flashing. Embedding the flashing in the bed joint of the inner wythe is another method.

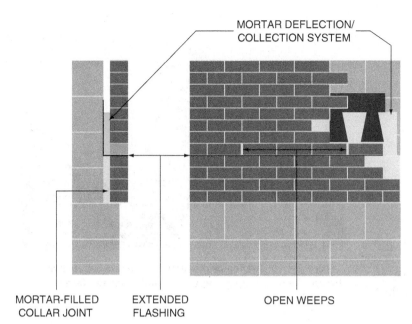

**Figure 16-19** This illustrates well-designed base flashing details for a cavity wall.

MORTAR DEFLECTION/
COLLECTION SYSTEM

MORTAR-FILLED
COLLAR JOINT

EXTENDED
FLASHING

OPEN WEEPS

## Weeps

In the course above the flashing, weep vents or open head joints spaced horizontally every 24" in the outer wythe direct moisture out of the wall (see Figure 16-19). A maximum spacing of 16" between weeps is permitted when using wick ropes. An open weave mesh, mortar-dropping collection material can be installed directly behind the weeps to prevent mortar droppings from blocking the weeps (see Figure 16-19). Open weeps and a mesh mortar-collection material can help reduce moisture-related problems by ventilating the cavity and equalizing the air pressure of the cavity with that of the outside air.

In addition to the previously described details at the base of cavity walls, through-wall flashing and weeps are needed at several other locations:

- Below and above wall openings
- At wall and roof intersections
- Through chimney walls above the roof line
- At shelf angles
- At the top of parapet walls

Flashing should be installed below windows and doorsills as well as other openings to prevent water leakage (see Figure 16-20).

Below sills, flashing should extend beyond each end of the opening to the head joints. Both ends of the flashing are turned up into the head joints a minimum of 1". This procedure forms a dam at each end for diverting the water outside through the weep vents. Open weeps should be included no more than 24" apart. These same procedures are used for installing flashing in walls continuing above adjacent roof lines.

Steel lintels spanning the top of openings require flashing also (see Figure 16-21). In addition to diverting moisture to the outside, flashing prevents corrosion of steel lintels. Corrosion decreases the load capacity that the lintel is designed to support.

FOLDING END
OF FLASHING
AT EACH END
FORMS A DAM

**Figure 16-20** Sill flashing diverts water entering at the sill to the outside.

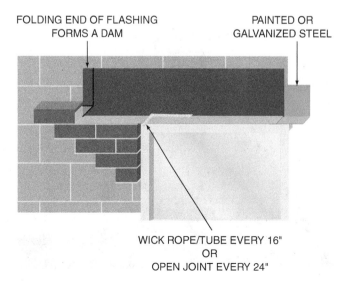

FOLDING END OF FLASHING
FORMS A DAM

PAINTED OR
GALVANIZED STEEL

WICK ROPE/TUBE EVERY 16"
OR
OPEN JOINT EVERY 24"

**Figure 16-21** Flashing and weeps prevent corrosion of the steel lintel, diverting water from the wall.

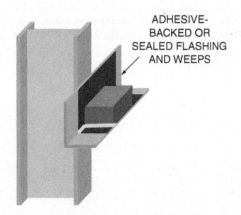

ADHESIVE-
BACKED OR
SEALED FLASHING
AND WEEPS

**Figure 16-22** Properly detailed shelf angles include flashing and weeps.

Flashing and weeps are required above shelf angles to prevent water from corroding and damaging the steel shelf angles (see Figure 16-22).

Because water seepage is likely between the flashing and shelf angle during wind-driven rains, self-adhering flashings or recommended sealants should be used to adhere the flashing to the shelf angle.

# Parapets, Copings, and Caps

A **parapet wall** is that part of a wall extending above the roofline (see Figure 16-23).

The cavity of a double-wythe wall continues to the top of the parapet, at which point it is covered with coping. **Coping** is the projecting top cap of a wall (see Figure 16-24).

Stone, precast concrete, and metal are three materials recommended for coping. Recognizing thermal movement of the wall system, the opportunity for moisture penetration is increased when using brickwork such as brick rowlocks

**Figure 16-23** This is a 12"-wide parapet wall.

**Figure 16-24** The top of this precast concrete coping slopes toward the roof, minimizing water runoff onto the face of the wall.

because of the numerous joints between the brick. Therefore, brick rowlocks are not recommended to be used as wall coping. Mortar joints between sections of precast concrete and stone coping should be raked out and filled with a baker rod and waterproof sealant.

Both sides of parapet walls can be exposed to severe weather conditions, including wind-driven rains and freezing temperatures. Through-wall flashing at the tops of parapet walls serves primarily to prevent water from entering the wall both below the coping and between the coping joints. It serves a different purpose than other flashings by keeping water from entering the wall rather than diverting water to the outside. Through-wall flashing should be installed at the top of the parapet wall and just below mortar bedding for the coping. Because the flashing prevents mortar from bonding to the wall below, the coping is secured to the wall using metal anchors. These anchors are embedded in the underside of the coping or in the joints between sections of coping. They penetrate the flashing and are embedded in mortar or grout in the wall cavity. To ensure a watertight seal, it is necessary to apply a sealant wherever the anchors penetrate the flashing.

Expansion joints are used to minimize cracking in parapet walls. Expansion joints placed in walls below parapets should continue to the top of the parapet wall (see Figure 16-25). In addition, expansion joints should be placed near the corners of parapet walls.

A **cap** is an architectural concrete or stone member for part of or for an entire wythe that is terminated below the top of an adjoining wall. Caps cover the exposed tops of masonry units where there is a reduction in the overall width of a wall (see Figure 16-26). A **capital** is an architectural concrete or stone top for piers and columns, protecting them from the weather and enhancing their appearances.

Like coping, caps require through-wall flashing. Caps and copings must be properly designed and installed. Their edges should extend beyond the face of the wall. The projecting

**Figure 16-25** **An expansion joint continues to the top of this parapet wall.**

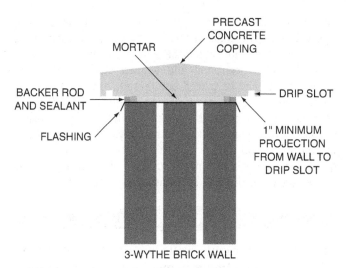

**Figure 16-27** **This coping installation includes the necessary details.**

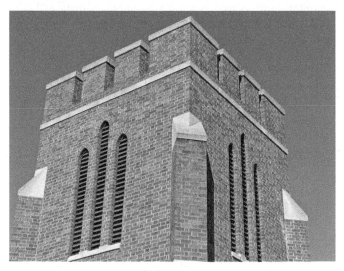

**Figure 16-26** **Architectural concrete caps protect the tops of brick walls ending below the tops of adjoining walls.**

sides of caps and copings should have a drip, a cutout in the underside preventing water from traveling back to and running down the face of the wall, to cause water to drip beyond the face of the wall. The inner lip of each drip should be a minimum of 1" from the face of the wall. The top of the caps and copings should slope downward a minimum of 15 degrees. Copings can slope in one or both directions. However, copings sloping in only one direction should slope toward the roof to minimize water runoff down the face of the wall. Figure 16-27 illustrates a properly detailed coping installation.

# Grout-Reinforced Walls

Composite and cavity walls can be reinforced with steel bar reinforcement and grout. Grout, a mixture of cement and small aggregates to which sufficient water is added to produce a pouring consistency without separation of the ingredients used to strengthen masonry walls, is placed either as low-lift or high-lift applications. Low-lift grouting, grouting done in multiple pours, is done within the air cavity of double-wythe walls as the walls are built. However, a multiwythe wall whose air cavity is filled with grout is considered as a multiwythe grouted masonry wall rather than a cavity wall. The narrow cavity of multiwythe grouted walls requires low-lift grout fills to ensure proper consolidation of the grout. When multiple lifts are poured, stopping each grout lift halfway between the courses is recommended because the separate lifts do not consolidate, creating a weak joint between the grout lifts. High-lift grouting is done once a wall is built to its story height or final height, whichever is less, where the grout is poured into and consolidated in the open cells of CMU walls with power equipment. There are advantages of high-lift grouting in open cell concrete masonry walls. The grout is well consolidated the total height of the wall, eliminating weak horizontal joints associated with low-lift grouting. Also, the bond between the rebar and grout is greater since the steel reinforcement rods do not become coated with cement, drying between lifts and reducing bond. High-lift grouting requires cleanouts at the bottom of the walls. Cleanouts are openings in the face-side of CMU walls, enabling the removal of mortar droppings before grouting. The openings are filled before grouting the walls.

Grout bonds to steel reinforcement rods, also called *rebar*, placed vertically in the open cells of CMUs and spaced according to engineered design specifications or code requirements. The bottom ends of the rebar, bent at right angles, are placed in the foundation footings as the concrete is placed. The steel reinforcement rods are placed in the wall in sections, permitting masons to lift block over the top of them. Sections of rebar are overlapped as much as 16" to 24" and joined using rebar tie wire or coil-shaped coupling assemblies as the walls are raised. Specially shaped wires can be placed in the cells of CMU walls as they are built to keep the rebars centered and aligned straight. High-lift grouts should not be placed before walls are given time to strengthen. Some sources claim that high-lift grouting can be done one day after the masonry is laid, whereas other sources recommend waiting three days before grouting. Placing grout too soon can result in "blowout" created by grout pressure at the bottom of the wall, causing mortar joints or even entire wall sections to come apart.

**CAUTION**

**To prevent blowout, walls grouted within three days of construction may require bracing by a competent person trained in the procedures of grouting and the operation of mechanical grouting equipment. Consult a structural engineer for recommended grouting procedures and precautions before grouting walls.**

The grout is consolidated to remove voids within the wall or cavity. Powered mechanical pencil vibrators and powered rebar shakers are two means of consolidating grout.

## Steel-Bearing Plates

Steel-bearing plates are embedded in the tops of masonry bearing walls to which structural steel beams or joists are secured. The sizes of the plates depend upon engineered design requirements and governing codes. Steel anchors welded to the bottom sides of the plates secure them to the tops of grouted masonry walls. Bond beams may be required to help distribute vertical loads imposed on steel bearing plates supporting structural steel members. Bond beams are constructed using specially formed bond beam CMUs, grout, and horizontally placed steel reinforcement bars. Properly designed bond beams can also be used to support loads above openings.

## Summary

- Cavity walls and composite walls are similar in that they are both multiwythe walls, but a 2" or wider air cavity is a distinguishing characteristic of cavity walls.
- Successful performance of cavity walls and composite walls depends on quality workmanship and attention to details such as wire joint reinforcement, control and expansion joints, and flashing and weeps.
- Ladder-type wire reinforcement is recommended for cavity walls, whereas truss-type wire reinforcement should be used for composite walls.
- To prevent the damaging effects of metal corrosion, wire reinforcement embedded in the mortar joints of exterior walls must have a hot-dipped galvanized finish.
- Grout and steel reinforcement is used to strengthen cavity and composite walls.

## Procedures

# Constructing a 12" Composite Jamb

**A** Lay two courses of 8" CMUs above a single course of 12" CMUs with a spacing for each course of 8". The 12"-block cross webs support the backside of the first course of 8" block. The faces of the 12" block and 8" block align plumb on the opposite side of the wall.

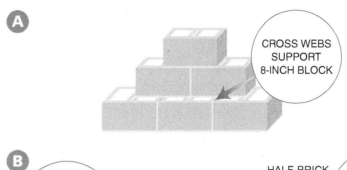

CROSS WEBS SUPPORT 8-INCH BLOCK

**B** Use Modular Scale 6 for the brick coursing. This permits six courses of standard size brick to be equivalent to 16", the same height as two courses of CMUs. The first course of brick projects beyond the face of the 12" CMUs, creating a 1" minimum air space between the backsides of the brick and the backup block wall. Lay a half-brick between the brick and block at the end of the jamb, repeating on every other course. Maximum brick projection should not exceed one-third the brick width or one-half the brick height.

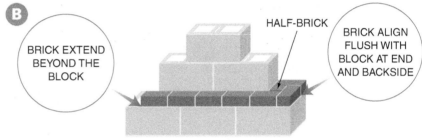

BRICK EXTEND BEYOND THE BLOCK

HALF-BRICK

BRICK ALIGN FLUSH WITH BLOCK AT END AND BACKSIDE

**C** Install through-wall base flashing above the first course.

**D** Lay the second course of brick above the first course, installing weeps at the head joints. (Note: Flashing detail between the brick and CMU backup wall is omitted in this illustration to better view the CMU wall.)

WEEPS

**E** Align the third course of brick level with the top of the first course of 8" block.

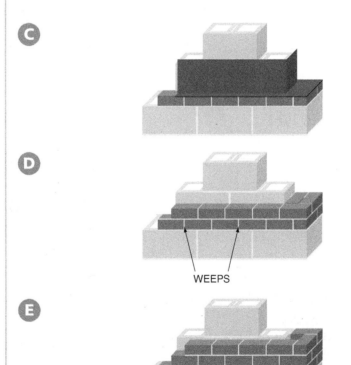

## Procedures

# Constructing a 12" Composite Jamb (continued)

**F** Lay the fourth course of brick.

**F**

**G** Lay the fifth course of brick.

**G**

**H** Lay the sixth course of brick aligned level with the top of the second course of 8" block. Embed truss-type or ladder-type wall reinforcement in the mortar joints of the block and brick to bond the masonry wythes together.

**H**

**I** Bed an 8" CMU in mortar above the wall below it. The course height is 8". The CMU should be aligned to create a uniform bond with the CMUs below it.

**I**

**J** Lay the next three courses of brick, using a course spacing like all other courses, Modular Scale 6. The brick and block should align level at the top of the lead.

**J**

## Review Questions

*Select the most appropriate answer.*

**1** The two masonry wythes of cavity walls are separated by an air space no less than

a. 1"
b. 1½"
c. 2"
d. 3"

**2** A fire-resistant wall built from the foundation and extending above the roof to restrict the spread of fire from one area of a building to another is called a

a. composite wall
b. fire wall
c. retaining wall
d. separating wall

**3** A rain screen wall design

a. contains protected openings and weeps permitting the passage of air to the cavity
b. permits air pressure equalization within the cavity to that of the outside air
c. cannot be constructed as a cavity wall
d. both a and b

**4** A type of wall joint reinforcement is

a. adjustable assemblies
b. ladder type
c. truss type
d. all of the above

**5** Which of the following statements are true?

a. Brick walls expand over time, requiring walls to have expansion joints
b. Block walls experience shrinkage over time, requiring them to have control joints
c. Both brick and block walls expand over time, requiring them to have both expansion and control joints
d. both a and b

**6** Horizontal expansion joints intended to minimize wall cracks by permitting both the expansion of brick masonry and the deflection of steel shelf angles are called

a. control joints
b. expansion joints
c. soft joints
d. both a and b

**7** The spacing scale that is used to align the bed joints of two dissimilar masonry materials at 16" intervals is the

a. brick mason's scale
b. modular spacing scale
c. oversize spacing scale
d. both a and b

**8** Composite or cavity wall joint reinforcements are placed as the wall is raised every

a. 8"
b. 16"
c. 24"
d. 32"

**9** Base flashing should be adhered to the outer side of the inner wythe, extending vertically on it a minimum of

a. 4"
b. 6"
c. 8"
d. 12"

10  Placed in the outer wythe just above the flashing, weep vents or open head joints should be spaced apart a maximum of

a. 8"

b. 16"

c. 24"

d. 32"

11  That part of a wall extending above a roof line is called a

a. cap wall

b. collar wall

c. fire wall

d. parapet wall

12  The cavity and top of double-wythe walls are covered with stone, precast concrete, or metal materials are called

a. capping

b. coping

c. plates

d. sills

13  A covering for part of or all of a wythe that is terminated below the top of a wall is called a

a. cap

b. coping

c. plate

d. sill

14  A cutout in the underside of a projecting stone or concrete unit intended to prevent water from traveling back to and running down the face of the wall is called a

a. drip

b. groove

c. relief cut

d. wash cap

# Chapter 17 | Brick Paving

B efore the development of concrete, brick was used to pave roads and walkways. Still today, the durability of brick makes it a popular choice for providing pedestrian walkways, patios, and vehicular driveways. Mortared brick paving systems are affordable alternatives to the bare look of concrete. Homeowners prize the appearance of brick paving because it's the "black tie" of home dress-up, giving a residence striking curb appeal.

## OBJECTIVES

After completing this chapter, the student should be able to:

- Describe different types of paving brick.
- List the elements of brick paving systems.
- Compare similarities and differences between the types of bases for brick paving systems.
- Describe the procedures for installing a rigid, brick paving system.

# Glossary of Terms

**basket weave pattern** a paving brick pattern where brick are paired perpendicular to adjoining pairs.

**brick density** a measure of a brick's weight as compared to its size or volume.

**brick pavers** brick intended for either pedestrian or vehicular traffic, also called paving brick.

**Classification MX paving brick** brick pavers used in climates where freezing does not occur.

**Classification NX paving brick** brick pavers designed for interior use only.

**Classification SX paving brick** paving brick designed to withstand freezing temperatures while saturated.

**control joint** a formed or sawed vertical separation of concrete that permits dimensional changes of the concrete, controlling the location of separation cracking.

**coping brick** specially shaped brick intended for the outer edge of a brick paving project.

**expansion joint** a separation in brickwork permitting the expansion of brickwork caused by thermal movements or the increasing volume of brick.

**field pattern** the brick pattern or configuration between the borders.

**flashed brick** brick showing color variations as a result of firing the brick with alternately too much or too little air.

**flexible base** a base below mortarless brick paving consisting of compacted sand above a compacted crushed stone base.

**grouting** the process of filling the mortar joints between the brick pavers.

**heavy traffic** brick paving designs intended for vehicular traffic such as roadways and loading docks.

**herringbone pattern** a brick paving pattern where each brick is aligned perpendicular to that brick beside it.

**light traffic** brick paving designs intended for residential pedestrian use only.

**medium traffic** brick paving designs for public pedestrian traffic and limited vehicular traffic.

**modular paving brick** paving brick measuring 3⅝" wide by 7⅝" long.

**mortared brick paving** brick paving bedded in mortar on a rigid base and having mortared joints between the paving brick.

**mortarless brick paving** brick paving without mortar joints.

**pattern bond** the visual pattern created by the arrangement of the brick.

**repressed paving brick** brick pavers first extruded and then compressed to increase density.

**rigid base** a base below brick pavers consisting of a concrete slab, compacted crushed stone sub-base, and a compacted subgrade.

**running bond pattern** brick rows having each brick offset half its length with the brick in adjacent rows.

**semi-rigid base** a base below brick pavers consisting of asphalt and compacted crushed stone.

**skid resistance** a measure of vehicular traction on a wet surface.

**slip resistance** a measure of pedestrian traction on a wet surface.

**spalling** the cracking and flaking of the bricks' surface due to freezing water.

**stacked bond pattern** a brick pattern where all joints between the brick form continuous straight lines perpendicular to each other.

# Design Considerations for Selecting Clay or Shale Brick Pavers

The durability of brick pavement provides an attractive option for walkways, steps, patios, pool decking, and even projects intended for vehicular traffic. Brick used for these applications are known as paving brick or brick pavers. Design considerations for brick paving include the following:

- Intended use
- Type of base
- Desired appearance

## Intended Use

The location and purpose of the brick paving are key factors in selecting an appropriate brick. Based on weather expectations, there are three classifications of paving brick. The classifications are SX, MX, and NX. Classification SX paving brick includes paving brick designed to withstand freezing temperatures while saturated with water. Class SX brick should be used for exterior paving wherever freezing may occur. Classification MX paving brick should only be used in climates where freezing does not occur. Classification NX paving brick are designed for interior use only.

Brick paving designs can be classified as light, medium, or heavy traffic applications. Light traffic identifies designs intended for residential pedestrian use, such as residential sidewalks and patios. Medium traffic includes designs for public pedestrian traffic and limited vehicular traffic, such as public sidewalks, residential driveways, commercial entrances, and parking lots. Heavy traffic identifies designs intended for vehicular traffic, such as roadways and loading docks. Medium and heavy traffic designs are beyond the scope of this unit.

The thickness of a paving brick is optional. A thickness of either 1⅜" or 2¼" is typical. Some paving brick are as thin as ½", whereas others may be 2½" thick. The thinner paving brick are used primarily for mortared, pedestrian traffic situations. The thicker paving brick are better for heavier load conditions such as vehicular traffic or for mortarless paving.

The density of the paving brick must meet or exceed the traffic requirement. Brick density is a measure of a brick's weight as compared to its size or volume. The process of extruding, molding, or repressing can be used to manufacture paving brick. While extruded or molded paving brick meet most requirements, repressed paving brick have a greater density. Repressed paving brick are first extruded and then compressed to increase density. They are intended for high traffic areas.

Paving brick are also rated according to slip and skid resistance. Slip resistance is a measure of pedestrian traction on a wet surface. Skid resistance is a measure of vehicular traction on a wet surface. Factors influencing slip and skid resistance include brick texture and edge design. A rough surface or wire-cut brick provides better traction. Also, a brick with a chamfered or rounded edge provides greater slip and skid resistance.

## Type of Base

The *base* is that part of the paving system over which the brick are laid. It supports the brick and distributes all weight placed upon the brick paving. Regardless of the type of base, a commercial or residential brick paving system can be designed for either pedestrian or vehicular traffic. Construction details should conform to local codes and recommended design specifications. Factors influencing the selection of base type include water drainage, function, mortared or mortarless joints, and costs. The three types of bases are 1) rigid bases, 2) semi-rigid bases, and 3) flexible bases.

### Rigid Base

A rigid base consists of a concrete slab, compacted crushed stone sub-base, and a compacted subgrade (see Figure 17-1). Compacted or undisturbed subsoil is necessary for supporting the weight of the concrete. Soils considered as top soils, loose soils promoting vegetation growth and not capable of compaction, must be removed. Soil is removed to a minimum depth of 8", permitting a 4" depth of #57 crushed stone sub-base and a 4" thick concrete slab. Placing 6-6-10-10 welded wire reinforcement or steel reinforcement rods approximately 1" from the bottom of the concrete strengthens it and helps it resist separating if cracks develop.

**FROM EXPERIENCE**

When placing concrete, consider ordering ready-mixed concrete, hauled to the jobsite and placed from the truck. With ready-mix concrete, all of the concrete can be placed in less time and as one consolidated pour. This eliminates poor bonding between batches when having to make several smaller pours mixed manually or in smaller capacity concrete mixers.

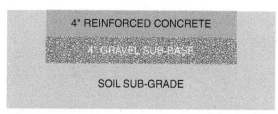

4" REINFORCED CONCRETE

4" GRAVEL SUB-BASE

SOIL SUB-GRADE

**Figure 17-1** A properly designed rigid-base system consists of a 4" or thicker concrete slab above a 4" crushed stone sub-base placed on compacted soil.

Exterior concrete slabs should have a minimum surface slope of ¼" per foot to allow water drainage. A broom finish rather than a smooth, trowel finish improves the bond between the mortar bedding for the brick and the concrete slab.

A rigid base is required for mortared brick paving, which is brick paving bedded in mortar on a rigid base with mortared joints between the paving brick A rigid base permits water drainage on a sloped surface. It is frequently used for residential walkways and patios.

**CAUTION**

**All underground utilities must be located and identified in and surrounding the area of construction. Governing regulations control the proximity of concrete slabs to buried utilities.**

### Semi-Rigid Base

A semi-rigid base consists of asphalt and compacted crushed stone below the paving brick. This system is limited to mortarless brick paving. Mortarless brick paving has no mortar joints between the brick. It is often used for commercial walkways.

### Flexible Base

A flexible base consists of compacted sand above a compacted crushed stone base (see Figure 17-2). Only mortarless brick paving should be laid over a flexible base.

The flexible base is used for both residential and commercial construction. Water is permitted to drain between the brick, through the setting bed and base, and into the soil. Flexible base systems permit easy removal and reuse of existing materials. This system is advantageous where access to underground utility and water lines is desired. Flexible base systems reflect a savings in material costs.

Regardless of the type of base, a commercial or residential brick paving system can be designed for either pedestrian or vehicular traffic. Construction details should conform to local codes and recommended design specifications. Since it is the scope of this chapter to address mortared brick paving on rigid bases only, the procedures for installing semi-rigid bases and flexible bases are not addressed. Mortarless paving systems are tasks typically undertaken by landscaping contractors.

### Desired Appearance

Options influencing the appearance of the finished project should be considered in the brick-selection process. These options include brick color, mortar color, brick size, and pattern bond.

### Brick Color

Brick manufacturers offer a variety of standard colors and custom colors upon request. Standard colors include a palette of grays, browns, oranges, reds, and pinks. The option of flashed brick increases the choices of standard colors. Flashed brick are brick showing color variations as a result of kiln-firing the brick with alternately too much or too little air.

### Mortar Color

The selection of standard mortar colors adds another option for the final appearance. Custom colors are available upon request.

## Brick Size

Modular paving brick measure 3⅝" wide and 7⅝" long. These dimensions assure pattern uniformity for mortared paving when ⅜"-wide mortar joints are used (see Figure 17-3).

Paving brick that measure 4" wide by 8" long are used for pattern uniformity for mortarless paving of the basket weave and herringbone patterns (see Figure 17-4). They can be used for mortared brick in the running bond pattern, but dimensions restrict them from being used for mortared brick in the basket weave and herringbone patterns.

Some manufacturers have 8"-square brick pavers, which create another appearance.

**Figure 17-3** The dimensions of a modular brick, 7⅝" × 3⅝", permit a uniformly aligned pattern when using ⅜" joints.

1"–1½" SAND

4" GRAVEL SUB-BASE

SOIL SUB-GRADE

**Figure 17-2** A leveling sand bed above a crushed stone sub-base forms a flexible base.

Figure 17-4 **With the brick edges touching adjoining brick, 4" × 8" brick are designed for a uniform pattern.**

Figure 17-5 **The profiles for two styles of coping brick intended for brick edges of steps and along pools are illustrated here.**

Special shapes offer added design possibilities. Half-radius or full-radius coping brick, specially shaped brick intended for the outer edge of a brick paving project, are available for the edge of a paving project (see Figure 17-5).

# Pattern Bonds

Pattern bond refers to the visual pattern created by the arrangement of the brick. Almost any pattern can be created, but there are four patterns frequently used. These include the running bond, the basket weave, the herringbone, and the stacked bond patterns. The intended use does not restrict the choice for the pattern bond. Technical Notes 14 of the Brick Industry Association reports that "bond pattern and unit orientation are not critical for load transfer in mortared paving."

## Running Bond Pattern

Each brick in the running bond pattern appears to be offset half its length with the brick in adjacent rows (see Figure 17-6). The rows can run either the length or width of the project. Aligning the rows parallel with the longer dimension of the project minimizes the need to cut brick at the ends. Another option is to lay the brick diagonally in relationship to the edge of the project.

## Basket Weave Patterns

Brick in the double basket weave pattern are paired perpendicular to adjoining pairs (see Figure 17-7).

Figure 17-6 **Brick in the running bond pattern are aligned in every other row and usually form a half-lap pattern with adjacent rows.**

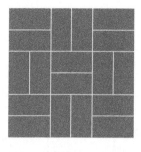

Figure 17-7 **Orienting pairs of brick perpendicular to adjoining pairs creates the double basket weave pattern.**

Figure 17-8 **This is one of many versions of the basket weave pattern.**

There are several versions of the basket weave pattern, one of which is shown in Figure 17-8.

## Herringbone Pattern

Aligning each brick perpendicular to the brick beside it creates the herringbone pattern. The herringbone pattern is typically laid at a 45-degree angle in relationship to the project's edge (see Figure 17-9).

Maintaining a uniform bond in the herringbone pattern requires more attention than patterns in which the mason's line or a straightedge makes it easier to align each row.

The 90-degree herringbone pattern is achieved by having the brick either parallel or perpendicular to the project's edge (see Figure 17-10).

Figure 17-9 **The herringbone pattern creates a "zig-zag" design.**

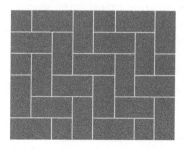

Figure 17-10 **The pattern is the same as in Figure 17-9, however, the brick are parallel and perpendicular to the borders.**

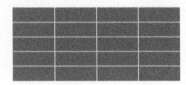

Figure 17-11 **Joints create continuous, straight lines with the stacked bond pattern.**

## Stacked Bond Pattern

The **stacked bond pattern** requires that all joints between the brick form continuous straight lines perpendicular to each other (see Figure 17-11).

# Laying Mortared Brick Paving

Brick paving that uses mortar requires masonry skills similar to the skills used for constructing masonry walls. Brick masons typically install rigid base systems requiring mortared joints. Because mortarless techniques do not require masonry skills, landscapers or homeowners typically install semi-rigid and flexible base systems. Semi-rigid and flexible base systems are beyond the scope of this unit.

Following are the procedures for laying mortared paving brick:

1. Provide the concrete base.
2. Lay the brick borders.
3. Lay the field pattern.
4. Grout the joints.
5. Tool the joints.
6. Clean the surface.

### Providing the Base

The first step for a mortared brick paving system is determining the dimensions for the concrete base. The dimensions depend on the selected brick and the pattern bond. Dry bonding the desired pattern using the specific brick intended for construction determines the dimensions for the base. Joints between the brick should be no less than ¼" or more than ½". The pattern may include a brick border. This border is typically equal to either the length or width of the brick. The total width for the project equals the sum of the borders and the field pattern, including the mortar joints.

A rigid base, as described earlier in this chapter, must be provided for mortared brick paving. A structurally sound, reinforced, and free of open cracks existing 4" or thicker concrete slab can be used as the rigid base. Existing slabs may not be the desired width the pattern requires. In this case, it may be desirable to project the brick slightly beyond the edges or set in from the edges of the concrete. Projecting the brick no more than ½" beyond the edges of a concrete slab is recommended, so that the projection of the

brick does not shear or break off. Existing slabs must be cleaned to remove soil, oil, coatings, organic matter, and other contaminates to ensure a good mortar bond.

A flat surface and rough texture improves good bonding of the mortar and concrete. A sloped surface is necessary for water runoff. A slope of ¼" per foot permits good drainage. Control joints in the concrete should be spaced no more than 16' apart. A **control joint** is a formed or sawed vertical separation of concrete that permits dimensional changes of the concrete base, controlling the location of separation cracking. The control joints in the concrete are to align with expansion joints in the brick paving. Technical Notes 14 of the Brick Industry Association states that "in most cases, an expansion joint spacing of 16 ft (5 m) in exterior mortared brick paving is adequate." Compliance with all local regulations governing the concrete mix, reinforcement, stone subbase, soil subgrade, and proximity to underground utilities is required. New concrete should be permitted to cure before laying the brick.

***For step-by-step instructions on constructing concrete slabs, see the Procedures section on pages 252–254.***

### Laying the Brick Borders

The borders and field pattern are dry bonded to determine the overall width for the brickwork. Borders on opposite sides are laid first because they serve to align the pattern and keep the surface flat (see Figure 17-12).

Dampening the concrete and the brick before laying improves the bond. The brick should be thoroughly wetted the day before or no later than several hours before being laid. Being low-absorption units, brick pavers easily move unintentionally if laid when their outer surfaces are saturated. Mortar is spread on the recently dampened concrete surface, and brick pavers are bedded in mortar within one minute or so to assure a good bond. Technical Notes 14A of the Brick Industry Association recommends "for exterior mortared brick paving on grade, Type M mortar is preferred.... Type S mortar may be used for exterior applications when the pavement is not in contact with the earth, such as suspended diaphragms or in interior applications." The brick along one border are bedded in mortar. A mason's line is used to ensure straight alignment of the outer edges of the border brick.

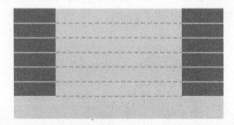

Figure 17-12 **The two borders are laid before beginning the field pattern. A uniform spacing keeps the brick on opposite borders equally aligned (broken lines).**

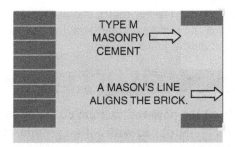

**Figure 17-13** A mason's line attached to the outside of the border aligns the brick.

Either the line or a mason's level aligns the top of the border brick flat. It should be noted that exterior brick paving surfaces are flat but not necessarily level. The paving surface follows the grade slope or is intentionally sloped to permit water drainage. The brick for the opposite border are bedded in mortar, maintaining the same spacing as for the first border (see Figure 17-13).

**FROM EXPERIENCE**

As each brick is bedded in mortar, apply light pressure to each brick and move it in a back and forth motion to improve the extent of bond between the mortar and the brick.

## Laying the Field Pattern

Attaching a mason's line to the borders on opposite sides permits laying a flat and uniform field pattern. The field pattern is the brick configuration between the borders. The mason's 4' level can be used instead of the mason's line for narrower field patterns. Each row of the field pattern is aligned with the brick uniformly spaced along each border (see Figure 17-14).

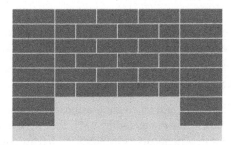

**Figure 17-14** The field pattern is aligned and kept flat with the border brick.

**FROM EXPERIENCE**

Moving the brick slightly in a back and forth motion while bedding it into the mortar increases bond strength. Covering the bedded brick with a plastic tarp prevents the bond-weakening effects of water evaporating from the mortar and keeps the brick clean.

Mortar protrusions appearing between the bedded brick that are less than ¾" from the surface of the brick are removed with a tuck pointing tool. This will permit adequate grouting of the joints between the brick later. Although mortar can be applied to the edges of brick as they are laid to fill the joints between them, filling these joints later is recommended. This procedure ensures fully mortared joints and is recommended because applying joints to the bricks' edges as they are bedded in mortar can misalign adjoining brick as the mortared brick is shoved against them. Unintentionally moving the brick already bedded in mortar also weakens bond strength.

Because dimensional changes of the concrete base and the brick paving are different, expansion joints, separations in brickwork permitting the expansion of brickwork caused by thermal movements or the increasing volume of brick, should align vertically with the control joints in the concrete base. Although expansion joints interrupt the brick pattern and may not be attractive, they should be spaced no more than 16' apart and aligned with the control joints placed in the concrete base. If there are no control joints in the concrete base, they can be created after completion of the project with a masonry saw capable of cutting the entire depth of both brickwork and concrete. Expansion joints should be filled with a compressible, weather-resistant sealant.

**CAUTION**

**Do not step or walk on mortared brick until at least 24 hours after bedding in mortar. Place barricades, caution cones, and yellow "caution" tape around the project to keep others away from it.**

## Grouting the Joints

The joints between the brick pavers are grouted, that is, filled with mortar using a tuck pointing tool, grout bag, or power grout gun, after the mortar-bedded brick have set for a minimum of 24 hours. Waiting at least 24 hours before grouting the joints prevents unintentionally breaking the bed-setting mortar bond between the bottom of the brick and the concrete surface. A power grout gun considerably reduces the time for filling joints and prevents a mason from experiencing wrist fatigue associated with squeezing mortar from a grout bag (see Figure 17-15).

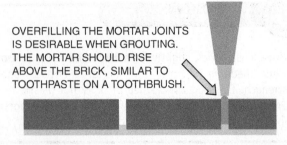

OVERFILLING THE MORTAR JOINTS IS DESIRABLE WHEN GROUTING. THE MORTAR SHOULD RISE ABOVE THE BRICK, SIMILAR TO TOOTHPASTE ON A TOOTHBRUSH.

**Figure 17-15** The joints are grouted, that is, filled with mortar.

**CAUTION**

Only those trained by a competent person and complying with governing regulations should operate a power grout gun. Electric motor-powered grout guns must be equipped with inline ground-fault circuit protection for operator safety.

**FROM EXPERIENCE**

It is best to overfill the joints with mortar so that the mortar rises above the brick. First, the protruding mortar shields the mortar between the joints, preventing water evaporation from the mortar between the joints. Additionally, some protruding mortar is compressed into the joints when tooling the joints, while at the same time, a shearing action of a jointer against the edges of the brick separates excess mortar from the joints.

## Tooling the Joints

Using a slightly concave or grapevine joint finish compresses the mortar, leaving a smooth-finish, water resistant joint (see Figure 17-16). Having joints as flat as possible minimizes standing water and accumulating dirt on the joints. Although the excess mortar can be removed from the surface using a mason's trowel, using a wet-dry vacuum cleaner or

TOOL JOINTS WHEN "THUMBPRINT HARD." TOOLING COMPACTS THE MORTAR FOR A MORE WEATHER-RESISTANT JOINT. EXCESS MORTAR IS EASILY REMOVED AFTER TOOLING.

**Figure 17-16** A concave-finished joint compresses the mortar and improves the joint's resistance to water.

powered blower is a better choice because the excess mortar is lifted from the surface without staining the brick or scarring the wet mortar joints.

**CAUTION**

Electric motor-powered vacuum cleaners and blowers must be equipped with inline ground-fault circuit protection for operator safety. Those exposed to the potential hazards of airborne debris must wear eye protection.

**FROM EXPERIENCE**

A ⅞" or 1" convex jointer forms a very slight concave joint surface for joints no wider than ½". The joints appear to be almost flush, but unlike flush cut joints never tooled, the mortar surface is compacted and less porous.

The joints are immediately tooled again. This improves the appearance of the joints and assures properly tooled joints. The brick surface is immediately covered with a tarp to prevent rapid evaporation of moisture from the mortar.

**FROM EXPERIENCE**

Using a power blower or shop-vacuum rather than a brush or broom to remove the excess mortar after tooling the joints prevents staining brick surfaces and scarring mortar joints when removing the remaining mortar. Mortar trapped between the bristles of brushes stains the brick, whereas blowing or vacuuming lifts the mortar from the surface.

*For step-by-step instructions for laying mortared brick paving, see the Procedures section on pages 255–256.*

## Cleaning the Surface

Brick pavers should be cleaned using only those materials and procedures recommended by both the brick and cement manufacturers. Chemicals considered as hazardous must be used with extreme caution and only by those trained by a competent person.

**CAUTION**

**Before using hazardous chemicals, you must be in compliance with all local, state, and federal regulations governing their use, storage, and disposal. Wear protective head gear, eye wear, face shield, clothing, gloves, and foot wear when using chemical cleaners.**

The brick manufacturer should be consulted before choosing to apply any water repellant. Although water repellants reduce a brick's rate of water absorption, some may prevent evaporation of ground moisture migrating to the surface from below the brick. This can cause brick **spalling**, the cracking and flaking of the bricks' surface, in regions experiencing freezing and thawing cycles.

## Summary

- Well planned and properly installed, mortared brick paving provides a low-maintenance surface for pedestrian and vehicular traffic, and transforms a no-frills concrete slab into an attractive and inviting area.
- Design considerations for mortared brick paving include intended use, brick and mortar color, border and field brick patterns, and joint finish.
- Compliance with all local, state, and federal regulations and using only recommended materials installed as recommended by industry standards provide safe and durable paving surfaces.
- As with other masonry projects, an apprentice should be supervised by an experienced mason when attempting mortared brick paving.

## Procedures | Constructing Concrete Slabs

**A** Remove soil to a minimum depth of 8" below grade, removing soils considered as top soils, those loose soils promoting vegetation growth and not capable of compaction. Compacted or undisturbed subsoil is necessary for supporting the weight of the concrete.

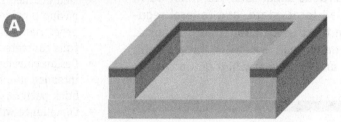

**A**

**B** Place a minimum 4" depth of #57 or #68 crushed stone into the excavated area, assuring that a 4"-thick concrete slab can be placed above it with its surface at the desired height.

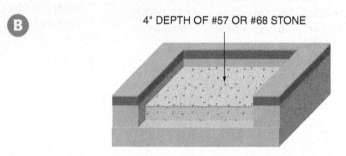

**B**

4" DEPTH OF #57 OR #68 STONE

**C** Using 2" framing lumber, form the sides for the concrete slab. Use screws to fasten the form boards to wooden stakes driven firmly into the ground. Be certain that the stakes are on the outsides of the form boards to prevent embedding the stakes in the concrete along its edges. Align the tops of the form boards level if a level concrete surface is intended, or slope the top of the form boards to follow the existing grade. A slope equivalent to ¼" per foot is needed to permit water runoff on exterior surfaces.

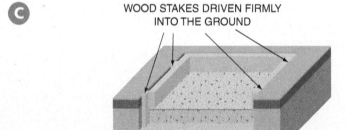

**C**

WOOD STAKES DRIVEN FIRMLY INTO THE GROUND

Place concrete with a strength rating of 3500 psi into the formed area, tapping the form boards with a rubber mallet to consolidate the concrete at its edges. Position either 6-6-10-10 welded wire or steel rod reinforcement meeting engineered specifications for the type of project and the geographic region in the concrete approximately 1" from the bottom of the slab as the concrete is placed. Keeping the reinforcement nearer the bottom of the concrete helps the concrete resist cracks from occurring at its bottom, the most likely place for such cracks to develop.

6-6-10-10 WELDED WIRE

$\frac{1}{2}$" STEEL REINFORCEMENT RODS
8" TO 12" ON CENTER

## Procedures

# Constructing Concrete Slabs (continued)

**E** Screed the surface of the concrete using a straight board or commercially available screed board. To screed concrete, hold the screed board against the top edge of the form boards on opposite sides of the concrete, and, while working it in a back-and-forth motion, move the screed board along the length of the concrete slab. The screed board consolidates the concrete and removes excess concrete at the surface. The screed board also creates a desirable surface for bedding brick in mortar, a flat surface with a rough texture permitting mortar to bond well.

- Use a hand-held cement finishing trowel for a smoother surface, removing surface ripples resulting from the screed board, where the exposed concrete is intended as the finished surface rather than a rigid base for mortared brick paving. Take a concrete edger along the top of concrete surfaces intended to remain exposed between its edges and the form boards to create a small radius at the top edges of the concrete. This prevents its edges from chipping. Pull the broom across the concrete surface to create a slip-resistant surface on exterior concrete, using an 18"-wide to 24"-wide floor broom.

**E**

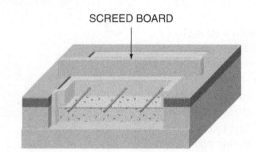

SCREED BOARD

## Procedures

# Laying Mortared Brick Paving

**A** Lay out the borders and the field pattern.

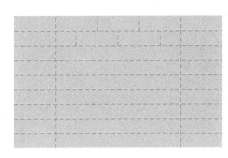

**B** Lay brick in mortar on opposite borders.

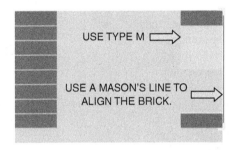

USE TYPE M ⟶

USE A MASON'S LINE TO ALIGN THE BRICK. ⟶

**C** Lay each row of the field pattern. Attaching a mason's line to the borders keeps each row aligned.

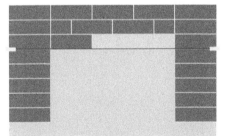

**D** Lay the final row as a border pattern if desired.

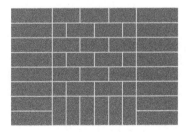

## Procedures

# Laying Mortared Brick Paving (continued)

**E** Overfill the mortar joints when grouting. The mortar should rise above the brick, similar to toothpaste on a toothbrush.

**F** Strike the mortar joints when they become "thumbprint hard." Remove excess mortar with a power blower or shop vacuum rather than a brush or broom to prevent mortar stains on the brick. Cover the surface with a plastic tarp to prevent excessive water evaporation from the mortar.

## Review Questions

*Select the most appropriate answer.*

1. **Brick used for exterior paving wherever freezing can occur should be classified as class**

   a. MX

   b. NX

   c. SW

   d. SX

2. **Paving brick intended for residential pedestrian use must meet applications for**

   a. light traffic

   b. medium traffic

   c. heavy traffic

   d. all of the above

3. **Paving brick that are first extruded and then compressed to increase their density for high traffic applications are called**

   a. commercial paving brick

   b. dense paving brick

   c. repressed paving brick

   d. vehicular paving brick

4. **The measure of pedestrian traction on a wet surface is called**

   a. fall resistance

   b. scuff resistance

   c. skid resistance

   d. slip resistance

5. **A properly designed rigid base for brick paving consists of**

   a. compacted subgrade

   b. crushed stone sub-base

   c. reinforced concrete

   d. all of the above

6. **Modular brick pavers measure**

   a. $3\frac{1}{2}" \times 7"$

   b. $3\frac{5}{8}" \times 7\frac{5}{8}"$

   c. $3\frac{1}{2}" \times 8"$

   d. $4" \times 8"$

7. **To permit water runoff, brick paving surfaces should have a minimum slope of**

   a. $\frac{1}{8}"$ per foot

   b. $\frac{1}{4}"$ per foot

   c. $\frac{3}{8}"$ per foot

   d. $\frac{1}{2}"$ per foot

8. **A procedure for improving the mortar bond for brick paving is**

   a. bed brick within 1 minute or so of spreading mortar

   b. dampen the brick pavers and the concrete base

   c. apply light pressure and move each brick in a back and forth motion

   d. all of the above

9. **The maximum spacing for control joints permitting differential movements of the brick and concrete base is**

   a. 10'

   b. 12'

   c. 16'

   d. 20'

10. **The type of mortar recommended for brick paving on grade in contact with the earth is**

    a. Type M

    b. Type S

    c. Type N

    d. both a and b

# Chapter 18

# Steps, Stoops, and Porches

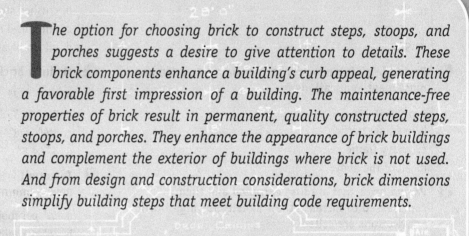

The option for choosing brick to construct steps, stoops, and porches suggests a desire to give attention to details. These brick components enhance a building's curb appeal, generating a favorable first impression of a building. The maintenance-free properties of brick result in permanent, quality constructed steps, stoops, and porches. They enhance the appearance of brick buildings and complement the exterior of buildings where brick is not used. And from design and construction considerations, brick dimensions simplify building steps that meet building code requirements.

## OBJECTIVES

**After completing this chapter, the student should be able to:**

- Identify and define the components of brick steps.
- List the procedures for constructing brick steps.
- Identify factors to consider when planning the dimensions for brick steps.
- Describe methods of providing foundation support below brick steps.
- Describe methods for constructing porches and stoops.

## Glossary of Terms

**loose fill** soil placed without the compaction necessary for structural support.

**porch** a sheltered area at the entrance of a building.

**rise** the vertical height between two adjoining horizontal step treads.

**riser** the vertical component of a step.

**slab footing** a concrete footing spanning the entire area below the structure it supports, continuing beyond its exterior walls a minimum of 6" on each side.

**stoop** an unsheltered platform at the entrance of a building.

**tread** the horizontal surface of a step supporting pedestrian traffic.

**tread depth** a measurement taken from the tread's front edge to its back edge adjoining the riser.

# Parts of a Step

The two parts of a step are the riser and the tread. The **riser** is the vertical component. The **rise** is the vertical height between two adjoining horizontal step treads. The **tread** is the horizontal surface of a step supporting pedestrian traffic. The **tread depth** is measured from the tread's front edge to its back edge between adjoining risers. Figure 18-1 identifies the parts of a step.

Local codes govern tread depth and riser height. Section R311.5.3.2 of the International Residential Code requires tread depth to be a minimum of 10" with no more than $^3/_8$" difference between the greatest and the smallest tread depth. Section R311.5.3.1 of the International Residential Code requires risers to be no higher than $7^3/_4$" and no more than a $^3/_8$" difference between the height of the lowest and highest riser for a flight of steps.

# Designing Steps

Following are the procedures for designing steps:

1. Determine the rise.
2. Determine the tread depth.
3. Determine the number and height of risers.
4. Determine the number of treads.

## Determining the Rise

The vertical rise of each step can be constructed with two courses of brick, one laid in the stretcher position and the other in the rowlock position. There are two reasons for this stretcher/rowlock design:

1. It creates an appropriate rise, approximately $6^3/_4$" when using standard size brick or approximately $7^1/_4$" with engineered or oversize brick.
2. Grade SW cored brick can be used for the tread when placed in the rowlock position.

The riser height can be changed by varying the width of the bed joints below the stretcher and rowlock courses within acceptable limits, $^1/_4$" to $^1/_2$". Since it is usually acceptable to vary bed joints as much as $^1/_8$" more or less than $^3/_8$", the height for the risers can be altered as much as $^1/_2$"

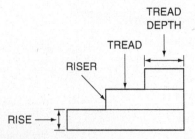

**Figure 18-1 This illustration identifies the parts of steps.**

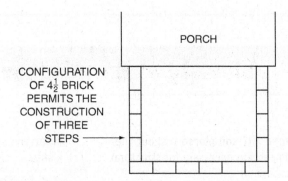

**Figure 18-2 This layout is designed for three steps.**

simply by changing the width of the two bed joints. The rise is equal to the sum of the following:

- The height of a brick bedded in the stretcher position
- The width of a brick bedded in the rowlock position
- Two, $^3/_8$" mortar joints (plus or minus $^1/_8$"), a total of $^3/_4$" (plus or minus $^1/_4$" each joint)

## Determining the Tread Depth

Depending on the size of the selected brick, the tread depth varies between $11^1/_2$" and 13" when the design for the tread depth is equivalent to $1^1/_2$ brick. Choosing to make the tread depth equivalent to the length of $1^1/_2$ brick simplifies the brick layout configuration. For example, if 4 treads are needed, then the brick layout configuration at the base extends a total of 6 brick from the stoop or porch. If 3 treads are needed, the brick layout configuration extends $4^1/_2$ brick at the base (see Figure 18-2).

The "rule of 25" is a guideline some use for determining riser height and tread depth. While complying with building code regulations, this guideline recommends that twice the rise added to the tread depth should equal 25". A tolerance of 1", more or less, is generally accepted. Therefore, the sum can range between 24" and 26". For example, if the height of the riser is $6^3/_4$", then an ideal tread depth may be $11^1/_2$", but a uniform tread depth between $10^1/_2$" and $12^1/_2$" is acceptable.

## Determining the Number and Height of Risers

Measurements are typically taken in increments of inches to the nearest $^1/_8$". The overall height from finished grade or walkway surface to the finished height of the stoop or porch is measured. Where applicable, the height from an existing or planned walkway to the finished height of the stoop or porch is measured (see Figure 18-3).

The number of risers required is determined by dividing the measured height by the desired rise. Increasing or decreasing the rise by changing the width of the bed joints within the allowable limits may eliminate a quotient remainder. The quotient represents the number of required risers. The following sample problem illustrates how to determine the number of needed risers.

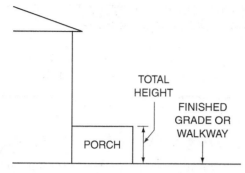

**Figure 18-3** In this illustration, total height is measured as the distance from finished grade to the top of a porch.

**Sample problem:**

If the measured height from the planned walkway surface to the finished height of a porch equals 36", and the step rise is planned to be 6¾", how many risers are needed?

$$\text{number of risers} = \text{total height"}/\text{rise"}$$
$$\text{number of risers} = 36 \text{ divided by } 6\tfrac{3}{4}$$
$$36/6\tfrac{3}{4} = 5\tfrac{1}{3}$$

Answer: number of risers = 5

Whenever the quotient has a remainder, either the riser height or the overall height from the walkway to the top of the stoop or porch must be changed. Choose to change the riser height only if the following are true:

- Each riser is to have the same height.
- Mortar bed joints are not less than ¼" or more than ½".
- Riser height does not exceed 7¾".

For the preceding sample problem, the height of each riser needs to be about 7³⁄₁₆" if overall height is to remain 36". Accomplishing this using standard-size brick requires bed joints to be more than ½" when each step is comprised of one stretcher course and one rowlock course. Therefore, the choice may be to have the riser height remain at 6¾" and raise the walkway surface at ground elevation or change the finished height of the porch or stoop. Sections R311.4.2 and R311.4.3 of the International Residential Code requires the floor or landing at an exit door, "a side-hinged door not less than 3' in width and 6'8" in height"... providing "for direct access from the habitable portions of the dwelling to the exterior without requiring travel through a garage," to be no more than 1½" lower than the top of the threshold, while the floor or landing at exterior doors other than the exit door can be no more than 7¾" lower than the top of the threshold.

Choosing to have five risers, each measuring 6¾" in height, requires changing the overall height to equal 33¾" (6¾" × 5). This requires a decision to either elevate the intended walkway at finished grade level or lower the proposed height of the porch. Lowering the porch level is not an option if it already exists. In some cases, lowering the porch or stoop level may not be an option if it causes the porch or

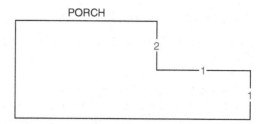

**Figure 18-4** There are two risers and one tread between finished grade level and the top of the porch.

stoop to be below the door sill more than the building code permits. It may therefore be necessary to raise the finished grade level at the base of the steps to have risers of equal height, constructed as one brick stretcher course and one brick rowlock course.

## Determining the Number of Treads

The number of treads for steps accessing a porch or stoop is one less than the number of risers between finished grade or the walkway and the top of the porch or stoop (see Figure 18-4).

If each tread is to be a depth of 1½ brick, the brick layout configuration from the existing porch or stoop is determined by the number of treads multiplied by 1½. Sometimes tread depth is designed to equal the length of 2 brick, approximately 16". In such cases, the base layout configuration is determined by multiplying the number of treads by 2.

The length between the two ends of steps varies. Factors influencing step length include the following:

- Appearance design considerations
- Expected volume of traffic
- Building codes and regulations

Steps are a design element that can influence the appearance of the building. Longer steps are usually found at a building's main entrances, giving it a more inviting appeal and allowing more traffic than narrower steps. Local codes may govern the minimum lengths for steps and the minimum space between installed hand railings. Typically, 36", measured from the inside of handrails where applicable, is considered as the minimum length for treads. Section R311.5.6 of the International Residential Code requires that "handrails shall be provided on at least one of each continuous run of treads or flight with four or more risers." A handrail should be at each side of open end steps for protection from falls. Always comply with local building codes when constructing steps.

## Constructing Brick Steps

Following is the procedure for constructing steps:

1. Place a concrete slab footing.
2. Complete below-grade masonry.
3. Lay mortared brick steps.

## Placing a Concrete Slab Footing

A concrete slab footing is placed below the frost line on undisturbed or compacted soil. A typical design consists of an 8"-thick concrete slab footing that extends a minimum of 6" beyond the outer sides of the steps, reinforced with ½" steel bar reinforcement, placed 8" to 12" on center. But for a site-specific design, a structural engineer should approve a soil's capacity for supporting a project's weight as well as the design of the supporting concrete footing.

> ⚠️ **CAUTION**
>
> **Before beginning to dig, utility companies or their agents must be notified so that all buried utilities are located and properly identified. Comply with all governing regulations pertaining to distances required between marked utilities and ground digs. Regulations generally prohibit construction above utilities and limit the proximity between concrete footings and buried utilities.**

The footing may be sloped ¼" per linear foot to permit the sloping of each tread for water runoff. A ¼" per foot of slope for treads is recommended for water runoff (see Figure 18-5).

Having a level footing and providing tread slope by varying the size of the bed joints of the brick for the treads is an alternative method of providing the necessary tread slope. The concrete slab may either be formed or placed in an open excavated area. Wet concrete is placed into temporary forms that are built using 2" framing lumber. Many times, concrete is placed directly in ground excavations, requiring no formwork. Typically, concrete for slab footings should have a minimum strength rating of 2500 psi and a slump not less than 3" or more than 4". It is recommended to use a cement float, a hand-held cast metal cement finishing tool, to finish the surface of the wet concrete, bringing cement paste to the top.

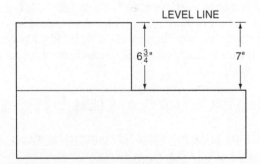

Figure 18-5 The tread is ¼" lower at its front edge to permit water drainage.

## Completing Below-Grade Masonry

With the top of the slab footing below finished grade, concrete masonry units (CMUs) rather than brick can be used as masonry fill below grade level and to the bottom of the first riser.

Structural bridging is needed wherever steps, stoops, and porches are built above loose fill. Loose fill is that earth placed without the compaction necessary for structural support. Backfilled against the exterior walls of foundations, loose fill is not suitable for supporting the weight of masonry. Unless the slab footing is placed below the backfill on soil tested capable of supporting it, bridging across the backfill from the foundation wall to undisturbed or compacted soil is necessary. If placed on loose fill, the steps shift from their original, intended position. Using concrete masonry lintels to bridge the backfill is one method. Concrete masonry lintels can bridge across loose fill, transferring the weight of the concrete slab footing and the brick steps to the foundation wall at one end and a concrete trench footing placed in the compacted or undisturbed soil at the opposite end (see Figure 18-6).

> ⚠️ **CAUTION**
>
> **Because of their weight, concrete masonry lintels should be placed by power-operated construction equipment operated by competent persons complying with OSHA safety standards. Comply with OSHA construction industry regulations 1926.651 and 1926.652 when working below grade in trenches created by foundation walls and soil embankments.**

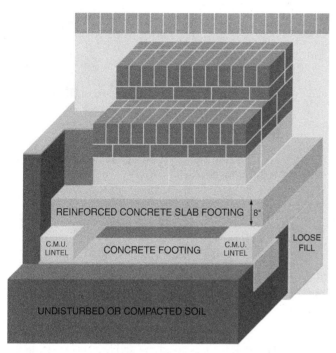

**Figure 18-6 Two concrete masonry lintels bridge the backfill and support the reinforced concrete slab footing. The lintels typically extend into the foundation wall a minimum of 4". For a specific job, the support for brick steps should be designed by a structural engineer.**

## Laying Mortared Brick Steps

**CAUTION**

**Do not step or walk on mortared brick until at least 24 hours after bedding in mortar. Place barricades, caution cones, and yellow "caution" tape, out from the project to keep others away from it.**

**CAUTION**

**A brick step built entirely of brick can be designed to accommodate the length of both the brick and mortar joints. Brick rated for "severe weather" applications should be used wherever freezing and thawing may occur. Type M mortar is recommended for brick at grade level and in contact with the earth. Begin by laying the outer perimeter of brick in the stretcher position, filling the inside of the perimeter brickwork with mortared brick. All mortar joints must be full to prevent water migration within the step (see Figure 18-7).**

**Figure 18-7 The first course is laid in the stretcher position. All joints are fully mortared. This 3-brick configuration permits constructing two steps, each having a 12" tread.**

**FROM EXPERIENCE**

After the brick for the outer course are bedded in mortar and aligned, bed the fill brick, leaving about ½" between adjacent brick. Fill the joints between them solid with wet mortar, ensuring no open voids for potential water migration and accumulation.

Next, a brick course is laid in the rowlock position. The front edge of the rowlock can either be aligned plumb with the face of the stretcher course below it or the rowlock can be projected ½" to ¾" beyond the face of the stretcher course. This completes the first step (see Figure 18-8). Grade SW cored brick can be used for the brick in the rowlock position, however solid brick are recommended and especially for the brick at ends where the cores should not be exposed. Mortar joints are tooled once the mortar becomes "thumbprint hard." As with mortared brick paving, a wide convex jointer forming a slight concave joint finish is recommended. The jointer compacts the mortar, creating a more water-resistant joint between the brick.

**FROM EXPERIENCE**

When cored brick are specified for building brick walls, request that matching solid brick be ordered for laying at the ends of rowlocks.

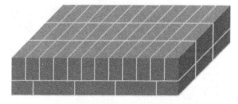

**Figure 18-8 The height of the riser is determined by the combined height of the stretcher and rowlock courses.**

Use well-tempered Type M mortar for steps, with a full extent of mortar bond to prewetted brick to strengthen the bond. Brick that are poorly bonded have a greater potential for coming loose, possibly causing an accident if one falls and becoming a liability for which the mason may be held responsible.

The front face of the next riser begins at a point 1½ brick from the front edge of the first tread. Laying brick in the stretcher position as done for the first step begins the next riser (see Figure 18-9).

**FROM EXPERIENCE**

Covering the completed tread with plastic sheathing, cardboard, or plywood protects it from mortar stains as the next riser is built. Cardboard and plywood can also protect the newly bedded brick from unintentional impacts and rapid water evaporation from the mortar.

Laying a course of brick in the rowlock position as previously done for the first step completes the second tread (see Figure 18-10).

For every 1½ additional brick from the porch or stoop in the layout configuration of the first course, an additional step can be constructed.

**Figure 18-9** The face of the second step begins at the center of the brick below it.

**Figure 18-10** Two steps having 6¾" risers and 12" treads are complete.

**Figure 18-11** Solid brick permit the option of laying the tread brick in the stretcher position.

**Figure 18-12** Combining two standard-size brick stretcher courses with a course of 1⅜"-thick pavers creates this step.

The brick pattern for a step can be laid in the reverse order if the brick for the tread are solid, meeting grade requirements for severe weather applications where applicable. One advantage of having the tread brick laid in the stretcher or header position rather than the rowlock position is the reduced number of exposed mortar joints, which are potential areas for water migration (see Figure 18-11).

Brick pavers can be used for step construction. Laying 1⅜"-thick brick pavers above two courses of standard-size brick results in a step that has an approximate 7" rise (see Figure 18-12).

***For step-by-step instructions on constructing brick steps, see the Procedures section on page 267.***

## Brick Paved Concrete Steps

On sloped terrains, concrete steps are often paved with mortared brick. The tread depth and riser height of a concrete step, generally formed and placed by construction crews other than the brick masons, should accommodate brick spacing and brick coursing (see Figure 18-13). Section R311.5.4 of the

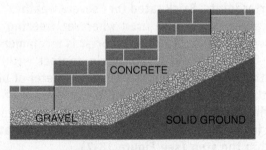

**Figure 18-13** Walkways at different elevations are connected by steps. The mortared brickwork is supported by reinforced concrete steps.

International Residential Code requires a flight of stairs to be no more than 12' between floor levels or landings.

# Porches and Stoops

A porch is a sheltered area at the entrance of a building. A stoop is an unsheltered platform at the entrance of a building. One or more steps may be needed at a porch or stoop. The outer perimeter walls for a masonry porch or stoop are constructed on a concrete foundation. Like steps, the concrete footing must be placed below the frost line and on undisturbed soil, or it must be supported, typically by a concrete masonry lintel or reinforced concrete bridging the loose fill, and always in compliance with building codes and engineered specifications. Crushed stone placed inside the walls supports a 4"- thick reinforced concrete slab atop the porch or stoop. The finished top surface may be mortared, brick pavers (see Figure 18-14).

## FROM EXPERIENCE

A stoop or porch should extend a minimum of 5' from the wall it adjoins, a distance permitting a person room to stand in front of an outward opening, side-hinged 36"-wide storm door or screen door.

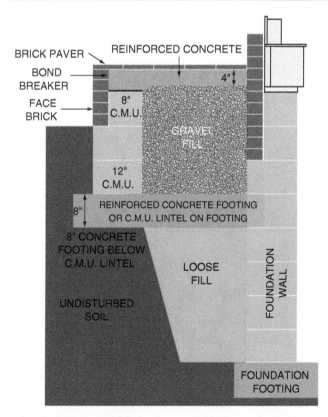

**Figure 18-14 This illustrates the components of a properly designed brick stoop or porch.**

Labels in figure:
BRICK PAVER
BOND BREAKER
FACE BRICK
REINFORCED CONCRETE
4"
8" C.M.U.
GRAVEL FILL
12" C.M.U.
8"
REINFORCED CONCRETE FOOTING OR C.M.U. LINTEL ON FOOTING
8" CONCRETE FOOTING BELOW C.M.U. LINTEL
LOOSE FILL
FOUNDATION WALL
UNDISTURBED SOIL
FOUNDATION FOOTING

**Figure 18-15 Placing a concrete slab inside the perimeter walls of a porch or stoop permits laying mortared brick paving, concealing the concrete slab.**

**Figure 18-16 This design is used for a brick porch or stoop that has an exposed, concrete surface. Having the concrete above and projecting beyond the brick walls of the porch or stoop caps the brick walls.**

A 4"-thick reinforced concrete slab may be placed inside the masonry walls, flush with the top (see Figure 18-15). This design permits laying mortared brick paving on the concrete slab, having no part of the concrete exposed. The slab is typically supported by crushed stone fill and the inside of the perimeter walls. Always comply with building codes and engineered specifications.

Where the concrete is to remain exposed, it may be placed above and projected beyond the masonry perimeter walls (see Figure 18-16). In either case, a bond break, material such as 30-pound construction felt paper, should be placed between the concrete and the brick walls to allow for differential movement.

## FROM EXPERIENCE

Before placing the concrete, cover the parts of the brick exposed to the concrete with 30-pound construction felt paper or PVC flashing. This acts as a bond break, permitting differential movements of the brick and concrete. Cover wet concrete with plastic tarps to prevent excessive evaporation and to permit hydration, the chemical reaction between cement and water.

# Summary

- Brick offers low-maintenance, attractive steps, stoops, and porches.
- Brick selected for steps and similar flatwork intended for pedestrian traffic must meet industry standards and governing regulations for such purposes.
- The performance of masonry steps, stoops, and porches depends not only on proper design and engineering but also on good quality workmanship.
- Steps, stoops, and porches require reinforced concrete footings not only capable of supporting the imposed

loads but also designed to span loose fill to prevent structural settlement.

- The illustrations and information in this chapter are for general guidance only. Prior to using this information, a registered professional engineer should be consulted to ensure that the details are appropriate for a particular project or geographical area in order to correct any errors or inaccuracies in the details and to ensure that the details comply with governing codes and regulations.

- As with other masonry projects, an apprentice should be supervised by a competent mason when attempting to construct steps, stoops, and porches.

## Procedures

# Constructing Brick Steps

**A** Lay the first course of brick on a properly constructed masonry foundation or footer. Use full bed joints and head joints. The brick in this illustration permits the construction of two treads. Each tread requires extending the first course 1½ brick from the existing structure. Slope the brickwork ¼" per foot from the back to the front to permit water runoff for the steps.

**A**

**B** Lay brick in the rowlock position aligned plumb with the stretcher course completed in Procedure A. The combined height of the stretcher course and the rowlock course is approximately 6¾" if using standard-size brick. Using oversize brick results in a riser height of approximately 7¼". Slope the brickwork ¼" per foot from the back to the front to permit water runoff for the steps.

**B**
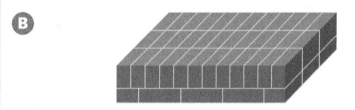

**C** Beginning at the center of the second row of rowlocks from the front, lay a stretcher course. Each tread should be equal to the length of 1½ brick. This is equal to approximately 12". Slope the brickwork ¼" per foot from the back to the front to permit water runoff for the steps.

**C**
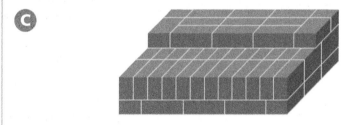

**D** Lay brick in the rowlock position and aligned plumb with the stretcher course below it. Half-brick are used on the last tread to permit a total tread depth of 12". The height between the last tread and the top of the stoop should be equal to the height of the risers. Slope the brickwork ¼" per foot from the back to the front to permit water runoff for the steps.

**D**
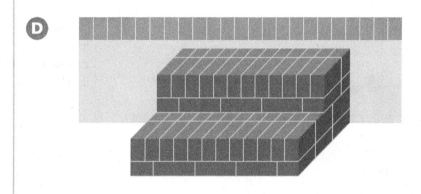

## Review Questions

*Select the most appropriate answer.*

**1** The vertical height between two steps is called the

a. depth

b. lift

c. rise

d. span

**2** The riser height between any two adjoining steps should be no more than

a. 7"

b. 7½"

c. 7¾"

d. 8"

**3** Tread depth should be a minimum of

a. 8"

b. 10"

c. 12"

d. 16"

**4** To permit water runoff, a tread should slope toward its front edge

a. ⅛" per foot

b. ¼" per foot

c. ⅜" per foot

d. ½" per foot

**5** Soil placed without compaction necessary for structural support is called

a. course aggregate

b. disturbed soil

c. loose fill

d. undisturbed soil

**6** A reinforced CMU sometimes used to support the weight of porches and steps is called a

a. beam

b. coping

c. header

d. lintel

**7** A sheltered entrance is called a

a. pavilion

b. platform

c. porch

d. stoop

**8** The type of mortar recommended for brick steps at grade level and in contact with the ground is

a. Type M

b. Type S

c. Type N

d. both a and b

**9** A concrete slab footing supporting brick steps should have a minimum thickness of

a. 4"

b. 6"

c. 8"

d. 12"

**10** The details for concrete slab footings that are to support steps, stoops, and porches are best designed by

a. a brick mason

b. a general contractor

c. a homeowner

d. a structural engineer

# Chapter 19

# Piers, Columns, Pilasters, and Chases

The design requirements for buildings can include piers, columns, pilasters, and chases. Standing alone, similar to posts, or as integral parts of masonry walls, masonry piers and columns can be purely decorative architectural details or serve as structural supports. Similarly, pilasters may be intended only as architectural design details, projecting from otherwise flat-appearing walls to create a three-dimensional wall face design. Or they may serve structural needs, enclosing vertical structural steel members or strengthening masonry walls. Masonry chases, however, typically serve only utilitarian purposes, primarily enclosing electrical panels and conduit, ductwork, and plumbing.

## OBJECTIVES

After completing this chapter, the student should be able to:

- Define the terms pier, pilaster, chase, and column.
- Identify and give uses for masonry piers, pilasters, chases, and columns.
- Lay out and build a pier, pilaster, chase, and column.

# Glossary of Terms

**actual size** the true dimensions of an object.

**cap, also called capping,** an architectural concrete or stone member for the top of a pilaster or part of or an entire wythe that is terminated below the top of an adjoining wall.

**capital** an architectural concrete or stone top for piers, columns, and pilasters that protects them from the weather and enhances their appearance.

**chase** a recessed area in a wall intended to provide space for electrical panels and conduit, heating and cooling ductwork, and plumbing.

**column** a masonry wall whose width does not exceed four times its thickness and whose height exceeds four times its least lateral dimension.

**compass brick** also called radial brick, curved brick that are used for curved brickwork.

**compressive strength** a measure of the vertical weight or force that a wall can support.

**concentrated load** the weight supported by another structural component.

**control joint** a joint controlling the location of separation in a concrete masonry wall resulting in the dimensional changes of the building materials.

**hollow masonry pier** a pier constructed of masonry units supporting no weight other than its own.

**lateral strength** a measure of the horizontal force that a wall is capable of resisting.

**masonry column** a masonry wall whose width does not exceed four times its thickness and whose height exceeds four times its least lateral dimension.

**nominal size** an expression of an object's approximate size rather than its actual size.

**pier** a support below another structural component.

**pilaster** a columnar projection from a masonry wall.

**pilaster block** a concrete masonry unit designed for constructing concrete masonry pilasters.

**radial pier** a round pier.

**reinforced masonry pier** a pier strengthened with concrete grout and steel reinforcing bars.

**structural pier** a pier designed to support a concentrated load.

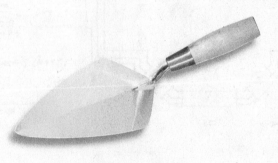

# Piers and Pilasters

A masonry **pier** is a support below another structural component. A pier can be freestanding, such as under a floor system, but a masonry wall between openings is also considered a pier if it supports the structure above it. A pier must be properly designed to carry the **concentrated load,** or the weight of other structural components. A **pilaster** is a masonry column projecting from a masonry wall. A pilaster may be used to enhance appearance or as a structural component of a reinforced masonry wall. Like a pier, a pilaster can serve as a support for a concentrated load.

# Nominal and Actual Sizes

The design for piers, pilasters, chases, and columns is often expressed in terms of nominal sizes. **Nominal size** is an expression of an object's approximate size rather than its actual size. The nominal dimensions for standard size brick and oversize brick considers the length of each brick to be 8" and its width 4". Figure 19-1 illustrates the layout for a rectangular brick pier having the nominal dimensions of 16" × 20".

By constructing piers, pilasters, chases, and columns using nominal sizes, dimensions can accommodate the actual lengths and widths of the brick.

For brickwork, the **actual size,** the true dimensions of an object, is expressed in feet, inches, and 1/16" increments. Using specific actual dimensions may require altering the length of the brick or the mortar joint widths. Whether given dimensions represent actual or nominal size should be clarified before construction begins. Piers and columns are typically supported by concrete slab footings. Their depth below grade, reinforcement, and concrete thickness and strength are site-specific and job-specific requirements that must comply with building codes and engineered specifications. The footings typically supporting hollow piers and columns for residential light posts and mailbox enclosures are 8"-thick reinforced concrete and extend 6" beyond the face of each wall, placed below the frost line on suitable soil.

# Types of Masonry Piers

Four types of masonry piers are radial piers, hollow piers, structural piers, and reinforced masonry piers.

**Figure 19-1** The nominal size of this brick pier is 16" × 20". Depending on brick dimensions and head joint width, the actual size can be more or less than the nominal size.

**Figure 19-2** Radial brick are used to build these radial brick piers.

## Radial Piers

A **radial pier** is a round pier (see Figure 19-2). The size of a radial pier can be expressed as its radius or its diameter. Compass brick are used to construct radial piers. **Compass brick,** also known as **radial brick,** are curved brick that are used for curved or radial brick walls.

## Hollow Piers

A **hollow masonry pier** is a pier constructed of masonry units supporting no weight other than its own. A hollow masonry pier should not be used as a structural support. To enhance appearance, a hollow pier is often used to enclose a structural steel column or similar column serving as a structural support (see Figure 19-3).

A freestanding, hollow pier can be used to enclose a mailbox or support a light fixture (see Figure 19-4).

**Figure 19-3** Each of the hollow brick piers encloses a steel structural column. The brick support no weight of the above structure.

Figure 19-4 This hollow brick pier supporting a light fixture creates an inviting entrance to the residence.

A brick column to which a light fixture or electrical receptacle is attached requires electrical wiring. A licensed electrician should install the wiring.

**CAUTION**

Masonry piers or columns such as those containing mailboxes and near roadways must be in compliance with Department of Transportation and utility company regulations and local ordinances governing easements, right of ways, and clear zones. Before beginning to dig, utility companies or their agents must be notified so that all buried utilities are located and properly identified. Comply with all governing regulations pertaining to distances required between marked utilities and ground digs. Regulations generally prohibit construction above utilities and limit the proximity between concrete footings and buried utilities.

## Structural Piers

A structural pier is designed to support a concentrated load. A single-wythe brick wall between openings and supporting a load is considered a pier, as shown in Figure 19-5.

## Reinforced Masonry Piers

A reinforced masonry pier is strengthened with concrete grout and steel reinforcing bars. Reinforced masonry piers support concentrated loads. The compressive strength of the masonry and the tensile strength of the steel are both factors governing the design of the pier. A structural engineer should design a reinforced masonry pier.

Figure 19-5 The brick wall between openings is considered a structural pier because it supports brick masonry above the openings.

Both brick and concrete masonry units (CMUs) are used for piers.

## Masonry Columns

The terms *column* and *pier* are often used to mean the same thing. Like a pier, a column can be designed to support a concentrated load. The distinction between a pier and a column is based on its dimensions. A masonry column is a masonry wall whose width does not exceed four times its thickness and whose height exceeds four times its least lateral dimension.

## Laying Out Piers and Columns

A column or pier is said to be square when:

- Each of the four corners are 90-degree angles.
- The front side and back side are equal lengths.
- The left side and right side are equal lengths.
- The diagonal measurements are equal.

The following procedures are recommended for laying out and constructing piers and columns.

**Step 1:** The layout of a column or pier begins with two lines perpendicular to each other. A framing square establishes a 90-degree angle. The length of two adjoining sides are measured and marked (see Figure 19-6).

**Step 2:** Using the framing square, a line perpendicular to either line established in Step 1 is drawn, overlapping the mark indicating its length. Equal to the length of the opposite side, the length of this line is measured and marked (see Figure 19-7).

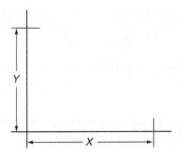

**Figure 19-6** Using a framing square, two lines at right angles are drawn. Measuring from their intersection, the length for the side and front are marked.

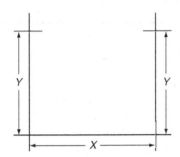

**Figure 19-7** Parallel with the left side, the right side is measured and marked equally in length.

 **FROM EXPERIENCE**

> Dry bonding brick to determine the length and width of the column or pier and then marking these lengths on the side of a stick or the mason's level enables you to lay out a square brick column or pier by gauging, an accurate method for laying out without relying upon measurements when exact dimensions are not specified.

**Step 3:** A line is drawn through the marks establishing the length for the left and right sides (see Figure 19-8).

**Step 4:** The diagonal measurements are taken. Equal measurements ensure a square pier (see Figure 19-9).

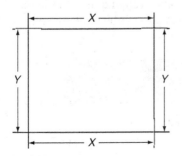

**Figure 19-8** With adjacent sides at right angles, opposite sides have equal lengths.

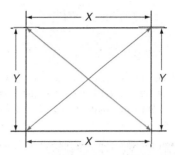

**Figure 19-9** The blue diagonal lines have equal lengths when the layout is square.

Repeating the first three steps are necessary if the diagonal measurements are unequal. Mistakes made when measuring or aligning the framing square cause the layout errors.

# Constructing Piers and Columns

All four corners of a pier or column are to be aligned plumb, and all four sides are to be aligned straight and level. Bed and head joints should be uniform and within the acceptable size limits, typically between ¼" and ½".

**Step 1:** For smaller projects, mortar is spread for an entire course of brickwork before laying and aligning the brick. Starting at one corner, brick are laid for the entire course, carefully aligning each brick with the layout lines. Right-handed masons hold the trowel with the right hand, and move forward along the outside faces of the walls in the same direction as if running the bases on a baseball field (see Figure 19-10). Left-handed masons may find it easier to work in the opposite direction.

By allowing brick to be bedded and aligned in mortar within one minute or so of spreading, the recommended time limitation, only one or two sides of larger projects are completed before proceeding with the adjoining sides.

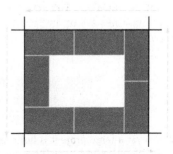

**Figure 19-10** Brick are aligned with the layout lines. Right-handed masons, those holding the trowel with the right hand, can work forwards by moving in the direction of the arrows.

Figure 19-11 **With the top of the brickwork level, the measurement at the left corner is less than the measurement at the right corner because the concrete slab or footing is not level.**

**FROM EXPERIENCE**

Spreading mortar uniformly for the bed joints of a course of brick helps to align the brick course level. Be sure adequate mortar is applied at the corners for bedding the brick. Being able to see any part of the top of the brick below the bed joint is evidence of inadequate mortar coverage.

**Step 2:** Using the masons' scales, at one corner establish the desired height for the course of brickwork. Wherever the length of the mason's level permits leveling from one corner to another, establish height or scale at only one corner of the brickwork. Constructing level coursing requires varying the bed joint widths when the base is not level (see Figure 19-11)

A project whose dimensions are greater than the length of the level requires establishing the course height at more than one corner. Where the base or footing is not level, the first or first few bed joints between courses are varied within acceptable limits, allowing the brick coursing to eventually align level. Height or scale is then measured from the top of the first course aligned level (see Figure 19-12).

**Step 3:** Referring to the corner establishing height or scale, all four sides are aligned level. Care must be taken not to alter the height of the brick at the corner referenced for scale.

**Step 4:** Each brick at the four corners is aligned plumb with the first course. For the first several courses, it may be easier to use a shorter, 2' level.

**Step 5:** The edge of the mason's level is used to align each side in a straight line with the two adjoining corners. Care must be taken not to misalign the brick at the corners establishing the plumb lines set up in Step 4.

Repeat these five steps for each course of brickwork.

Figure 19-13 **A pilaster projects from the wall line. The masonry units are bonded to the wall.**

**FROM EXPERIENCE**

At the four corners, when brick permit, be selective and use only brick whose face and end align square to one another. This enables the sides to be aligned square and straight as possible.

# Masonry Pilasters

A masonry **pilaster** is a columnar projection from the face of a wall. The projection and width of a pilaster depends on the arrangement of the masonry units. The size of a pilaster is expressed in terms of its projection from the original wall line and also by its width. The pilaster in Figure 19-13 projects the length of one brick from the wall line, and its width is equivalent to the length of two brick. The nominal size for this pilaster is described as having an 8" projection and a 16" width.

Pilasters are popular design elements for masonry walls. Bonded to walls by the brick pattern, they strengthen walls and add appealing architectural detailing to them. Pilasters can increase both the compressive and lateral strength of a wall. **Compressive strength** is a measure of the vertical weight or force that a wall can support. It is expressed in pounds per square inch, or psi. **Lateral strength** is a measure of the horizontal force that a wall is capable of resisting. Increasing a wall's compressive strength rating permits it to support more weight. Increasing its lateral strength helps a wall resist high wind velocities and other horizontally imposed forces.

*For step-by-step instructions for constructing a corbelled brick pilaster, see the Procedures section on pages 278–280.*

A pilaster alters the appearance of a wall. Pilasters add architectural detailing to walls (see Figure 19-14).

A cap, also called capping, is often found at the top of a pilaster. A cap is an architectural concrete or stone top for

Figure 19-12 **Whenever a large project is built on a base that is not level, the first bed joint is used to level the brickwork, providing bed joints comply with specified size limitations, typically ¼"–½". Height or scale is taken from the top of the first level course (arrows).**

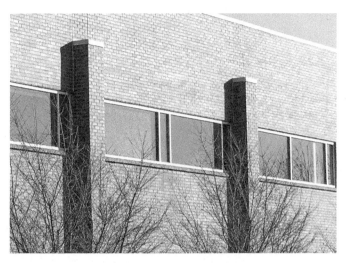

Figure 19-14 **These brick pilasters enhance the appearance of the wall. Precast concrete caps at the top of each pilaster protect them from water intrusion.**

Figure 19-15 **Brick columns along the walkway and pilasters projecting from the brick wall support the arches.**

pilasters that protects their tops from the weather and enhances their appearances. Caps or capping can also be an architectural concrete or stone member for part of or an entire wythe that is terminated below the top of an adjoining wall.

Both columns and pilasters are used as support for the arches in Figure 19-15.

Both CMUs and brick are used for pilaster construction. As with load-bearing columns and piers, a structural engineer should design a load-bearing pilaster.

## Laying Out Pilasters

A pilaster creates a separate wall line, and a wall that has one or more pilasters is typically raised as two wall lines for each course of brickwork (see Figure 19-16).

Figure 19-16 **The pilaster creates two parallel wall lines (arrows).**

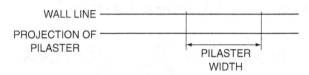

Figure 19-17 **The line representing the face of the wall is parallel to the line representing the face of the pilaster. The two shorter lines indicating the width of the pilaster's face are perpendicular to the wall lines.**

**Step 1:** First, the width of the pilaster is determined. Pilasters are usually designed to accommodate the dimensions of the brick. For example, an 8"-wide pilaster is equal to the length of one brick and a 12"-wide pilaster is equivalent to the length of one and one-half brick. The width of the pilaster in Figure 19-16 is considered to be 16" because it is two-brick wide.

**Step 2:** The projection of the pilaster is determined. The projection of a pilaster is the distance measured from the primary wall line to the face of the pilaster. The projection of the pilaster in Figure 19-16 is equivalent to the sum of the width of the brick and the width of a head joint, or 4".

**Step 3:** Layout lines for the wall and the pilaster are established. Because the face of the wall and the pilaster are usually intended to be parallel, the layout lines representing their faces should be parallel. Lines representing the width of the pilaster are perpendicular to the lines representing its face (see Figure 19-17).

The length and width of most brick and accompanying head joints are accommodated by having the layout lines in 4" increments. Pilasters with a projection of 4" or 8" are common. The pilaster in Figure 19-15 projects 8" from the wall.

## Constructing Pilasters

For each course, the brickwork for the wall is laid ahead of the brickwork for the pilaster. In this procedure, a mason's line is used to accurately align each course of brick along the wall without interference between the line and the brick forming the pilaster. Walls 4' in length or shorter can be aligned using the mason's 4' level.

**Step 1:** The course of brick in the main wall line is laid ahead of the brick forming the pilaster. Omitting all brick projecting from this wall line forming the pilaster permits straight and level brickwork for the entire length of the wall (see Figure 19-18).

**Step 2:** Brick forming the pilaster are laid after the same course of the main wall line is completed. The pilaster brick

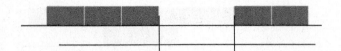

**Figure 19-18** Only the brick in the face of the wall are laid. For a longer wall, brickwork is laid to the line. The mason's level is used as a straight-edge to align a short wall.

**Figure 19-19** The brick forming the pilaster project perpendicular to and are aligned parallel with the main wall.

are aligned level with the brick coursing of the wall line. Observing the layout lines for the first course assures accurate projection of the pilaster, its sides perpendicular to the face of the wall, and its width parallel with the face of the wall (see Figure 19-19). The end and the face of the brick at the ends of the pilaster are aligned plumb before aligning its face straight.

**FROM EXPERIENCE**

Raising the mason's line to the height of the next course after the brick for a wall are laid and before the pilaster is constructed keeps the line from interfering with the process of laying the brick for the pilaster.

Attaching a mason's line to previously constructed pilasters at opposite ends of a wall allows the mason to align straight the faces of multiple pilasters. Steps 1 and 2 are repeated for each course of brickwork. Piers, columns, and pilasters are classified as plain or reinforced. Those classified as plain have no reinforcement. Those classified as reinforced are strengthened with steel rods embedded in concrete or cement grout.

## CMU Piers and Pilasters

CMUs are used for columns, piers and pilasters. A **pilaster block** is a CMU designed for constructing plain or reinforced concrete masonry pilasters (see Figure 19-20). The pilaster block stabilizes load-bearing walls and provides an ideal arrangement for a control joint. A **control joint** regulates the location of separation in a concrete masonry wall resulting in dimensional changes of the building materials. Control joints reduce wall cracking by minimizing the stress forces in a wall.

**Figure 19-20** A recess on the two opposite sides of the pilaster block bonds it with the wall yet allows for dimensional changes of the wall.

**Figure 19-21** An 8" CMU is used to construct this pilaster.

**Figure 19-22** Combining a 4" block and a 12" block creates a pilaster and bonds it with the wall.

Standard CMUs can be used to construct a pilaster in concrete masonry walls (see Figure 19-21).

Using different widths or sizes of CMUs on every other course bonds the pilaster with the wall (see Figure 19-22).

In Figure 19-22, the 4" block is laid as the course is built. Afterwards, the 12" block is laid level with the course. The face of the 12" block is in a plumb line with the 8" block on the course below it (see Figure 19-21). Alternating course patterns, as illustrated in Figures 19-21 and 19-22, permits the pilaster to be tied into the wall. The pilaster in these two illustrations has a projection of 8" and a nominal width of 16".

The combination of various block sizes permits different projections and widths for CMU pilasters. One example is illustrated in Figure 19-23.

Alternate courses for the wall illustrated in Figure 19-23 are constructed as shown in Figure 19-24.

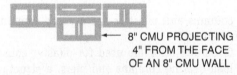

← 8" CMU PROJECTING
4" FROM THE FACE
OF AN 8" CMU WALL

**Figure 19-23** The combination of a 4" and an 8" block forms a pilaster having a 4" projection for this 8" block wall.

**Figure 19-24** A 4" CMU forms the pilaster on alternate courses.

# Masonry Chases

A masonry chase is a recessed area in a wall typically intended to provide space for electrical panels and conduit, heating and cooling ductwork, or plumbing. Chases are often included in commercial CMU walls (see Figure 19-25).

Figure 19-26 shows alternate courses laid in the running bond pattern for the chase illustrated in Figure 19-25.

Chases can be formed in a CMU wall by sawing the block as in Figure 19-27.

Figure 19-28 illustrates alternate courses of the CMU wall with the chase shown in Figure 19-27.

**Figure 19-25** A chase having a depth of 4" is formed using a 4" block in an 8" CMU wall.

**Figure 19-26** For an 8" wall laid in the running bond, a chase is formed using two 4" CMUs and two half 4" CMUs (outlined in black). Figure 19-25 illustrates alternating courses.

**Figure 19-27** Using a masonry saw to cut block creates a chase.

**Figure 19-28** For a wall laid in a running bond, a chase is formed with CMUs laid in this pattern and alternating courses as illustrated in Figure 19-27.

## CAUTION

Only those trained by a competent person and complying with governing regulations and safety standards should operate a masonry saw.

## FROM EXPERIENCE

CMUs that have been cut to form a chase must be handled with care to prevent them from breaking.

A chase in a multiwythe brick wall can be formed simply by omitting brick in a wythe. The sides of a chase should be plumb and the depth uniform for both brick and CMU walls.

# Summary

- Masonry piers, columns, pilasters, and chases are design elements involving precise layout and quality workmanship.
- Besides performing as structural support elements, piers, columns, and pilasters add attractive architectural detailing to a structure.
- A structural engineer should design walls having piers, columns, pilasters, or chases to ensure adequate strength for supporting structural loads.

| Procedures | Constructing a Corbelled Brick Pilaster |

**A** Corbel the first course of the pilaster no more than ¾" from the wall line. The corbelled brick must be level with others in the brick course and plumb on both sides and the front. Corbelled brick forming the pilaster are cut shorter. The face of the corbelled brick should be parallel with the wall line.

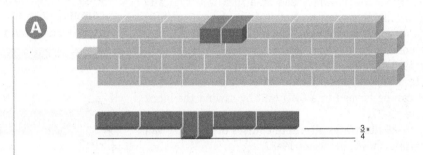

**B** The pilaster interrupts the next course of brick on the main wall. Two half-bricks or bats allow for the adjoining pilaster.

**C** The corbelled brick is turned opposite those below it. A split brick is laid behind the corbelled brick. The brick must be level with the course and aligned plumb on both sides and the front. The corbelling for the second course should be equal to that of the first course, with ¾" being the maximum corbel for any one course and 1½" overall.

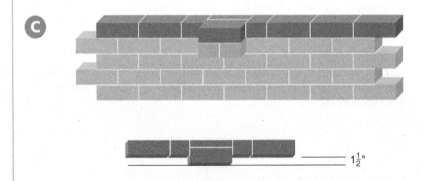

**D** For each course, brick of the original wall line are laid before adding the corbelled brick. Aligning the course of brick straight with the edge of the mason's level or line is possible only without obstruction of the corbelled brick.

**E** The corbelling of the brick for the third course is equal to that of each previous course and extends no more than 2¼" from the original wall line. The brick should be level with the wall and aligned plumb on each side.

$2\frac{1}{4}$"

**F** The fourth course of the corbelled pilaster should be equal to the corbelling for the previous courses and no more than 3" from the original wall line. Only by having the brick on the fourth course of the corbelled pilaster equal to the length of the brick on the second course can plumb alignment for both sides of the pilaster be maintained. As illustrated, either clipping the backside of the corbelled brick or the two in the wall line behind it permits the proper corbel.

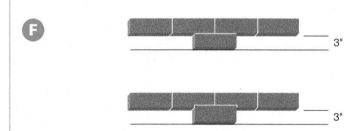

3"

3"

**G** The next course of the wall is laid ahead of the brick for the pilaster.

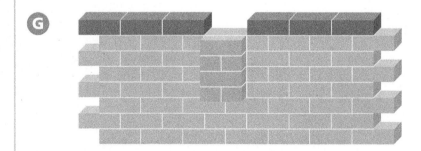

## Procedures

# Constructing a Corbelled Brick Pilaster (continued)

**H** The pilaster must be level with the wall and aligned plumb on both sides and the front. The total projection of the pilaster is equivalent to the sum of the width of the brick plus allowance for a head joint between the corbelled brick and the wall. Having brick measuring 3⅜" wide, the overall projection of this corbelled pilaster is 3¾".

**H**

$3\frac{3}{4}$"

**I** The brick within the wall line securely bonds the corbelled pilaster.

**I**

**J** The pilaster is completed before proceeding to the next course. It should be level with the wall and aligned plumb. When oriented with the width of the pilaster, each brick must be of equal length.

**J**

## Review Questions

*Select the most appropriate answer.*

**1** A structural support bearing a concentrated load is called a

a. structural column

b. structural pier

c. hollow pilaster

d. both a and b

**2** A decorative brick post on top of which a lighting fixture is likely to be installed is called a

a. hollow pier

b. hollow column

c. structural pier

d. both a and b

**3** The permitted location for a masonry column near a roadway, such as one intended as a light post or mailbox enclosure, may be regulated by the

a. local Department of Transportation

b. local government

c. utility companies

d. all of the above

**4** In relationship to its width or lateral dimensions, a column is

a. at least two times taller than its least lateral dimension

b. at least four times taller than its least lateral dimension

c. shorter than a pier

d. wider than a pier

**5** A rectangular layout is considered to have square corners when

a. the length of the front and back sides are equal

b. the length of the left and right sides are equal

c. the diagonal measurements are equal

d. all of the above

**6** To build as one walks forward, right-handed masons build four-sided piers and columns by

a. walking in a counter-clockwise rotation around the outside of the project

b. walking in the same pattern as walking the bases on a baseball field

c. walking in a clockwise rotation around the outside of the project

d. both a and b

**7** A columnar projection of masonry from a wall is called

a. an abutment

b. a column

c. a pier

d. a pilaster

**8** An architectural concrete or stone member for the top of a pilaster is called a

a. cap

b. capital

c. coping

d. header

**9** An architectural concrete or stone member for the top of a pier or column is called a

a. capital

b. coping

c. finial

d. header

**10** A recessed area in a masonry wall intended to provide space for an electrical panel, ductwork, and plumbing is called

a. an alcove

b. a chase

c. a flue

d. a groove

# Chapter 20

# Appliance Chimneys

*T*he primary function of a chimney is to vent combustion byproducts to the outside while protecting the structure and its occupants from fire and the life-threatening effects of breathing carbon monoxide. The two categories of chimneys most often constructed by masons are those that vent appliances and those that vent fireplaces. Appliance chimneys are designed to vent the combustion byproducts of a variety of appliances. Although many of today's appliances are vented by means other than masonry chimneys, there remains a demand for them because of the architectural styling they add to a structure.

## OBJECTIVES

After completing this chapter, the student should be able to:

- ⊗ Identify the parts of brick masonry chimneys.
- ⊗ Explain important regulations and codes governing the construction of masonry chimneys.
- ⊗ Construct an appliance chimney.

# Glossary of Terms

**appliance chimney** a chimney designed to vent solid fuel appliances, those fired with coal and wood, and appliances fired with oil, natural gas, or liquid petroleum propane gas.

**chimney base flashing** noncorrosive metal flashing placed between rows of shingles and turned upward against the outside of the chimney walls.

**chimney cap** the masonry element atop a chimney designed to eliminate water penetrating the top of the chimney, made of reinforced concrete, either prefabricated or cast-in-place.

**cleanout** a door-enclosed opening below the thimble permitting access within the flue lining for removing combustion by-products and other accumulated debris.

**corbelling** the process of setting the face of brick out beyond the face of the brick below it.

**counter-flashing** a noncorrosive metal having the shape of an inverted "L," placed in the chimney walls and turned down over base flashing.

**cricket** a sloped section of roof designed to divert water from behind the upper side of a chimney.

**cross-sectional area** the area as calculated from inside the walls of the flue lining.

**fireblocking** also called fire stopping, a noncombustible material preventing air drafts and the potential spread of smoke and fire between floors or ceilings.

**fireclay** clay capable of withstanding high temperatures, used for chimney flue liners and chimney thimbles.

**flue lining** a rectangular or round, hollow chimney lining intended to contain the combustion by-products and protect the chimney masonry walls from heat and corrosion.

**thimble** a round, fired clay, chimney component used as a chimney inlet and attached horizontally from the flue lining to the face of the wall into which the appliance connector is inserted.

# Chimney Support

An **appliance chimney** is a chimney designed to vent solid fuel appliances, those fired with coal and wood, and appliances fired with oil, natural gas, or liquid petroleum propane gas. Their tall, columnar shape places a concentrated load on a comparatively small surface area. This weight requires a properly sized and supported footing. Section R1001.1.1 of the International Residential Code requires that a concrete footing supporting a chimney be "at least 12" thick and at least 6" beyond each side of the exterior dimensions of the chimney. Footings shall be founded on natural, undisturbed earth below the frostline. In areas not subject to freezing, footings shall be located a minimum of 12" below finished grade" (see Figure 20-1). Steel rod reinforcement may be a requirement of the structural engineer designing some chimney footings. In all cases, however, footings must be in compliance with all federal, state, and local laws and regulations. Having a licensed structural engineer design a footing for a specific job is recommended because requirements are different depending upon the intended geographic location for the chimney.

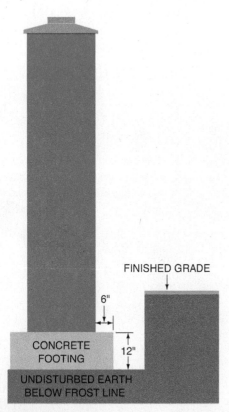

FINISHED GRADE

6"

CONCRETE FOOTING

12"

UNDISTURBED EARTH BELOW FROST LINE

**Figure 20-1** A properly sized footing for supporting an appliance chimney is placed on approved, undisturbed soil below the frost line.

# Chimney Walls

According to Section R1001.7 of the International Residential Code, the walls for a masonry chimney "shall be constructed of solid masonry units or hollow masonry units grouted solid with not less than 4" nominal thickness." Solid masonry units are those with a cross-sectional area of 75% or more of the bed area. Many cored brick are considered as solid units. The cores of hollow masonry units, those whose cores are more than 25% of the bed area, must be filled with grout. *Grout* is a mixture of cement, aggregates, and water mixed to a consistency that can be poured but that is not so wet that the aggregates separate. Enough water is added to permit pouring the grout into the cores, yet not too much as to allow settling of the heavier aggregates. Grade SW brick are recommended for the chimney walls. The Brick Industry Association Technical Notes 19B recommends that "brick should conform to ASTMC 216, Grade SW, or ASTM C 62, Grade SW to assure sufficient durability. Paving brick should conform to ASTM C 902, Class SX."

Brick Industry Association Technical Notes 19B recommends "To allow for both weathering and thermal considerations, Type N Portland cement-lime mortar is recommended for the chimney. Type S Portland cement-lime mortar is acceptable and may be necessary when the chimney is subjected to high lateral forces such as wind loads in excess of 25 psf (1.2 kPa) or seismic loads. Where the chimney is in contact with earth, Type M Portland cement-lime mortar is recommended."

# Chimney Linings

Section R1001.8 of the International Residential Code states that "masonry chimneys shall be lined. The lining material shall be appropriate for the type of appliance connected, according to the terms of the appliance listing and manufacturer's instructions." A **flue lining** is a rectangular or round, hollow chimney lining intended to contain the combustion products and protect the chimney walls from heat and corrosion. Flue linings are made of **fireclay,** clay capable of withstanding high temperatures, with walls being a minimum of ⅝" thick. Although clay linings can crack from thermal shock, a result of starting a very rapid hot fire in a cool chimney or the result of a chimney fire, tests show that they do not crack from normal use.

Providing a ½" minimum air space between the flue lining and the masonry outer walls reduces the likelihood of cracked liners caused by thermal expansion (see Figure 20-2). Section R1001.9 of the International Residential Code requires one building a chimney "to maintain an air space or insulation not to exceed the thickness of the flue liner separating the flue liners from the interior face of the chimney masonry walls."

The 2' sections of flue linings should be joined with tight mortar joints using medium-duty refractory mortar conforming to ASTM C 199, a requirement of the International Residential

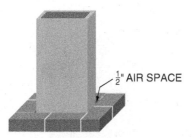

½" AIR SPACE

**Figure 20-2** A ½" minimum air space separates the outer brick walls from the cast and fired clay, terra-cotta flue lining.

Code, Section R1001.9. The size of a rectangular flue liner refers to the outside dimensions, whereas the size of a round flue liner refers to its inside diameter. Three sizes of rectangular flue linings frequently used are 8" × 8", 8" × 12", and 12" × 12".

**FROM EXPERIENCE**

Flue linings may have structural cracks unnoticeable to the eye. Tapping lightly with a hammer on the sides of a liner as it stands on end produces a ringing sound on liners that have the structural integrity required for chimney construction. A dull, crackling sound indicates a cracked flue liner. Use only flue liners that have no structural cracks when constructing chimneys.

## Sizing the Chimney Liner

Building code regulations specify the required size of a flue liner, based on the intended appliance and its connector. Section R1001.11 of the International Residential Code prohibits chimney flues from being smaller in cross-sectional area than that of the connector size for the appliance. The cross-sectional area of a rectangular flue lining is the area derived from multiplying the length and width as measured from inside the walls of the flue lining. An example is illustrated here.

Cross-Sectional Area of a Rectangle
Area = width × length

Determine the cross-sectional area of an 8" × 12" flue lining. The dimensions, 8" × 12", represents the nominal outside dimensions of the flue lining, typically measuring 9" × 13". The actual inside dimensions, from which area is calculated, typically measure approximately 7" × 11".

Area = width × length
Area = 7 × 11
Area = 77 square inches

Calculating the area of a circle is necessary for determining the cross-sectional area of a round flue lining or a round appliance connector. An example is illustrated here.

Area of a Circle
Area = diameter² × 0.7854

Determine the cross-sectional area of a woodstove connector having an 8" diameter.

Area = diameter$^2$ × 0.7854
Area = $8^2$ × 0.7854
Area = (8 × 8) × 0.7854
Area = 64 × 0.7854
Area = 50.27 square inches

This information is needed for determining the size of flue lining required for any given appliance that has a round connector. For example, is an 8" × 8" rectangular flue lining adequate for venting a woodstove that has an 8" diameter connector? The following example answers this question.

Find the area of the flue lining:

The dimensions, 8" × 8", are nominal outside dimensions. The actual opening measures approximately 6¾" × 6¾".

Area = width × length
Area = 6¾ × 6¾
Area = 6.75 × 6.75
Area = 45.6 square inches

Since it is determined that an 8"-diameter connector has a cross-sectional area of 50.27", it can be concluded that an 8" × 8" flue lining with a cross-sectional opening measuring 45.6" is too small. However, an 8" × 12" flue lining with a cross-sectional area of approximately 77 square inches is adequate.

## Cleanout Openings

A chimney cleanout is a door-enclosed opening below the thimble that permits access within the flue lining for removing combustion by-products and other accumulated debris below the thimble. A cleanout opening with a minimum height of 6" is required. Section R1001.14 of the International Residential Code requires that "cleanout openings shall be provided within 6" of the base of each flue within every masonry chimney. The upper edge of the cleanout shall be located at least 6" below the lowest chimney inlet opening. The height of the opening shall be at least 6". The cleanout shall be provided with a noncombustible cover."

## Chimney Thimbles

A fireclay thimble is a round, fired clay, chimney component used as a chimney inlet and attached horizontally from the flue lining to the face of the wall into which the appliance connector is inserted. Where passing through walls of combustible material, the thimble must be surrounded by a minimum of 12" of solid masonry framed into the combustible wall.

The thimble cannot be smaller than the appliance connector pipe. To comply with International Residential Code Section R1001.9, flue linings must be fully supported on all sides and begin at least 8" below the lowest inlet thimble (see Figure 20-3).

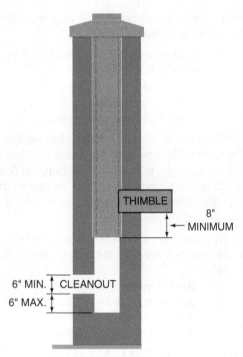

**Figure 20-3** A properly positioned thimble and cleanout access are illustrated.

# Chimney Height

According to Section R1001.6 of the International Residential Code, "chimneys shall extend at least 2' higher than any portion of a building within 10' but shall not be less than 3' above the highest point where the chimney passes through the roof." A chimney cap eliminates moisture penetration at the top of the chimney. The cap, made of reinforced concrete, can be prefabricated or cast-in-place. The following recommendations, illustrated in Figure 20-4, should be considered when providing a cast concrete chimney cap:

- Extend the cap a minimum of 2½" beyond the chimney on all sides and have a drip cutout on the underside to prevent water migration from the underside edges to the masonry walls.
- Make the cap a minimum of 4" thick adjacent to the lining and sloping away from the flue on all sides to improve water drainage and reduce wind turbulence at the flue.
- Install a bond break between the brickwork and the concrete cap to permit differential movement of the cap and the brick masonry, helping to eliminate open cracks in the cap.
- Anticipate thermal expansion of the flue lining and prevent water penetration between the lining and cap with a silicone-sealed, ³/₈" gap between the lining and the cap.
- Make the cap no less than 2" or more than 6" below the flue lining.

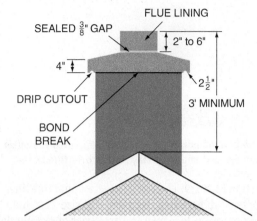

**Figure 20-4** The chimney top illustrated here includes design details for a weather-resistant chimney cap.

# Constructing an Appliance Chimney

Appliance chimneys must be constructed in compliance with all governing regulations and building codes. The mason should review local building code regulations addressing the design and construction of chimneys before beginning construction. A professional mason experienced in chimney construction should supervise apprentice masons and those inexperienced with chimney construction.

## Placing the Footing

Because of the varying depth at which frost penetrates, the bearing capacity of different soils, and the different seismic zones, local building code regulations govern requirements for placing and reinforcing the concrete footing. Covering fresh concrete to prevent water evaporation and allowing the footing to cure for a minimum of seven days before constructing the chimney permits the concrete to gain strength for carrying the load of the chimney.

**CAUTION**

**Working in ground excavations where chimneys are likely to begin pose safety hazards that must be addressed. The surrounding ground must be examined by a competent person for indications of potential cave-ins. Where there is a potential for cave-ins and in excavations 5' or more in depth, protective systems meeting the intent of OSHA Standards 1926.652 must be used.**

**Figure 20-5** This 2 × 2½ brick pattern having nominal dimensions of 16" × 20" accommodates an 8" × 12" flue lining.

## Laying Out the Chimney Walls

Single-wythe, 4" brick walls are acceptable for most chimneys. The outside dimensions of the brick walls must permit using the required size flue lining and the ½" minimum air space between the lining and the brick walls. For example, a 2 × 2½ brick pattern can be used for single-wythe brick walls enclosing an 8" × 12" flue lining (see Figure 20-5).

The actual dimensions of the length and the width depend on the brick dimensions and head joint widths. Joint width should be no less than ¼" or more than ½".

A chimney is square when

- each of the four corners are 90-degree angles.
- the front side and back side are equal length.
- the left side and right side are equal length.
- the diagonal measurements are equal.

To ensure a square chimney, the following four procedures are recommended.

**Step 1:** The layout of a chimney begins with two lines perpendicular to each other. A framing square establishes a 90-degree angle. Based on the layout pattern for the brick outer walls, the mason measures and marks the lengths of the two adjacent sides (see Figure 20-6).

**Step 2:** Aligning the framing square with the line draws a line perpendicular to one of the two lines established in Step 1. The length of this side is measured and marked on the line (see Figure 20-7).

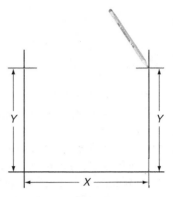

**Figure 20-7** Parallel with the left side, the right side is measured and marked equally in length.

**Step 3:** A line drawn through the marks establishing the length for the left and right sides completes the layout (see Figure 20-8).

**Step 4:** Measuring the lengths of the diagonals determines accurate layout. Equal measurements assure a square chimney (see Figure 20-9).

Repeating the first three steps is necessary if the diagonal measurements are unequal. Mistakes made when measuring or aligning the framing square cause layout errors.

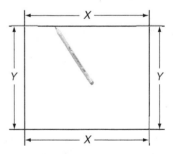

**Figure 20-8** This is the completed layout.

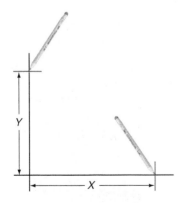

**Figure 20-6** Using a framing square, two lines at right angles are drawn. Measuring from their intersection, the length for the side and front are marked.

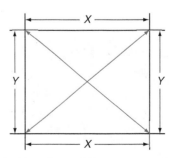

**Figure 20-9** The blue diagonal lines are equal lengths when the layout is square.

## Building the Brick Walls

Constructing chimneys to their intended height requires scaffolds and possibly working on roofs. Only those trained by a competent person should build or access scaffolds. The 5'-wide scaffold frames must be adequately braced or tied to restrain and prevent them from tipping at heights above 20'. Section 1926.451(g)(1) Of the OSHA Construction Industry Regulations requires that "each employee on a scaffold more than 10' above a lower level shall be protected from falling to that lower level." Section 1926.501(b)(1) requires that when on roofs, for example, "each employee on a walking/working surface (horizontal and vertical surface) with an unprotected side or edge which is 6' or more above a lower level shall be protected from falling by the use of guardrail systems, safety net systems, or personal fall arrest systems." A different OSHA regulation applies when employees stand on roofs rather than working from scaffolds. With few exceptions, a minimum clearance of 10' must be kept between scaffolds and power lines, per Section 1926.451(f)(6). Comply with OSHA construction standards, Subpart L "Scaffolds" and Subpart M "Fall Protection," when working on scaffolds or roofs.

When brick can be bedded within 1 minute or so of spreading mortar, mortar is spread for the entire course of brickwork before laying any brick. Starting at one corner, brick are laid for the entire course, carefully aligning each brick with the layout lines. If right-handed, hold the trowel with the right hand, moving forwards and in a counterclockwise direction as if running the bases on a baseball field (see Figure 20-10). Left-handed masons may find it easier to work in the opposite direction. Permitting brick to be bedded and aligned within 1 minute or so of placing the mortar as recommended by the Brick Industry Association Technical Notes 7B, only one or two sides of larger chimneys should be bedded and aligned before proceeding with the adjoining sides.

For every course, the four corners are aligned plumb, and each of the four sides is aligned straight and level. Widths of bed joints and head joints are kept uniform, at approximately ⅜" wide.

## Providing a Cleanout Opening

A cleanout opening is provided on one side of the chimney, located within 6" of the base of the flue, its upper edge at least 6" below the thimble. A noncombustible cover is required for the cleanout. Cleanout doors fitting an 8" × 8" opening, a space equivalent to the length of one brick and three standard size brick courses high, are available from masonry suppliers.

## Installing Clay Flue Linings

The first flue lining must be installed ensuring at least an 8" extension of the lining below the bottom edge of the thimble. All four sides of the flue lining must be supported from below with masonry. A procedure frequently used to support the first flue lining is to corbel the brick walls of the chimney, laying brick in the header position so that the corbelled header brick support the first flue lining (see Figure 20-11). A brick can be corbelled no more than half its height or one-third its width, whichever is less. Installing the first flue lining after several courses of brick for the walls of the chimney are laid above the corbelled headers prevents the weight of the flue liner from tilting the headers intended to support its weight.

Installing the first flue lining at the base of the chimney on solid masonry permits a fully lined chimney and the best support for each additional flue lining section. This procedure requires a cutout in the flue lining at the cleanout opening (see Figure 20-12).

Figure 20-11 Supporting the first lining are corbelled brick laid on all four sides in the header position.

Figure 20-12 A cutout in the side of the flue lining, supported by masonry below it, permits access for cleaning.

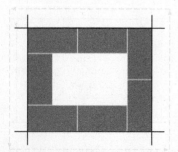

Figure 20-10 Moving in the direction of the arrows, a right-handed mason aligns the brick with the layout lines.

FROM EXPERIENCE

An electric hand-held 7¼" circular saw equipped with a diamond masonry-cutting blade makes cutting flue linings easy for those not having access to stationary power masonry saws.

CAUTION

**Always wear approved eye, face, hearing, and respiratory protection when cutting masonry materials with power saws and be trained by a competent person before using the saws. Electric-powered tools must be equipped with inline ground-fault circuit interrupters for operator safety and protection from electric shock and electrocution.**

An air space no less than ½" or greater than the thickness of the walls of the flue lining should be maintained between the lining and the backsides of the exterior chimney walls. Minimizing mortar protrusions and eliminating mortar bridges assures a good air space. Spreading no more mortar for the bed joints than needed minimizes mortar protrusions.

FROM EXPERIENCE

Laying a section of flue lining ahead of the outer brick walls as recommended makes it difficult to remove mortar protrusions and mortar bridges. Holding the level's edge against the face of the brick to prevent misalignment, plaster any mortar protrusions against the backside of the brick with the trowel blade to allow for an open air space.

The 2' sections of flue lining are bonded with medium-duty refractory mortar, forming tight joints that are smooth on the inside of the flue linings.

A chimney may have multiple flues. This permits more than one appliance and the option of having a masonry fireplace vented from separate flues within the same chimney walls. Section R1001.10 of the International Residential Code requires that "when two or more flues are located in the same chimney, masonry wythes shall be built between adjacent flue linings. The masonry wythes shall be at least 4" thick and bonded into the walls of the chimney. Exception: When venting only one appliance, two flues may adjoin each other in the same chimney with only the flue lining separation between them. The joints of the adjacent flue linings shall be staggered at least 4" (see Figure 20-13).

WALL TIES EVERY 3 COURSES

**Figure 20-13** A 4" brick wythe tied to the outside walls separates a 12" × 12" flue from an 8" × 12" flue.

## Installing Clay Thimbles

To connect the fireclay thimble with the flue lining, an opening in the side of the flue lining is needed. The opening is cut slightly larger than the outside diameter of the thimble, allowing for a ¼" mortared joint using medium-duty refractory mortar between the thimble and the liner. Drawing a circle onto the side of the flue liner, using the outside of the thimble as a pattern, becomes a reference for making the opening.

FROM EXPERIENCE

Use a power drill and ¼" masonry drill bit to drill a series of holes close together through the lining along the outside of the line marked for the location of the thimble. Then, using a masonry saw, make two cuts perpendicular to each other through the center of the intended cutout circle (see Figure 20-14). Lightly tap the area of the intended cutout to remove the pieces.

CAUTION

**Always wear approved eye, face, and hearing protection when using electric saws and impact "hammer" drills and be trained by a competent person before using the drills. Electric-powered tools must be equipped with inline ground-fault circuit interrupters for operator safety and protection from electric shock and electrocution.**

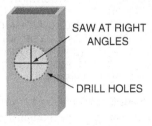
SAW AT RIGHT ANGLES
DRILL HOLES

**Figure 20-14** Drilling holes and sawing permits removing the four sections without cracking the flue lining.

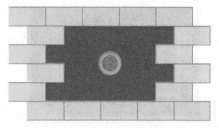

**Figure 20-15** Solid masonry brickwork ensures safe enclosure for the thimble where it passes through the hollow concrete masonry wall. Three courses of standard size brick are bonded to every course of CMUs.

The thimble should be aligned flush with the inside of the lining, bonding the two with medium-duty refractory cement. The thimble must extend to the face of the wall where the appliance connector is inserted. Fireclay thimbles passing through masonry walls do not pose the fire hazards of thimbles passing through walls of combustible materials. Brick Industry Association Technical Notes 19B recommends that the minimum distance between a thimble and the room ceiling above it be 18". A minimum of 12" of solid masonry is recommended to surround thimbles passing through CMU masonry walls (see Figure 20-15).

Thimbles passing through walls of combustible material such as wood framing must be enclosed in a minimum of 12" of solid masonry completely surrounding the outer walls of the thimble. Building codes typically allow properly designed wall framing to support the solid masonry (see Figure 20-16).

Only one thimble should be connected to a flue lining passageway. But to permit changing the location for connecting an appliance to a different floor level, some local governing regulations may permit more than one thimble connected to the same flue lining at different heights of the chimney. However, connecting only one appliance to a single flue is permitted. When this condition is permitted, all other thimbles should be sealed with mortared, solid masonry to prevent air leaks and potential fire risks. If relocating an appliance connection to a second thimble is desired, then the thimble from which the appliance is moved must be sealed solid with mortared brick masonry.

## Providing Adequate Clearances

Section R1001.15 of the International Residential Code requires a minimum 2" air space clearance to combustibles between the exterior chimney walls and nearby interior combustible building materials, that is, chimneys built within the exterior wall of the building.

Placing a minimum 1" layer of noncombustible material to serve as fireblocking or fire stopping between chimneys and floor or ceiling framing is required. **Fireblocking** or **fire stopping** is a noncombustible material preventing air drafts and the potential spread of smoke and fire between floors or ceilings. Unfaced fiberglass batt insulation may be used as fire stopping. To comply with Section R1001.16 of the International Residential Code, it should "only be placed on strips of metal or metal lath laid across the space between combustible material and the chimney." When wall framing is used to support the brickwork surrounding the thimble, building codes typically permit the wood framing to touch and support the brickwork, bearing directly upon the framing and providing that a minimum of 12" of solid masonry surrounds the outside of the thimble (see Figure 20-17). However, one should be sure that the supporting wood framing and other details comply with local building codes.

Section R1001.15 of the International Residential Code requires that "chimneys located entirely outside the exterior walls of the building, including chimneys that pass through the soffit or cornice, shall have a minimum air space clearance of 1 inch (see Figure 20-18)".

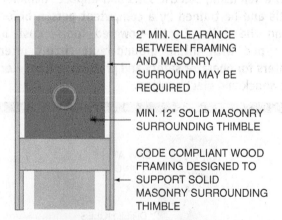

2" MIN. CLEARANCE BETWEEN FRAMING AND MASONRY SURROUND MAY BE REQUIRED

MIN. 12" SOLID MASONRY SURROUNDING THIMBLE

CODE COMPLIANT WOOD FRAMING DESIGNED TO SUPPORT SOLID MASONRY SURROUNDING THIMBLE

**Figure 20-16** Mechanical codes require 12" of solid masonry surrounding the outside of the thimble, supported by properly designed, code-compliant framing. Check local building codes for specific requirements.

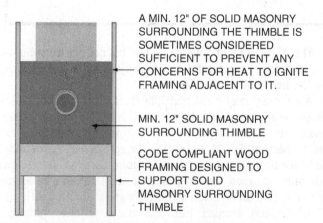

A MIN. 12" OF SOLID MASONRY SURROUNDING THE THIMBLE IS SOMETIMES CONSIDERED SUFFICIENT TO PREVENT ANY CONCERNS FOR HEAT TO IGNITE FRAMING ADJACENT TO IT.

MIN. 12" SOLID MASONRY SURROUNDING THIMBLE

CODE COMPLIANT WOOD FRAMING DESIGNED TO SUPPORT SOLID MASONRY SURROUNDING THIMBLE

**Figure 20-17** The 12" of solid masonry surrounding the thimble is sometimes considered as sufficient thermal insulation between the thimble and combustible framework. Always comply with local building codes.

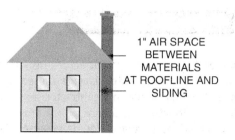

**Figure 20-18** A 1" air space between combustible materials and masonry chimneys built entirely outside a structure's exterior walls is required.

## Anchoring a Chimney

Single-wythe chimney walls are to be secured using corrosion-resistant wall ties to the structure's framework. Brick Industry Association Technical Notes 19B states that "single-wythe chimneys should be attached to the structure. This is generally accomplished by using corrosion-resistant metal ties spaced at a maximum of 24" (600 mm) on center. Multi-wythe chimneys that are not masonry bonded should be bonded together using metal wire ties." Depending upon the assigned seismic design category for the region where a chimney is to be constructed, additional reinforcement may be required. Refer to Section R1003.3 of the International Residential Code for specific seismic reinforcing requirements. Consulting local governing regulations for specific regional requirements is advised.

## Topping Out a Chimney

The top of a chimney must be at least 3' above the highest point where it penetrates the roofline or a minimum of 2' higher than any portion of the structure within 10' (see Figure 20-19).

Brick Industry Association Technical Notes 19B recommends extending the last flue lining above the finished brickwork to permit a minimum 2" reveal above the top of the concrete cap. Placing PVC flashing or 30-pound construction felt paper on top of the brickwork serves as a bond break, isolating the concrete cap from the brickwork and

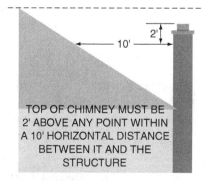

**Figure 20-19** Although the chimney is lower than the roof peak, it meets requirements of being a minimum of 2' above any part of the structure within a 10' horizontal distance.

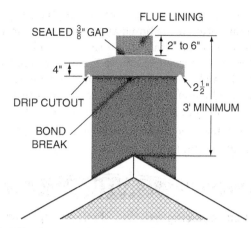

**Figure 20-20** These details are observed to improve weather resistance and fire safety.

providing for differential movements of brick and concrete (see Figure 20-20).

A precast or cast-in-place concrete cap is recommended rather than a mortar cap. Mortar is intended for bonding masonry units together. It is not intended to be used to form chimney caps or any other structural members. The following procedures are recommended for setting a precast concrete cap.

1. Provide a bond break between the brickwork and the cap.
2. Bed the cap in Type N mortar.
3. Fill the gap between the lining and cap with a noncombustible, compressible material such as fiberglass insulation.
4. Seal the filled gap between the lining and cap with a butyl or silicone caulk.

Cast-in-place procedures permit forming a cap for a specific job as the chimney is built. A cast-in-place cap is formed using a removable form while following these procedures.

1. Assemble and align the sides of the form with the sides of the chimney. Use commercially available forms with a drip edge and 2" framing lumber to make the form. Use a masonry saw to create a drip edge cutout after the cap is cast and the form is removed.
2. Secure a ½"-thick layer of noncombustible, compressible material such as fiberglass insulation around the outside of the flue liner.
3. Place concrete above the brickwork onto the flashing material as a bond break, and reinforce it with embedded steel reinforcement rods or 6-6-10-10 welded wire mesh.
4. Trowel the concrete surface smooth, and tap the edges of the concrete form with a hammer to consolidate the concrete and remove air pockets entrapped between the form and concrete.
5. Cover fresh concrete with a lightweight, moisture-proof tarp to prevent excessive water evaporation.
6. After the concrete cures for a minimum of 48 hours, remove the form.

7. Remove the excess material forming the compressible joint between the lining and cap near its surface, and seal the joint with butyl or silicone caulk.
8. Install a removable rain cap to prevent rain, leaves, birds, and rodents from entering the chimney. Depending on the model, it may be attached to the flue lining or anchored to the chimney cap.

Brick Industry Association Technical Notes 19B, Figure 3, illustrates chimney cap details.

### CAUTION

**When replacing existing chimney caps, never block the chimney opening itself. The building's occupants may suffer carbon monoxide poisoning if gases are not vented through the flue to the outside.**

## Installing Chimney Flashing

Corrosion-resistant sheet metal base flashing and counter-flashing prevent water from migrating between the exterior chimney walls and the adjacent roofing materials. Chimney **counter-flashing,** in the form of an inverted L, is built into the chimney walls (see Figure 20-21). The counter-flashing overlaps the noncorrosive metal **chimney base flashing,** flashing placed between rows of shingles and turned upward against the side of the chimney. There are several similar procedures for installing base flashing, however, chimney base flashing should be installed in accordance with governing codes and the shingle manufacturer's recommendations as the shingles are installed. The resulting combination of properly installed counter-flashing and base flashing is a waterproof joint between the chimney walls and the roof.

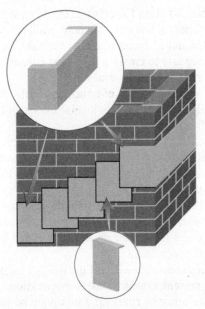

**Figure 20-21 Counter-flashing is embedded in the mortar joints. Each stepped piece laps the lower section at least 2".**

### FROM EXPERIENCE

It may be that a masonry contractor's insurance coverage does not include water damage caused by leaking chimney flashing. A masonry contractor should consider having a contractor's insurance policy providing completed operations coverage, for claims after the job is complete, for water damage to a building caused by leaking flashing.

The following chimney flashing details are recommended by the Brick Industry Association Technical Notes 19B:

- Counter-flashing should extend a minimum of 4" above the roofline.
- Counter-flashing should be inserted into mortar joints or a groove cut into the wall to a depth of ¾" to 1".
- Counter-flashing should be mortared solidly into the joint or inserted into a sawed groove, after which it is sealed with butyl or silicone caulk.
- Sections of stepped counter flashing higher up the roof line should lap over the lower sections a minimum of 2".
- Counter flashing should lap the base flashing by at least 3".
- All flashing joints should be sealed with a butyl or silicone caulk.

## Providing a Chimney Cricket

Section R1001.17 of the International Residential Code requires chimneys not intersecting the ridgeline and having a length parallel to the ridgeline greater than 30" to be provided with crickets. A **cricket** is a sloped section of roof designed to divert water from behind the upper side of a chimney (see Figure 20-22). The side of the chimney adjoining the cricket requires the same flashing details as do the other sides of the chimney at a roofline. Construction details of a cricket are considered as a roof framing detail and are not addressed in this chapter. Crickets must be constructed in compliance with International Residential Code Figure R1001.17 and Table R1001.17.

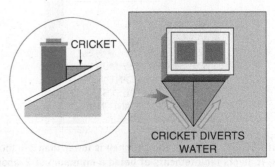

**Figure 20-22 A cricket diverts water from the upper side of the chimney toward the sides.**

## Summary

- Masonry chimneys require specific design details and quality workmanship. Faulty design and construction can endanger lives and properties.
- A chimney flue is intended to vent only one appliance. No other connections are allowed.
- Only competent, experienced masons who comply with governing building code regulations and recommended industry practices should construct chimneys.

- In add those c̲ꞈaving construction liability insurance, contractor̲. g chimneys should be licensed coverage, ins̲ mpleted operations insurance property dama̲ꞈ contractor in the event of future result of complet̲ ꞈrsonal injuries sustained as a
- Providing a waterpr̲ roofs requires knowle̲ꞈ between chimney walls and procedures for the inter̲flashing details and roofing ꞈofing materials.

# Review Questions

Select the most appropriate answer.

**1** An appliance chimney is used to vent

a. natural gas and oil furnaces and hot water heaters

b. oil-burning furnaces

c. wood-burning stoves

d. all of the above

**2** The minimum thickness of a concrete footing for an appliance chimney is

a. 8"

b. 12"

c. 16"

d. 20"

**3** The minimum nominal width of masonry units used for masonry appliance chimneys walls is

a. 4"

b. 6"

c. 8"

d. 12"

**4** A rectangular or round, hollow chimney lining intended to contain the combustion products and protect the chimney walls from heat and corrosion is called a

a. chimney casement

b. chimney pot

c. flue lining

d. thimble

**5** The space between the brick walls of the chimney and the chimney liner should be

a. filled with mortar

b. filled with grout

c. a minimum ½" air space

d. a minimum 1" air space

**6** Sections of chimney lining should be joined using

a. latex-based caulk

b. medium-duty refractory mortar

c. Type N mortar

d. Portland cement

**7** The minimum size of the chimney liner to which a wood-burning stove is to be vented is

a. 8" × 8"

b. 8" × 12"

c. 12" × 12"

d. dependent upon the cross-sectional area of the required stove connector

**8** The chimney component attached horizontally from the chimney lining to the face of the wall into which the appliance connector is inserted is called a

a. chimney casement

b. chimney pot

c. flue lining

d. thimble

**9** Beyond the highest point where a chimney penetrates a roofline, it should extend a minimum height of

a. 1'

b. 2'

c. 3'

d. 4'

**10** A chimney cleanout should be located no closer than

a. 6" below the thimble

b. 12" below the thimble

c. 16" below the thimble

d. 24" below the thimble

⑪ **A masonry wall between chimney liners should be**

a. 4" solid masonry

b. bonded to the chimney walls

c. poured grout

d. both a and b

⑫ **The minimum distance between a chimney thimble and a room ceiling is**

a. 12"

b. 18"

c. 24"

d. 32"

⑬ **Masonry surrounding a thimble passing through a wall of combustible materials is to be**

a. a minimum of 4" beyond the outer wall of the thimble

b. a minimum of 6" beyond the outer wall of the thimble

c. a minimum of 8" beyond the outer wall of the thimble

d. a minimum of 12" beyond the outer walls of the thimble

⑭ **The minimum air space required between the exterior walls of masonry chimneys and interior combustible building materials, that being chimneys built within exterior walls, is**

a. ½"

b. 1"

c. 1½"

d. 2"

⑮ **For a chimney built entirely outside a building, there must be a minimum air space between the exterior walls of the chimney and combustible exterior building materials of**

a. ½"

b. 1"

c. 1½"

d. 2"

⑯ **The minimum thickness of noncombustible material placed between a chimney built within a building's exterior walls and combustible-material ceiling framing, intended to block air drafts, is**

a. ½"          c. 2"

b. 1"          d. 3"

⑰ **The maximum recommended vertical spacing for which a chimney is to be anchored to structural framing with metal ties is every**

a. 8"          c. 24"

b. 16"         d. 32"

⑱ **A chimney cap should be made of**

a. cast-in-place concrete

b. precast concrete

c. mortar

d. both a and b

⑲ **When placing a chimney cap, it is recommended to**

a. provide a bond break between the brickwork and the cap

b. have a ⅜"-wide, water-sealed, caulked joint between the flue liner and the cap

c. have a mortar-filled joint between the flue lining and the cap

d. both a and b

⑳ **Preventing water from traveling down the outside of chimney walls, beyond the roof, and into the structure is**

a. base flashing installed between rows of roofing shingles

b. counter flashing installed in the chimney walls

c. tar roof coating between the chimney and roof

d. both a and b

㉑ **A double-sloped roof behind the upper side of the chimney parallel with the ridgeline diverting water from behind the chimney toward the sides is called a**

a. cricket

b. dam

c. saddle

d. valley

# Chapter 21 | Masonry Fireplaces

Today's masonry fireplaces are just as popular for the atmosphere they create as for the heat they generate. Many homeowners request fireplaces not so much as a supplemental source of heat but as an architectural focal point. Safe and efficient operations require constructing fireplaces in accordance with governing building code regulations and recommended, time-proven, masonry construction practices. The many designs for fireplaces give homeowners options for selecting a style compatible to their needs and the structure's architectural styling.

## OBJECTIVES

After completing this chapter, the student should be able to:

- Identify the components of a wood-burning fireplace.
- Explain basic features of the five types of masonry fireplaces.
- Describe factors governing the performance of a fireplace.
- List building code requirements for a single-face masonry fireplace.
- Explain procedures for constructing a single-face masonry fireplace.

# Glossary of Terms

**air-circulating fireplace** a fireplace circulating room air through baffled, heat-exchange chambers behind the firebox walls.

**air intake** a passageway for bringing outside air for combustion into the firebox.

**ash pit** a hollow space below the inner hearth for the deposit of ashes.

**base** those components of a fireplace, including concrete footings, masonry foundation walls, ash pit, air inlet, and reinforced concrete hearth base.

**chimney** that part of a fireplace venting combustion byproducts to the outside air while protecting the house and its occupants from fire and the life-threatening effects of carbon monoxide poisoning.

**combustion chamber** also called the firebox, the enclosed space beyond the fireplace face where combustion occurs.

**draft** a measure of the passage of air from the firebox to the smoke chamber.

**fire stopping** also called fireblocking, a noncombustible material preventing air drafts and the potential spread of smoke and fire between floors or ceilings.

**firebox** also called the combustion chamber, the enclosed space beyond the fireplace face where combustion occurs.

**fireplace brick** also called firebrick, refractory brick that has thermal resistance and thermal stability for lining the fireplace firebox.

**fireplace surround** the face of the fireplace immediately surrounding the opening.

**hearth** the floor of the firebox.

**hearth base** the structural support for the firebox floor.

**inner hearth** the floor of the combustion chamber.

**multi-face fireplace** a fireplace with openings on two or more sides.

**outer hearth** also called hearth extension, that part of the fireplace extending beyond the fireplace face in front of the firebox floor.

**Rosin fireplace** a single-face fireplace with a curved, back-wall design.

**Rumford fireplace** a single-face fireplace design with a tall and shallow firebox with significantly flared sides and a vertical back for radiating more heat than a conventional fireplace.

**single-face fireplace** a fireplace with an opening on only one side.

**smoke chamber** an inverted funnel-shaped area extending from the level of the throat damper to the beginning of the chimney flue lining.

**smoke shelf** the area at the base of the smoke chamber, directly behind the throat damper opening.

**throat** the narrow passage between the combustion chamber and the smoke chamber.

**throat damper** a cast iron or steel frame and valve plate control, controlling the burning rate by managing the passage of air from the firebox to the smoke chamber.

# Styles of Fireplaces

The most recognized styles of fireplaces include single-face fireplaces, multi-face fireplaces, air-circulating fireplaces, Rumford fireplaces, and Rosin fireplaces. They differ primarily in firebox design.

## Single-Face Fireplaces

A conventional single-face fireplace has an opening on only one side (see Figure 21-1). It is the fireplace design typically found in homes built since the mid-twentieth century.

## Multi-face Fireplaces

A multi-face fireplace is open on two or more sides (see Figure 21-2). Multi-face fireplaces are less energy-efficient than single-face fireplaces because there is less firebox surface area radiating heat into the room.

**FROM EXPERIENCE**

Multi-face fireplace openings placed in the paths of potential air drafts such as those resulting from opening exterior doors can cause the unintentional diversion of smoke into room areas. The potential for air drafts should be considered when choosing to build multi-face fireplaces.

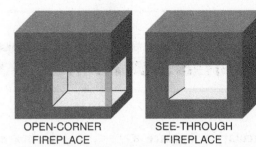

OPEN-CORNER
FIREPLACE

SEE-THROUGH
FIREPLACE

**Figure 21-2** Multi-face fireplaces can be open on the front and side or on the front and back as illustrated.

## Air-Circulating Fireplaces

An air-circulating fireplace circulates room air through baffled, heat-exchange chambers behind the firebox walls (see Figure 21-3). Electric-operated fans draw room air through inlet vents. The airflow slows as it passes a series of baffles. The air is heated and then forced into the room through one or more vents. Popular during the last half of the twentieth century, double-wall, heat-circulating fireplace forms perform similarly. Fans near the bottom pull room air into the double steel firebox walls, circulating it between a series of baffles and exhausting heated air above the firebox opening.

**Figure 21-1** The conventional single-face fireplace is often requested in finer homes.

**Figure 21-3** The vent near the bottom on the right side, concealing a fan drawing cooler room air into an air chamber behind the steel firebox walls, and the one above the opening, returning heated forced air into the room, are evidence of an air-circulating fireplace. Not shown is a second vent drawing cooler room air into the chamber on the opposite side.

**Figure 21-4** **A shallow, widely flared firebox is an obvious characteristic of the Rumford fireplace.** *(Courtesy of Buckley Rumford Company)*

## Rumford Fireplaces

A Rumford fireplace is a single-face fireplace having a tall and shallow firebox with significantly flared sides and a vertical back for radiating more heat than a conventional fireplace (see Figure 21-4). Its smaller throat is designed to reduce air turbulence, removing smoke while conserving heat. Count Rumford is credited with the design of the Rumford fireplace in the latter years of the eighteenth century. Thomas Jefferson used the Rumford fireplace design for the fireplaces in his home, Monticello. They were used in homes through the 1800s until the conventional, single-face fireplace of the twentieth century gained popularity. Section R1003.6 of the International Residential Code permits the construction of Rumford fireplaces "provided that the depth of the fireplace is at least 12" and at least one-third of the width of the fireplace opening, that the throat is at least 12" above the lintel and is at least $\frac{1}{20}$ the cross-sectional area of the fireplace opening."

## Rosin Fireplaces

Professor P.O. Rosin developed the Rosin fireplace in the earlier part of the twentieth century. The Rosin fireplace is a single-face fireplace with a curved, back-wall design (see Figure 21-5). The Rosin firebox is available as a cast-refractory component that can be installed in an existing fireplace or used in a new fireplace.

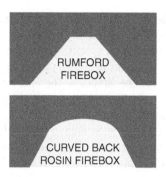

**Figure 21-5** **Compare the curved-back design of a Rosin fireplace with the widely splayed sides of a Rumford fireplace.**

# Fireplace Parts

Fireplaces have four major parts: the base, firebox, smoke chamber, and chimney.

## Fireplace Base

The fireplace base consists of concrete footings, masonry foundation walls, an ash pit, an air inlet, and a reinforced concrete hearth base (see Figure 21-6).

To comply with Section R1003.2 of the International Residential Code, a concrete footing supporting a fireplace should be a minimum of 12" thick and extend a minimum of 6" beyond the exterior walls. Section R1003.2 of the International Residential Code also requires that footings must be placed on natural, undisturbed earth or engineered fill below the frost depth or 12" below finished grade in areas not subject to freezing. Regional and local building code regulations govern specific requirements. Steel rod reinforcement may be a requirement of the structural engineer designing some chimney footings. A fireplace footing should be level with and bonded to the concrete footings supporting the structure's foundation walls. In all cases, however,

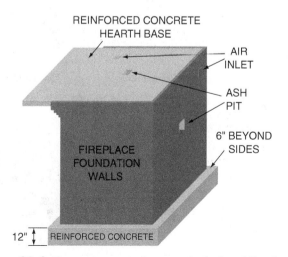

**Figure 21-6** **The components of a properly designed fireplace base are illustrated here.**

footings must be in compliance with all federal, state, and local laws and regulations. Having a licensed structural engineer design a footing for a specific job is recommended since requirements are different depending upon the intended geographic location for the fireplace. The following information (see Figure 21-7 and Table 21-1) is provided by the Brick Industry Association for sizing single-face fireplace openings.

Fireplace foundation walls should be constructed of solid masonry, meaning materials are cored no more than 25%. The Brick Industry Association Technical Notes 19 recommends the foundation walls to be a minimum of 8" thick with no voids except for the ash pit and an air inlet to the combustion chamber.

**Table 21-1 Single-Face Fireplace Dimensions[a], Inches[b]**

| Finished Fireplace Opening | | | | | | | Rough Brick Work[c] | | | | Steel Angle[d] |
|---|---|---|---|---|---|---|---|---|---|---|---|
| A | B | C | D | E | F | G | H | I | J | K | N |
| 24 | 24 | 16 | 11 | 14 | 18 | 8¾ | 32 | 21 | 19 | 10 | A-36 |
| 26 | 24 | 16 | 13 | 14 | 18 | 8¾ | 34 | 21 | 21 | 11 | A-36 |
| 28 | 24 | 16 | 15 | 14 | 18 | 8¾ | 36 | 21 | 21 | 12 | A-36 |
| 30 | 29 | 16 | 17 | 14 | 23 | 8¾ | 38 | 21 | 24 | 13 | A-42 |
| 32 | 29 | 16 | 19 | 14 | 23 | 8¾ | 40 | 21 | 24 | 14 | A-42 |
| 36 | 29 | 16 | 23 | 14 | 23 | 8¾ | 44 | 21 | 27 | 16 | A-48 |
| 40 | 29 | 16 | 27 | 14 | 23 | 8¾ | 48 | 21 | 29 | 16 | A-48 |
| 42 | 32 | 16 | 29 | 16 | 24 | 8¾ | 50 | 21 | 32 | 17 | B-54 |
| 48 | 32 | 18 | 33 | 16 | 24 | 8¾ | 56 | 23 | 37 | 20 | B-60 |
| 54 | 37 | 20 | 37 | 16 | 29 | 13 | 68 | 25 | 45 | 26 | B-66 |
| 60 | 37 | 22 | 42 | 16 | 29 | 13 | 72 | 27 | 45 | 26 | B-72 |
| 60 | 40 | 22 | 42 | 18 | 30 | 13 | 72 | 27 | 45 | 26 | B-72 |
| 72 | 40 | 22 | 54 | 18 | 30 | 13 | 84 | 27 | 56 | 32 | C-84 |

[a]Adapted from *Book of Successful Fireplaces*, 20th Edition.

[b]SI conversion: mm = in. × 25.4.

[c]L and M are equal to outside dimensions of flue lining plus at least 1" (25 mm). L is greater than or equal to M.

[d]Angle sizes: A — 3 × 3 × ¼", B — 3½ × 3 × ¼", C — 5 × 3½ × ⁵⁄₁₆".

The **hearth base** supports the firebox floor and the outer hearth extending beyond the front of the firebox. Section R1003.9 of the International Residential Code states that "masonry fireplace hearths and hearth extensions shall be constructed of concrete or masonry, supported by

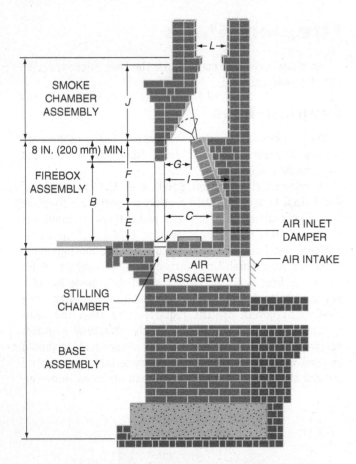

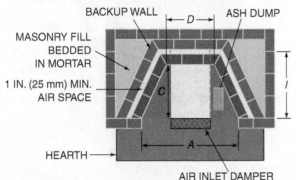

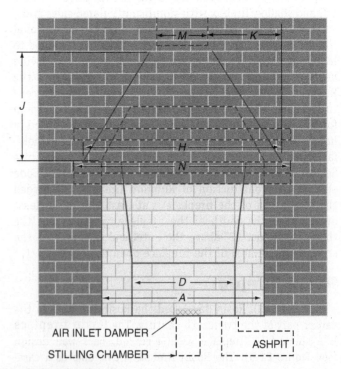

**Figure 21-7** The preceding three illustrations and table are provided by the Brick Industry Association for sizing single-face fireplaces. *(Courtesy of the Brick Industry Association)*

noncombustible materials, and reinforced to carry their own weight and all imposed loads. No combustible material shall remain against the underside of hearths and hearth extensions after construction." Section R1003.9.1 of the International Residential Code requires the minimum thickness of fireplace hearth to be 4". A steel-reinforced concrete slab cantilevered from the fireplace foundation walls typically serves as both the inner and outer hearth support.

A properly engineered, structural concrete slab placed on approved soil can be used as both a base and hearth support if foundation walls are unnecessary.

## Firebox

The firebox, also called the combustion chamber, is where combustion occurs. The firebox includes the fireplace opening, the hearth, the combustion chamber, and the throat (see Figure 21-8).

### Opening

Firebox openings vary in size. However, the dimensions of the firebox opening determine other design details, including the combustion chamber dimensions and the flue size.

### Hearth

The hearth is the floor of the firebox. It consists of both the inner and outer hearths. The hearth can be level with the adjacent flooring or it can be elevated to create a raised hearth.

The inner hearth serves as the floor of the combustion chamber. It is usually lined with heat-resistant refractory brick known as fireplace brick (also called firebrick). Fireplace brick have thermal stability unlike standard brick. They show no evidence of cracking or chipping after consistent cycles of heating and cooling. Firebrick are slightly larger than standard brick, usually measuring 4" wide, 9" long, and $2\frac{1}{4}$" thick. The inner hearth has provisions for an ash pit and an air intake. The ash pit is a hollow space below the inner hearth for the deposit of ashes. A pivoted metal door located flush with the hearth surface permits sweeping the ashes into the ash pit. For a hearth constructed on a concrete slab where the design does not permit an ash pit, ashes are swept into a removable pail below a cast

iron grating. To comply with Section R1005.2 of the International Residential Code, fireplaces must have an exterior air intake capable of providing all combustion air. The air intake is a passageway for bringing outside air for combustion into the firebox. Energy-efficient homes that have little air infiltration require outside air for proper combustion. Otherwise, as Brick Industry Association Technical Notes 19 explains, a noticeable drop in room air pressure results from room air loss through the chimney. This results in efficiency reduction of the fireplace, as the lower room pressure promotes cold outside air infiltration in addition to the removal of heated room air out the chimney. The air intake consists of a screened, closeable louver on an outside wall connected to a noncombustible, insulated air duct. The air duct leads to an inlet located on or near the floor of the firebox. The inlet should have a damper to regulate the airflow. According to Brick Industry Association Technical Notes 19, "a damper is required to control the volume and direction of the air flow. This is necessary because cold outside air channeled into the fireplace expands and could possibly result in more air than is needed for draft and combustion. This can create a spillover effect into the room prior to the air being warmed. The inlet can be located in the sides or the floor of the combustion chamber, preferably in front of the grate for best performance. If the inlet is located toward the back of the combustion chamber, ashes may be blown into the room by drafts for the inlet. As an option, the inlet can be located on or near the floor within 24" (600 mm) of the firebox opening. Any inlet should be closeable and designed to prevent burning material from dropping into concealed combustible spaces." Section R1005.4 of the International Residential Code requires the air passageway to be "a minimum of 6 square inches and no more than 55 square inches, except that combustion air systems for listed fireplaces shall be constructed according to the fireplace manufacturer's instructions." To reduce the risk of fire, air intakes cannot be located in a garage, attic, basement, or crawl space.

The outer hearth, or hearth extension, extends beyond the fireplace opening and into the room. To comply with Section R1003.10 of the International Residential Code, it must extend a minimum of 8" beyond the sides of the opening and 16" in front of the opening and for openings larger than 6 square feet, the outer hearth must extend at least 12" beyond the sides and 20" in front of the opening. It too must be constructed of noncombustible materials. A raised hearth extension is generally built when the firebox floor is higher than the height of the room floor. Having the hearth extension level with the raised firebox floor conceals the front edges of the firebrick for the inner hearth. Although a raised hearth extension reduces floor area, it provides a fireside seating area for the room's occupants.

### Firebox

The firebox (combustion chamber) is the enclosed space beyond the fireplace face where combustion occurs. A larger fire in the combustion chamber improves efficiency. Brick

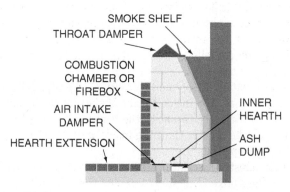

**Figure 21-8** This cross-sectional illustration identifies the parts of the firebox, also called the combustion chamber.

Industry Association Technical Notes 19 states that "firebox dimensions should be selected so that the fire fills the combustion chamber during operation". Careful consideration should be given to the size of the opening when designing a fireplace. Building code regulations require larger openings to have larger combustion chambers and flue area. As a result, more heated room air can escape out the chimney. Choosing a larger opening for appearance alone can result in lowered efficiency if smaller fires are sustained in the larger size combustion chamber. A 36"-wide firebox is typical for room areas up to about 300 square feet.

The walls of the combustion chamber are typically lined with firebrick. The sidewalls are aligned plumb and angled to create a narrower chamber at the back. This angled design permits better radiation and reflection of heat into the room. The first few courses of the back wall are aligned plumb. It then slopes toward the front, creating a narrow opening at the top known as the throat.

### Throat

The throat is the narrow passage between the combustion chamber and the smoke chamber. The throat extends the full width of the combustion chamber to enable the fire to burn more evenly across the back wall.

A high form throat damper consists of either a cast iron or steel frame and a valve plate control. It controls the burning rate by managing the draft, a measure of the passage of air from the firebox to the smoke chamber. With no fire in the combustion chamber, a closed damper valve plate prevents room air from escaping out the chimney. A cast iron damper is recommended over a steel damper because cast iron is more resistant to heat.

## Smoke Chamber

The smoke chamber extends from the level of the throat damper to the beginning of the chimney flue lining (see Figure 21-9). Its purpose is to pass the byproducts of

**Figure 21-9** Heated air and combustion byproducts rise through the open throat damper and into the smoke chamber. Heavier, outside cooler air settles into the smoke chamber from the open top of the flue, and heated air and gases rise up the flue.

combustion, smoke and gases, into the chimney. It functions like a funnel. Smoke and gases exiting the combustion chamber and passing through the narrow throat collect in the smoke chamber. Heavier, cooler outside air enters the flue from above and creates an air draft as it forces the heated air upwards through the chimney flue.

At the base of the smoke chamber directly behind the damper opening is the smoke shelf. Sloping the back wall of the combustion chamber forward until it meets the rear frame of the fireplace damper forms the smoke shelf. The smoke shelf is a design element helping to prevent air currents and downdrafts from entering the combustion chamber and causing smoke to enter the room. Being directly below the chimney flue, the smoke shelf collects ashes and soot that gravitate to the base of the smoke chamber. With no fire in the fireplace and cool surfaces, the valve plate of some dampers can be removed to access the smoke shelf, allowing the removal of combustion byproducts accumulating on the smoke shelf.

Considerations for having a properly functioning smoke chamber include the following:

- The smoke chamber should extend the full width of the throat.
- To comply with Section R1003.8 of the International Residential Code, the minimum thickness of the front-, back-, and side walls must be 8" of solid masonry where no lining is provided and a minimum of 6" where a lining of firebrick at least 2" thick or a lining of vitrified clay at least ⅝" thick is used.
- To comply with Section R1003.8.1 of the International Residential Code, "the inside surface of the smoke chamber shall be inclined no more than 45 degrees from vertical when prefabricated smoke chamber linings are used or when the smoke chamber walls are rolled or sloped rather than corbelled. When the inside of the smoke chamber is formed by corbelled masonry, the walls shall not be corbelled more than 30 degrees from vertical."
- The back wall of the chamber is aligned plumb with the back of the smoke shelf.
- To comply with Section R1003.8 of the International Residential Code, "corbelling of masonry shall not leave unit cores exposed to the inside of the smoke chamber." And "when the inside surface of the smoke chamber is formed by corbelled masonry, the inside surface shall be parged smooth."
- Section R1003.8.1 of the International Residential Code requires that "the inside height of the smoke chamber from the fireplace throat to the beginning of the flue shall not be greater than the inside width of the fireplace opening."
- Brick Industry Association Technical Notes 19 recommends that the smoke shelf should extend the full width of the smoke chamber, level with the damper frame, either flat or curved downwards at the rear wall.

Preformed, smoke chamber and firebox components are available. Manufacturers of these products claim the components to be engineered for time saving, correctly designed, and more efficient fireplaces.

## Chimney

The fireplace **chimney** vents combustion byproducts to the outside air while protecting the house and its occupants from fire and the life-threatening effects of carbon monoxide poisoning. The chimney begins at the bottom of the first flue lining atop the smoke chamber and ends at the top of the last flue lining or any rain cap above it (see Figure 21-10).

A flue lining is a rectangular or round, hollow chimney lining intended to contain the combustion products and protect the chimney walls from heat and corrosion. Flue linings are made of fireclay. Brick Industry Association Technical Notes 19B states that "flue liners should conform to ASTM C 315. They should be thoroughly inspected just prior to installation for cracks or other damage that might contribute to smoke and flue gas leakage."

Although clay linings crack from thermal shock, meaning sudden and extreme temperature changes, tests show that they do not crack from normal use. Thermal shock can be a result of starting a very rapid, hot fire in a cool chimney using such highly combustible materials as paper and kindling wood, or it can be the result of a chimney fire.

Most suppliers refer to the size of a rectangular flue liner as being its overall outside dimensions, the actual opening dimensions being approximately 1" less. The size of a round liner typically refers to its inside diameter, the approximate diameter of the opening. Most liners are available in 2' sections. Like firebrick, a flue lining withstands the continuous

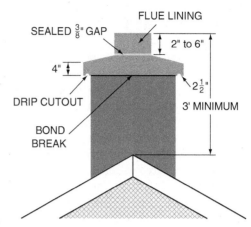

**Figure 21-11 The chimney top illustrated here includes design details for a weather-resistant, safe operating chimney.**

heating and cooling cycles of the fireplace. The flue lining is enclosed in solid masonry walls whose nominal thickness is not less than 4". Providing a minimum ½" air space while complying with Section R1001.9 of the International Residential Code, requiring "...to maintain an air space or insulation not to exceed the thickness of the flue liners separating the flue liners from the interior face of the chimney masonry walls" reduces the likelihood of cracked liners caused by thermal expansion. The size of the flue lining depends on the size of the fireplace opening. Larger openings require larger flues. Rumford and Rosin fireplaces typically require smaller flues, a result of higher combustion temperatures and emitting fewer combustion byproducts. Building code regulations specify the required size of flue lining based on the fireplace design and opening size.

Section R1001.6 of the International Residential Code requires that "chimneys shall extend at least 2' higher than any portion of a building within 10' but shall not be less than 3' above the highest point where the chimney passes through the roof" (see Figure 21-11). A chimney cap eliminates moisture from penetrating the top of the chimney brickwork. The cap, made of reinforced concrete, can be prefabricated or cast-in-place. A cap should have the following details:

- It should extend a minimum of 2½" beyond the chimney on all sides and have a drip cutout on the underside to prevent water from working its way back to the walls.
- It should be a minimum of 4" thick and slope away from the flue on all sides for the purposes of shedding water and reducing wind turbulence at the flue.
- There should be a bond break between the brickwork and the concrete cap to allow for the differential movement of it and the brick masonry.
- Thermal expansion of the flue lining should be accommodated by a silicone-sealed, ⅜" gap between the lining and the cap.
- It should be no less than 2" or more than 6" below the flue lining.

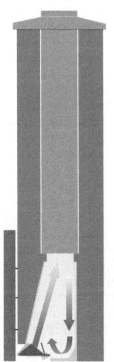

**Figure 21-10 Beginning at the top of the smoke chamber, the chimney vents smoke and deadly gases to the outside. An air space, no less than ½" or more than the thickness of the flue lining walls, between the brickwork and the flue lining extends the height of the chimney to permit thermal expansion of the flue lining.**

# Fireplace Surround

The **fireplace surround** is the face of the fireplace immediately surrounding the opening. The appearance of the fireplace is influenced greatly by the fireplace surround. It can be stone, brick, tile, marble, or a combination of these materials. Combustible materials such as wood mantles are often included. Section R1003.12 of the International Residential Code requires that "woodwork or other combustible materials shall not be placed within 6" of a fireplace opening. Combustible material within 12" of the fireplace opening shall not project more than $\frac{1}{8}$" for each 1" distance from such opening." Such is the case for a wooden mantle. The masonry face above the opening is supported by a steel lintel, a cast stone or natural stone lintel, a reinforced brick lintel, or an approved design brick arch. Masonry should be secured to the front backup walls of the combustion chamber and the smoke chamber with corrosion-resistant metal ties embedded in the mortar joints at vertical and horizontal spacing not exceeding 24" and allowing a tie to support no more than 2.67 square feet of wall area. Typical installation of wall ties is found at a 16" spacing, both vertically and horizontally.

# Constructing a Single-Face Fireplace

Fireplaces must be constructed in compliance with all building code regulations. One should review current building code regulations addressing the design and construction of fireplaces before beginning construction. Referring to the International Residential Code, Chapter 10, "Chimneys and Fireplaces" is recommended. A professional mason experienced in fireplace construction should supervise apprentice masons and those inexperienced with fireplace construction.

## Placing the Footing

Building plans showing the overall dimensions for the outer walls of the fireplace are necessary for providing a footing capable of supporting the masonry walls. The concrete footing is reinforced with steel reinforcing rods. Because of the varying depth at which frost penetrates, the bearing capacity of different soils, and the different seismic zones, concrete should be placed in accordance with current local building code regulations.

**CAUTION**

**Before beginning to dig, all underground utilities must be located and identified in and surrounding the area of construction by the utility companies or their agent. Governing regulations control the proximity of concrete slabs to buried utilities.**

A newly placed concrete footing should be covered with a waterproof tarp to prevent excessive water evaporation and should be allowed to cure for a minimum of seven days before constructing the masonry base.

**CAUTION**

**Working in ground excavations where fireplace bases are likely to begin poses safety hazards that must be addressed. The surrounding ground must be examined by a competent person for indications of potential cave-ins. Where there is a potential for cave-ins and in excavations 5' or more in depth, protective systems meeting the intent of OSHA Standards 1926.652 must be used.**

## Constructing the Base

The base of the fireplace is constructed of solid masonry, or brick with a cross-sectional area of 75% or more of the bed area. Grade SW brick are recommended. Local regulations may permit grouted and reinforced hollow concrete masonry units (CMUs). The base should be tied into adjoining masonry foundation walls.

A through-wall opening for accessing the ash pit and another opening for a fresh air supply are needed. A ferrous metal, tightly closing, clean-out door and frame are installed for access to the ash pit. A commercially available fresh air supply kit is installed in the base. These kits include an outside louver, air duct, firebox vent, and installation instructions. If a commercially available air duct is not used, a masonry passageway for airflow leading to the firebox is formed using brick. Installing a screened louver, preferably one that can be operated from inside, in the outside wall of the base may be approved as an outside air supply enclosure when commercially available kits are not used. The Brick Industry Association Technical Notes 19 recommends "to help decrease the velocity of the air through the inlet, a space before the inlet should be constructed as a stilling chamber." Failure to do so can result in higher temperatures that can damage metal inlet dampers and wood-holding metal fire grates.

Projecting the brickwork from a brick-masonry base provides additional support for the outer hearth. The projection for each course is limited to no more than $\frac{1}{3}$ the brick's bed depth or more than $\frac{1}{2}$ the brick's face height, which is approximately 1". The Brick Industry Association Technical Notes 19 recommends that "the overall horizontal projection should be limited to one-half of the wall thickness unless the corbel is reinforced." The height of the finished hearth in relationship to the finished floor must be known for proper height of the base. The fireplace base should stop at a point below the intended finished height of the hearth equal to the sum of the thickness for the mortared fireplace brick and the concrete base for the hearth.

## Constructing the Hearth Support

A minimum 4" thick, reinforced structural 3500 psi concrete slab is placed on top of the fireplace base. In most cases, the concrete is cantilevered from the base so that the outer and inner hearth can be constructed on one, steel-reinforced, continuous concrete slab. Forms are needed for the construction of the concrete hearth base. A cutout for the rectangular opening of the ash pit is needed in the inner hearth form. The air inlet included in the base continues through the hearth support. It is routed to the back or sidewalls of the firebox, or within 24" of the firebox opening on or near the floor. By laying out the firebrick pattern for the inner hearth, the location of the ash pit can be determined and its opening formed, permitting the substitution of the ash dump for a firebrick. Likewise, if the air inlet is to be in the firebox floor, it is helpful to route the air passage where only one firebrick needs cutting. These two steps should be taken for the firebox floor layout pattern:

1. Lay the firebrick so that the front edge of the firebox floor aligns with the fireplace surround or face.
2. Improve the appearance of the firebox floor with a firebrick pattern centered with the width of the firebox opening.

Because the inner hearth form cannot be removed from below once the concrete cures, it should be constructed of an approved noncombustible forming material. Combustible materials create a fire hazard as heat conducted through the hearth over a period of time to the form creates a potential for spontaneous combustion. In addition, hot ashes and embers diverted to the ash pit can ignite a combustible form. Section R1003.9 of the International Residential Code states that "no combustible material shall remain against the underside of hearths and hearth extensions after construction."

A properly engineered floor joist design can be used to support the outer hearth form and fresh concrete, but the form should be removed once the concrete cures. Floor framing should not be used to permanently support the outer hearth. If so, the outer hearth may crack and separate from the inner hearth due to differential movements of the framing materials and the masonry.

⚠️ **CAUTION**

**An open area between floor framing and the hearth base before the concrete hearth support is placed is a hole through which a person might fall. Persons must be prevented from falling through such open holes, OSHA 1926.501 (4). Coverings or guard rail systems meeting OSHA construction industry standards 1926.502 must be used when the unprotected area is 6' or more above lower levels.**

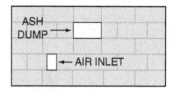

**Figure 21-12** The ash dump and air inlet are aligned between the inner hearth's fireplace brick.

 **FROM EXPERIENCE**

Having sample materials for the finished floor and for the outer hearth permits constructing the hearth support at the intended level when the outer hearth and adjoining floor are to be aligned level with one another.

## Constructing the Inner Hearth

The ash dump opening and air inlet vent are aligned between the fireplace brick (see Figure 21-12). These openings must align with openings previously formed in the hearth's concrete base support.

Proper fireplace brick moisture content increases bond strength. Dampening the fireplace brick before bedding them in mortar will increase bond strength. Section R1003.5 of the International Residential Code requires that firebrick be bedded in medium duty refractory mortar, limiting joint width to no more than ¼". Such joints are less likely to deteriorate and crack. Ensuring a level firebox floor helps to construct level firebox walls. The ash dump door and air inlet vent must fit their respective positions in the firebox floor.

 **FROM EXPERIENCE**

The fireplace brick pattern for the inner hearth is aligned from the center of the opening's width. This procedure ensures pattern symmetry in relationship of the inner hearth and the firebox sidewalls.

## Constructing the Firebox Walls

The dimensions of the firebox opening determine the design and dimensions for the firebox sidewalls and back wall. Measured from the face of the fireplace surround, the inner hearth must have a minimum depth of 20" in order to comply with Section R1003.6 of the International Residential Code. Technical information provided by the Brick Industry Association provides dimensions proven successful for proper fireplace performance. Refer to the Brick Industry Association's Table of Standard Fireplace Dimensions, page 300, for their recommended dimensions. The sidewalls are equally angled,

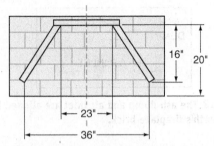

Figure 21-13 Lines representing the front and back widths are laid out from the center of the hearth. The exposed face of the back wall is 20" from the front edge of the inner hearth. Allowing for the face surround to be flush with the front of the inner hearth, the layout for the firebox sidewalls begins 4" behind the inner hearth's front edge.

creating a narrower inner hearth at the back than at the front, a design intended to help reflect heat into the room. The back wall eventually slopes toward the fireplace front opening. This design helps reflect heat into the room also.

Layout lines are drawn onto the firebox floor representing the sidewalls and the back wall. Figure 21-13 illustrates the dimensions for a 36" firebox.

Fireplace brick should be thoroughly wetted several hours before bedding in mortar. Fireplace brick should be bedded in the position recommended by the brick producer. At least one brick company recommends bedding them on edge in the shiner position. They do not accept responsibility for firebrick damaged by heat buildup resulting in brick whose 4" sides are bedded in mortar. Each of the first three courses of the sidewalls and back wall are aligned level, plumb, and straight as they are bedded in joints no more than ¼" wide, using medium-duty refractory mortar. The Brick Industry Association Technical Notes 19A recommends that "fireclay mortar joints should be ¹⁄₁₆" to ³⁄₁₆" (1.6 to 4.8 mm) thick to reduce thermal movements and mortar joint deterioration. When using fireclay mortar, extremely thin mortar joints may be obtained by using the Pick and Dip method. This consists of dipping the unit into a soupy mix of fireclay mortar and immediately placing it in its final position. The mortar joints need be only thick enough to provide for dimensional irregularities in the unit being laid."

*For step-by-step procedures on laying out the sidewalls and back wall for a 36" single-face fireplace, see the Procedures section on page 314.*

To comply with Section R1003.5 of the International Residential Code, including a 2"-thick firebrick lining, the total thickness of the back wall and sidewalls must be a minimum of 8" of solid masonry. Where firebrick is not used to line the firebox, the total thickness of the back- and side walls must be at least 10" of solid masonry. Brick Industry Association Technical Notes 19A recommends a 1" minimum air space between the firebrick and the surrounding brickwork, left open or filled with a compressible, noncombustible material such as fibrous insulation to reduce

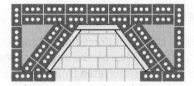

Figure 21-14 A 1" air space separates the double-wythe, solid masonry backup wall from the firebrick.

the stress from thermal movements of the firebrick (see Figure 21-14). Brick Industry Association Technical Notes 19A also recommends that "the wall behind the firebrick at the rear of the firebox should be at least 8" (200 mm) thick. A greater thickness may be required to support higher chimneys. With the exception of the combustion chamber walls, wall ties should be used at all intersections where the wall is not masonry bonded. These ties should be spaced a maximum of 16" (400 mm) vertically, and embedded at least 2" (50 mm) into bed joints of the brick masonry."

With few exceptions, Section R1003.11 of the International Residential Code requires that "all wood beams, joists, studs, and other combustible material shall have a clearance of not less than 2" from the front faces and sides of masonry fireplaces and not less than 4" from the back faces of masonry fireplaces."

Veneer wall ties are used to bond the double-wythe backup walls and to anchor the face surround to the backup walls. Although a maximum spacing of 24" is allowed, no more that 2.67 square feet supported by a single tie, a 16" vertical and horizontal spacing is recommended for veneer wall ties.

The sloping of the back wall towards the front of the firebox typically begins on the fourth course. Sloping the firebrick and maintaining ⅛" to ¼" bed joints prevents the back wall coursing from remaining level with the sidewalls. As a result, continuing from the fourth course, the brick for the sidewalls are no longer cross-lapped, bonding with the back wall. Building the sidewalls ahead of the back wall allows the mason to establish the intended slope for the back wall. Making a full-size drawing showing the profile of the firebox helps determine the slope of the sidewalls (see Figure 21-15).

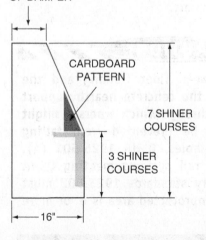

Figure 21-15 Drawing the diagonal red line from the top to the back wall shows the slope or angle for continuing the back wall beyond the third course. Firebrick are laid in the shiner position.

**Figure 21-16** The top course of fireplace brick on the back wall are cut, permitting the back to be level with the sides.

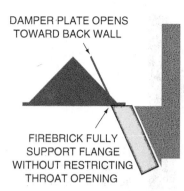

**Figure 21-17** The rear damper flange rests entirely on the fireplace brick, protecting it from intense heat. The damper plate opens toward the back wall.

A cardboard pattern can be made to represent the angle for cutting the sidewall brick.

As a result of sloping the firebrick for the back wall, coursing becomes lower for the back wall than the sidewalls. Using a masonry saw, brick are cut to the required height so that the top course of the back wall is level with the sidewalls (see Figure 21-16).

### CAUTION

Always wear approved eye, face, hearing, and respiratory protection when cutting masonry materials with power saws, and be trained by a competent person before using the saws. Electric-powered tools must be equipped with inline ground-fault circuit interrupters for operator safety and protection from electric shock and electrocution.

### CAUTION

To permit proper curing of the mortar, the Brick Industry Association Technical Notes 19A recommends not building a fire in a firebox within 30 days of construction.

### FROM EXPERIENCE

Placing a 1" layer of sand on the inner hearth after it is built protects the firebrick from mortar droppings and brick from becoming chipped as construction above continues.

## Setting the Throat Damper

The throat damper should set flat and level on the completed firebox walls. To protect the back flange of the damper from intense heat, it should be fully supported above the back wall (see Figure 21-17).

Ensuring a tight seal between the damper and the fireplace walls, the bottom sides of the damper flanges are embedded in medium duty refractory mortar. Preventing the mortar from coming above the edges of the perimeter flanges allows for thermal expansion of the damper. The damper is positioned with the damper plate opening toward the back wall of the fireplace (see Figure 21-17). When open, the damper plate helps deflect any downdraft that may otherwise force smoke into the room. Because the damper expands when heated, mortar and masonry must not be permitted to contact it.

### FROM EXPERIENCE

If thermal expansion is restricted, either the damper or the brickwork may crack. Placing a noncombustible, compressible fibrous insulation between the throat damper and adjacent masonry ensures unobstructed thermal expansion of the damper.

## Constructing the Smoke Chamber

Where no lining is provided, Section R1003.8 of the International Residential Code requires a minimum of 8" thick, solid masonry for the front-, back-, and sidewalls of the smoke chamber. Remember that in most cases, a minimum 2" air space is required between the face of the masonry walls and combustible material at the front and sides and a minimum 4" air space at the back (International Residential Code Section R1003.11). Sloping the back wall of the firebox forward creates the smoke shelf (see Figure 21-18).

It should be no higher than the damper flange. As stated in Brick Industry Association Technical Notes 19A, the smoke shelf may be flat and level with the damper flange (refer to Figure 21-18) or curved and slightly below it (see Figure 21-19).

The smoke shelf extends the full width of the throat. Until above the top of the damper, usually three or four courses depending on the damper, the sidewalls of the smoke chamber are aligned plumb and parallel with the

Figure 21-18 The smoke shelf is directly behind and level with the throat damper.

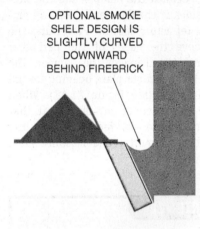

Figure 21-19 Having a curved profile, the smoke shelf is slightly lower than the damper.

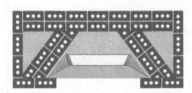

Figure 21-20 8" brick walls form the sides of the smoke chamber walls.

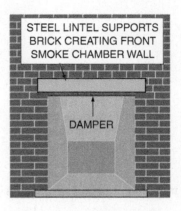

Figure 21-21 A steel lintel supports the brickwork for the front smoke chamber wall above the damper.

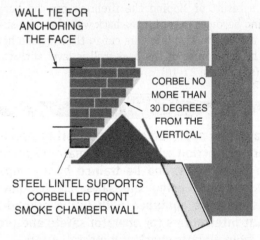

Figure 21-22 Brick courses laid in the header position are corbelled toward the back, reducing the smoke chamber to a width equal to the flue lining opening.

sides of the damper (see Figure 21-20). Laying the walls in the header position strengthens them.

Care must be taken to prevent mortar or brick from contacting the metal damper. If thermal expansion is restricted, either the damper or the brickwork can crack.

Once above the damper, the sidewalls and front wall are sloped uniformly toward the center of the smoke chamber by corbelling each course of brickwork. Not only to permit thermal expansion but also because the damper is not designed to support the weight of brickwork, a steel lintel having a minimum bearing of 4" on solid masonry at each side supports the front wall of the smoke chamber (see Figure 21-21 and Figure 21-22).

Corrosion-resistant wall ties are embedded in the bed joints vertically and horizontally no more than 24" apart for eventually anchoring the face surround (see Figure 21-22).

Both the sidewalls and front wall are to be sloped no more than 30 degrees from the vertical and parged smooth when walls are constructed from corbelled masonry. Laying the corbelled brick in a header position and backing it up with solid masonry provides stronger walls. The walls of the smoke chamber can be inclined as much as 45 degrees from the vertical when approved prefabricated smoke chamber linings are used (International Residential Code Section R1003.8.1). Each course of brickwork is corbelled uniformly until the opening at the top of the smoke chamber is equivalent to that of the flue lining. Because the sidewalls of the smoke chamber are originally parallel to the sides of the throat damper, creating a rectangular opening requires corbelling more towards the front than at the back (see Figure 21-23).

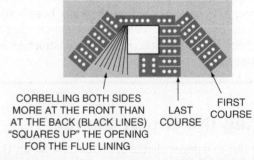

Figure 21-23 Each course of the smoke chamber sides is corbelled more toward the front of the fireplace to create the rectangular opening supporting the flue lining.

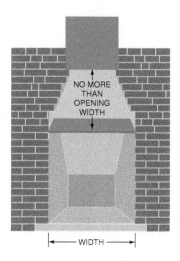

Figure 21-24 Smoke chamber height is not more than the opening width. Corbelled brick are parged, creating a single, uniform slope.

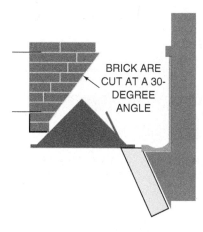

Figure 21-25 The corbelled brick are cut at an angle, permitting a uniform layer of parging for the smoke chamber. Brick on the sidewalls are cut likewise.

To comply with Section R1003.8.1 of the International Residential Code, the completed smoke chamber must be no higher from the fireplace throat to the beginning of the flue than the inside width of the fireplace opening (see Figure 21-24). Parging or plastering the exposed brick with refractory cement strengthens the walls and improves the airflow.

Cutting solid brick at the appropriate angles creates a uniformly sloped smoke chamber. The result is a more uniform layer of parging, which is less likely to separate from the brick (see Figure 21-25).

### FROM EXPERIENCE

Placing used paper cement bags above the damper keeps mortar droppings from bonding to the smoke shelf. Reaching through the open damper from below, the bags are removed along with accumulated mortar droppings after the chimney is completed.

## Constructing the Chimney

### CAUTION

Constructing fireplace chimneys to their intended height requires scaffolds and possibly working on roofs. Only those trained by a competent person should build or access scaffolds. 5'-wide scaffold frames must be adequately braced or tied to restrain and prevent them from tipping at heights above 20'. Section 1926.451(g)(1) Of the OSHA Construction Industry Regulations requires that "each employee on a scaffold more than 10' above a lower level shall be protected from falling to that lower level."

Section 1926.501(b)(1) requires that when on roofs for example, "each employee on a walking/working surface (horizontal and vertical surface) with an unprotected side or edge which is 6' or more above a lower level shall be protected from falling by the use of guardrail systems, safety net systems, or personal fall arrest systems." A different OSHA regulation applies when employees stand on roofs rather than working from scaffolds. With few exceptions, a minimum clearance of 10' must be kept between scaffolds and power lines, according to Section 1926.451(f)(6). Comply with OSHA construction standards, Subpart L "Scaffolds" and Subpart M "Fall Protection," when working on scaffolds or roofs.

The smoke chamber walls should completely support all sides of the flue lining without obstructing the flue opening. The first flue lining above the smoke chamber walls should begin below the height of the combustible ceiling joist structural framing. Each liner is bedded in a medium-duty refractory mortar to keep the joints tight and smooth on the inside. Setting one section of the flue liner ahead of the brickwork permits parging the outside walls of the liner at the joint, which is a recommended practice.

### FROM EXPERIENCE

Flue linings may have structural cracks unnoticeable to the eye. Tapping lightly with a hammer on the sides of a liner as it stands on end produces a ringing sound on liners that have the structural integrity required for chimney construction. A dull, crackling sound indicates a cracked flue liner. Use only flue liners without structural cracks when constructing chimneys.

Grade SW brick are recommended for the chimney walls. Regional code regulations specify the mortar type for chimney construction. Type N mortar permits thermal movement of the brickwork. However, Type S mortar is necessary where wind loads exceed 25 psf or where required for seismic loads.

Although a single-wythe solid masonry wall meets requirements for enclosing the flue lining, fireplace chimneys are typically larger. This may be to enclose an additional fireplace or appliance flues. But appearance may be the only reason for having a larger chimney. The areas between the flue lining and exterior chimney walls are filled with solid masonry bedded in mortar, leaving an air space no less than ½" wide or more than the thickness of the flue lining wall between the outside walls of the flue liner and the brick fill.

### Installing Chimney Fire Stopping

Maintaining the minimum 2" space between combustible materials and the chimneys' outer walls is observed for those built within interior walls or through exterior walls. A 1" layer of fire stopping or fireblocking, noncombustible material is required between chimneys and floor or ceiling joists to prevent air drafts and the potential spread of smoke and fire between floors or ceilings. To comply with Section R1001.16 of the International Residential Code, it should "only be placed on strips of metal or metal lath laid across the space between combustible material and the chimney." Unfaced fiberglass batt insulation may be used as fire stopping. Section R1001.15 of the International Residential Code requires that "chimneys located entirely outside the exterior walls of the building, including chimneys that pass through the soffit or cornice, shall have a minimum air space clearance of 1"." Also, a minimum of 6" is required between the inside wall of flue linings and combustible materials.

### Anchoring the Chimney

Depending on regional seismic design regulations, the fireplace chimney may require anchorage using corrosion-resistant, ³/₁₆" × 1" strap metal to floor and ceiling joists. Brick Industry Association Technical Notes 19 states that "fireplaces and chimneys are typically attached to the structure by steel straps located at each floor or ceiling line. Consult the local building code for design loads and prescriptive requirements." Brick Industry Association Technical Notes 19B states that "single-wythe chimneys should be attached to the structure. This is generally accomplished by using corrosion-resistant metal ties spaced at a maximum of 24" (600 mm) on center. Multi-wythe chimneys that are not masonry bonded should be bonded together using metal wire ties."

### Racking the Chimney

Racking the brickwork for the outer walls reduces the chimney size. Not only does a smaller chimney save materials and time, but racking also creates a stepped, sloping design (see Figure 21-26).

Each course of the racked brickwork is capped with a mortar wash to prevent water entering the brick (see Figure 21-27).

**Figure 21-26** Above the smoke chamber, brick are racked or set in, narrowing the chimney with each course racked. This procedure forms stepped shoulders.

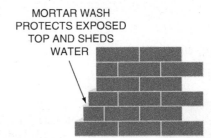

MORTAR WASH PROTECTS EXPOSED TOP AND SHEDS WATER

**Figure 21-27** A separate mortar wash cap above each course prevents water from entering the top of the brick.

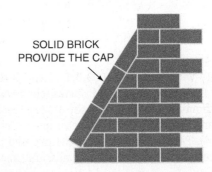

SOLID BRICK PROVIDE THE CAP

**Figure 21-28** The uniform slope, capped with solid brick, reduces the width of the chimney walls.

Another design requires solid brick bedded in mortar over a uniformly sloped surface (see Figure 21-28). Such chimneys have sloped shoulders.

Racking of one or both sides are design options when the full chimney width is not needed for multiple flues. Racking one side permits or may require sloping the flue lining. Racking allows for relocating the centerline of the flue. However, the flue can be sloped no more than 30 degrees from the vertical (see Figure 21-29).

### Topping Out the Chimney

Chimneys must be at least 3' above the highest point where the chimney passes through the roofline (see Figure 21-30).

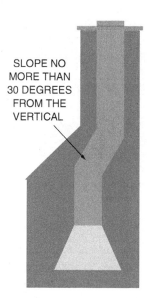

Figure 21-29 The flue lining is sloped, repositioning the centerline of the chimney. In addition, racking the brickwork reduces construction materials and time and adds a sloping design element.

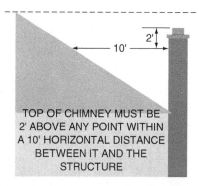

Figure 21-31 Although the chimney is lower than the roof peak, it meets requirements of being a minimum of 2' above any part of the structure within a 10' horizontal distance.

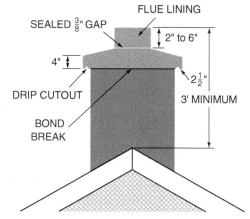

Figure 21-30 The details illustrated here are for weather-resistant chimney tops.

The flue lining should extend above the finished brickwork, permitting it to extend 2 to 6" above the finished cap. Flashing placed on top of the brickwork serves as a bond break, isolating the cap from the brickwork (see Figure 21-30).

In some instances, chimneys are not necessarily required to extend 3' above the roofline. Where design permits, the chimney must only be a minimum of 2' higher than any portion of the structure within 10' (see Figure 21-31).

### Installing the Chimney Cap

Although there is much evidence that mortar is used to make chimney caps, concrete caps are recommended. Concrete is cast to form a structural element. Mortar is intended as a bonding agent only. Concrete caps are more durable than mortar caps. The following procedures are recommended for setting a prefabricated concrete cap:

1. Provide a bond break between the brickwork and the cap.
2. Bed the cap in Type N mortar.

3. Fill the gap between the lining and cap with a noncombustible, compressible material such as fiberglass insulation.
4. Seal the filled gap between the lining and cap with a butyl or silicone caulk.

The following procedures are used to construct a cast-in-place cap using a removable form:

1. Create a drip edge for the underside of the cap as part of the form. An alternative method is to saw the cutout after removing the form.
2. Secure a ½" thick layer of noncombustible, compressible material such as fiberglass insulation around the outside of the flue liner.
3. Place concrete on the flashing material used as a bond break and reinforce with embedded steel reinforcement rods or reinforcement wire.
4. Trowel the concrete surface smooth, and vibrate the edges of the concrete form using a hammer or electric pad sander to remove air pockets from the concrete's surface.
5. After the concrete cures for a minimum of 48 hours, remove the form.
6. Remove the excess material forming the compressible joint between the lining and cap near its surface, and seal the joint with butyl or silicone caulk.

Brick Industry Association Technical Notes 19B, Figure 3, illustrates chimney cap details.

Installing a removable rain cap prevents rain, leaves, birds, and rodents from entering the chimney. Depending on the model, it may be attached to the flu lining or anchored to the chimney cap.

### Installing Chimney Flashing

Corrosion-resistant sheet metal base flashing and counter-flashing prevent water from migrating between the exterior chimney walls and the adjacent roofing materials. Counter-flashing, in the form of an inverted L, is built into the chimney walls (see Figure 21-32). The counter-flashing overlaps the chimney base flashing, which is noncorrosive

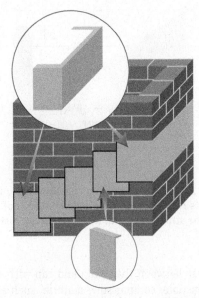

Figure 21-32 Counter-flashing is embedded in the mortar joints. Each stepped piece laps the lower section at least 2".

metal flashing placed between rows of shingles and turned upward against the side of the chimney. There are several similar procedures for installing base flashing, however, chimney base flashing should be installed in accordance with governing codes and the shingle manufacturer's recommendations as the shingles are installed. The resulting combination of properly installed counter-flashing and base flashing is a waterproof joint between the chimney walls and the roof.

The following procedures are necessary to ensure proper performance of the counter-flashing:

- Counter-flashing should extend a minimum of 4" above the roofline.
- Counter-flashing should be inserted into mortar joints to a depth of ¾" to 1".
- Counter-flashing should be mortared solidly into the joint.
- Sections of stepped counter flashing should lap over the lower sections a minimum of 2".
- All flashing joints should be sealed with a butyl or silicone caulk.

 **FROM EXPERIENCE**

It may be that a masonry contractor's insurance coverage does not include water damage caused by leaking chimney flashing. A masonry contractor should consider having a contractor's insurance policy providing completed operations coverage, for claims after the job is complete, for water damage to a building caused by leaking flashing.

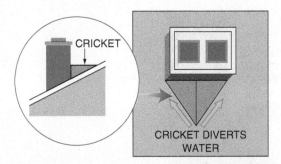

Figure 21-33 A cricket diverts water from the upper side of the chimney toward the sides.

### Constructing a Chimney Cricket

Chimneys not intersecting the ridgeline and having a length parallel to the ridgeline greater than 30" require a cricket. A chimney *cricket* is a sloped section of roof designed to divert water from behind the upper side of a chimney (see Figure 21-33). The side of the chimney adjoining the cricket requires the same flashing details as the other sides of the chimney. Construction details of a cricket are considered a roof framing detail and are not addressed in this chapter.

## Constructing the Fireplace Surround

The fireplace surround or face can be constructed as the firebox is being constructed or anytime afterward. Masons often leave the surround as a provision for working inside on days when the weather makes it impractical to work outside. The brickwork should be bonded to ensure a uniform bond with appropriate head joints once above the opening. Laying out the bond for the face by working from the centerline of the firebox opening assures proper spacing of brick above the opening. A layout with ⅜" head joints may result in having the actual opening width of the surround face slightly narrower than the firebox sidewalls (see Figure 21-34).

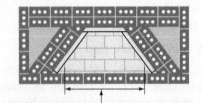

ACTUAL OPENING IS SLIGHTLY NARROWER THAN FIREBOX. THIS PERMITS HAVING APPROPRIATE HEAD JOINT WIDTH FOR THE BRICK SURROUND.

Figure 21-34 Laying out the face surround from the center of the firebox opening assures proper spacing of brick above the opening.

The front of the face surround is aligned with the forward-most edge of the fireplace brick. This ensures that hot coals and embers are exposed to only the fireplace brick after a fireplace screen or glass door enclosure is installed.

The firebox opening width determines the height of the fireplace opening. Section R1003.7 of the International Residential Code requires that "masonry over a fireplace opening shall be supported by a lintel of noncombustible material. The minimum required bearing length on each side of the fireplace opening shall be 4". The fireplace throat or damper shall be located a minimum of 8" above the lintel." A compressible, noncombustible material such as fiberglass insulation should be placed at each end of the steel lintel. Its purpose is to reduce the possibility of having the masonry face crack by accommodating the thermal expansion of the steel. Engineered specifications may or may not require a steel lintel for supporting brickwork above arched openings, reinforced brick masonry lintels, or precast concrete and stone lintels.

## Constructing the Hearth Extension

The hearth extension is aligned level with the firebox floor. Brick, marble, slate, tile, stone, or other noncombustible material may be used for the hearth extension. For openings less than 6 square feet, it must extend 16" in front of and 8" beyond the sides of the opening. Openings 6 square feet and larger require extensions 20" in front of and 12" beyond the sides. If desired, the hearth can extend the full length of the surround and farther than required into the room. However, adequate hearth support as described earlier is required.

## Prefabricated Fireplace Kits

Fireplace kits are manufactured providing the major fireplace components for conventional fireplaces, Rumford fireplaces, and Rosin fireplaces, including prefabricated fireboxes, throat sections, and smoke chambers. These kits include components that simplify the construction process and ensure the correct design. There are several manufacturers of masonry fireplace kits and components.

**CAUTION**

**Allow newly constructed fireplaces to dry out for a minimum of 28 days before starting fires. Build small break-in fires at first. Always avoid very hot, rapid burning fires. Sudden temperature changes of the fireplace materials can fracture them. Install a fire screen in front of the firebox as fire protection to the surrounding area. Never leave a burning fire unattended. Provide adult supervision for children, keeping them a safe distance from the firebox at all times.**

## Summary

- This chapter reveals the many details that must be addressed when building masonry fireplaces. In addition to possessing skills necessary to construct masonry walls, constructing masonry fireplaces requires knowledge of the technical design details and building code regulations governing their construction.

- The different fireplace sizes and designs provide options for most any room area or architectural styling.

- Only competent, experienced masons who comply with current building code regulations should construct fireplaces. Masons constructing fireplaces should be licensed contractors with completed operations insurance coverage. This insures the contractor in the event of future damages or injuries sustained as a result of completed work.

- An apprentice mason should not attempt to build a fireplace without the supervision of a competent mason who is knowledgeable of fireplace construction.

## Procedures

# Laying Out the Sidewalls and Back Wall for a 36" Single-Face Fireplace

**A** On the inner hearth, draw a line 20" from and parallel with the front edge of the inner hearth to represent the alignment for the face of the back wall of the firebox.

**A**
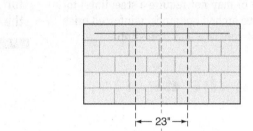

**B** Locate points 23" apart on the line drawn in Procedure A by measuring 11½" from each side of the inner hearth's centerline. These two points represent the intersections of the face sides of the firebox sidewalls with the back wall.

**B**
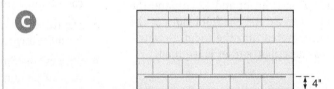

**C** So that the front side of the face surround will be aligned with the front of the inner hearth, draw a line 4" from and parallel to the front edge of the inner hearth. The intersection of this line with the lines representing the sidewalls identifies the front of the firebox sidewalls.

**C**

**D** Locate points 36" apart by measuring 18" from each side of the inner hearth's centerline onto the line drawn in Procedure C, and mark these two points representing the width of the firebox at its front, completing the layout for the firebox sidewalls.

**D**
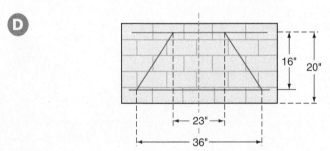

## Review Questions

*Select the most appropriate answer.*

**1** The minimum width for the walls of a fireplace base is

a. 4"
b. 8"
c. 12"
d. 16"

**2** The hearth base of a fireplace supports the

a. firebox inner hearth
b. outer hearth
c. adjoining structural framing
d. both a and b

**3** Combustion or burning is intended in that part of a fireplace called the

a. air intake
b. firebox
c. smoke chamber
d. throat

**4** A passageway for providing combustion air from outside is called the

a. air intake
b. firebox
c. smoke chamber
d. throat

**5** A fireplace throat

a. is the narrow passage between the combustion chamber and smoke chamber
b. extends the full width of the combustion chamber
c. is lined with rectangular or round flue linings
d. both a and b

**6** A smoke shelf

a. is formed by sloping the back walls of the firebox toward the front
b. can be flat or curved
c. is believed to help prevent downdrafts causing smoke to spill out into the room
d. all of the above

**7** A fireplace chimney

a. begins above the smoke chamber
b. should extend a minimum of 3' above the highest point where it penetrates the roofline
c. should have walls 12" or thicker
d. both a and b

**8** The masonry face of a fireplace is called the fireplace

a. frontage
b. hearth
c. surround
d. veneer

**9** The form supporting wet concrete placed for the inner hearth should

a. permit concrete that has a 4" minimum thickness
b. be an approved noncombustible material
c. provide for an air inlet and ash dump
d. all of the above

**10** To comply with the International Residential Code, and including a minimum 2"-thick firebrick lining, the total thickness of the back and sidewalls of the firebox should be

a. 6"
b. 8"
c. 10"
d. 12"

**11** The damper door plate on a throat damper opens

a. toward the back of the smoke chamber, opposite the fireplace opening
b. toward the front of the smoke chamber, toward the firebox opening
c. down into the firebox
d. straight up toward the chimney

12 **A fireplace smoke shelf should not be**

a. higher than the support flange of the throat damper
b. lower than the support flange of the throat damper
c. level with the support flange of the throat damper
d. included in the construction of a conventional fireplace

13 **Brick laid to construct the smoke chamber should be**

a. laid in the header position
b. corbelled no more than 30 degrees from the vertical
c. parged with refractory cement where exposed to the smoke chamber
d. all of the above

14 **The minimum space required between the outer walls of chimneys and combustible materials where chimneys are built within interior walls or through exterior walls is**

a. 1"
b. 1½"
c. 2"
d. 3"

15 **The minimum space required between the inside walls of flue linings and combustible materials is**

a. 4"
b. 6"
c. 8"
d. 10"

16 **Approved fireblocking should be**

a. an approved noncombustible material such as unfaced fiberglass insulation
b. supported by metal placed between the combustible materials and the chimney
c. a minimum of 1" thick
d. all of the above

17 **Measured from the vertical, the flue lining of a chimney should slope no more than**

a. 15 degrees
b. 20 degrees
c. 30 degrees
d. 45 degrees

18 **The minimum vertical distance between the head or top of a masonry fireplace opening and the fireplace throat is**

a. 4"
b. 6"
c. 8"
d. 12"

19 **The steel lintel at the head or top of a masonry fireplace face should**

a. have a 4" minimum bearing on the masonry to each side of the opening
b. have a compressible, noncombustible material at each end to accommodate thermal expansion of the steel
c. be ⅛" thick steel
d. both a and b

20 **Fireplace kits**

a. provide fireplace components, including preformed fireboxes, throat sections, and smoke chambers
b. simplify the construction of fireplaces
c. ensure correct component designs
d. all of the above

# Chapter 22 | Brick Masonry Arches

hey may be design elements credited to the empires of early civilizations, but masonry arches continue to charm people today. Their gracious curves capture the attention of onlookers as they become fascinated by the arch's mysterious capability to support whatever is above it. Although no longer considered as essential components in construction since the advent of engineered steel and wood structural components, brick masonry arches remain popular because they uniquely address both form and function. With a diversity of shapes, brick arches are able to complement a variety of styles, from European gothic to colonial American architecture.

## OBJECTIVES

After completing this chapter, the student should be able to:

- ⊗ Identify brick arches by their shapes.
- ⊗ Identify and define the parts of an arch.
- ⊗ Construct a semicircular brick arch.

# Glossary of Terms

**abutments** walls or piers supporting an arch.

**bonded arch** an arch consisting of two or more rows of voussoirs forming the depth.

**camber** the rise above the spring line for a jack arch.

**circular arch** also called a bull's eye arch, a 360-degree round opening framed with voussoirs.

**compression** the weight or forces above an arch that it must be designed to resist.

**creepers** the brick adjacent to the arch voussoirs.

**depth** the height of the brickwork forming the arch ring.

**extrados** the imaginary curved line at the top edge of an arch's voussoirs.

**gauged brick** a brick that has been tapered to be used as a voussoir.

**Gothic arch** an arch having a rise equal to or greater than its span formed by two overlapping circles of equal length radii coming to a point at its center.

**horseshoe arch** a circular arch greater than the 180-degree design of a semicircular arch but less than the 360-degree design forming a circular arch.

**intrados** the imaginary curved line at the bottom edge of an arch, the edge between the vertical face and horizontal soffit.

**jack arch** an arch design whose rise above the spring line is very little or nonexistent.

**keystone** the voussoir at the center of an arch.

**major arch** an arch having a span greater than 6' or rise-to-span ratio greater than 0.15.

**minor arch** an arch whose span does not exceed 6' with maximum rise-to-span ratios of 0.15.

**multicentered arch** also called an elliptical arch, an arch having more than one radius.

**rise** the measure of the height of an arch from its spring line, the imaginary line at which an arch begins to curve, to its center.

**segmental arch** an arch formed by one radius line less than a semicircle.

**semicircular arch** an arch appearing as half of a circle, formed by one radius line and 180 degrees of rotation.

**skewback** that part of an abutment supporting the voussoir at the end of an arch.

**soffit** the horizontal underside of an arch, perpendicular to the face of a wall, below an arched opening.

**span** the horizontal distance between the supporting abutments at each end of an arch opening.

**spring line** the imaginary line at which each end of an arch begins.

**triangular arch** an arched opening formed by two straight inclined sides.

**Tudor arch** an arch having a rise less than its span formed by two overlapping circles of equal length radii coming to a point at its center.

**unbonded arch** an arch consisting of a single row of voussoirs.

**Venetian arch** a semicircular arch flanked by a narrower horizontal line of brickwork on both sides.

**voussoir** the name given each masonry unit forming an arch ring.

**wood centering** a temporary arch form.

# Arch Terminology

*Arches* are both attractive and functional elements of brick masonry walls. Before structural steel or engineered wood products, brick arches were necessary for spanning openings and supported their imposing loads. Centuries-old arches seen surviving in many parts of the world today testify to their capability to endure both time and nature (see Figure 22-1). Arches are recognized by their shapes, spans, and depths.

## Shapes

There are several arch shapes. The more familiar shapes include the semicircular, segmental, multicentered, Gothic, Tudor, and jack. Additional shapes are the bull's eye or circular, horseshoe, triangular, and the Venetian.

## Spans

Depending upon the span, the horizontal distance between the supporting abutments at each end of an opening, arches are classified as either minor arches or major arches (see Figure 22-2).

The Brick Industry Association Technical Notes 31A defines a minor arch as one whose span does not exceed 6' and with maximum rise-to-span ratio of 0.15. Brick Industry Association Technical Notes 31A defines a major arch as having a span greater than 6' or rise-to-span ratio greater than 0.15.

## Depth

Voussoirs, the masonry units forming the arch ring, create arch depth. Arch depth is measured as the height of the voussoirs (see Figure 22-3).

Arch depth should increase as span increases. Brick Industry Association Technical Notes 31 recommends that both semicircular and segmental arches have a 1" minimum depth per foot of span or 4", whichever is greater. Technical Notes 31 also recommends that the depth for a jack arch be equal to or more than 4" plus 1" per foot of span or 8", whichever is greater.

Arch depth is classified either as bonded or unbonded. A bonded arch consists of two or more rows of voussoirs forming the depth (see Figure 22-4).

An unbonded arch consists of a single row of voussoirs (see Figure 22-5).

Figures 22-6 through 22-10 show examples of various types of arches. Arch depths and optional sizes of voussoirs permit a multitude of pattern designs.

**Figure 22-1 Evidence of the structural bearing strength created with an arch is seen here.**

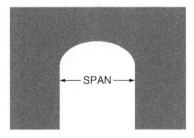

**Figure 22-2 Arches are classified according to their span, the distance between the sides of an arch.**

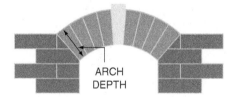

**Figure 22-3 Arch depth refers to the total height of the voussoirs forming the arch.**

**Figure 22-4 This arch is classified as a bonded arch.**

**Figure 22-5** This arch is classified as an unbonded arch.

**Figure 22-6** This bull's eye or circular arch is formed with three rings of voussoirs laid in the rowlock position. Four contrasting colored keystones complement the brick.

**Figure 22-7** Brick in the soldier position form an unbonded, semicircular arch surrounding an unbonded, jack arch above each window. A contrasting color keystone is seen on both semicircular arches.

**Figure 22-8** An arch with a depth of 8" is constructed with brick laid in the soldier position. A keystone is placed at the center.

**Figure 22-9** A brick and a half brick configuration permit the design of this bonded arch with a 12" depth.

**Figure 22-10** Having a 2½ brick configuration, this bonded jack arch has a 20" depth.

**Figure 22-11 Either a tapered voussoir brick (left) or a standard brick (right) are used to construct arches.**

## Brick Shapes for Arches

Brick arches are constructed using either standard rectangular-shape brick or tapered voussoir brick (see Figure 22-11).

Using standard rectangular-shape brick results in tapered head joints. Brick Industry Association Technical Notes 31 recommends "the thickness of mortar joints between arch brick should be a maximum of ¾" (19 mm) and a minimum of ⅛" (3 mm). When using mortar joints thinner than ¼" (6 mm), consideration should be given to the use of very uniform brick that meet the dimensional tolerance limits of ASTM C 216, Type FBX, or the use of gauged brickwork" (see Figure 22-12).

Tapered brick permit having uniform width head joints (see Figure 22-13). Tapered brick, also called **gauged brick,** are produced by brick manufacturers to accurate dimensions using a grinding process or made using a masonry saw at the job site. Tapered brick are recommended for small-radius arches, to limit the width variation of the mortar joint size between the top and bottom of adjacent voussoirs.

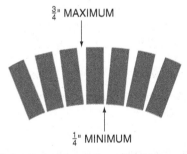

$\frac{3}{4}$" MAXIMUM

$\frac{1}{4}$" MINIMUM

**Figure 22-12 Typical mortar joints should be no wider than ¾" or narrower than ¼". ⅛" joints are allowed only when very uniform size brick are used.**

TAPERED BRICK PERMIT UNIFORM JOINT WIDTH

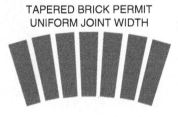

**Figure 22-13 Equal width mortar joints between tapered voussoirs create a more uniform appearance.**

**Figure 22-14 Three rings of brick laid in the rowlock position form this arch. An arch's structural strength and resistance to compression rely on full mortar joints. An arch must resist compression, which is the weight or forces above it. Compressive forces try to push an arch's voussoirs closer together. Without full mortar joints, the forces can move the voussoirs.**

Laying brick that are not tapered and in the soldier position is not recommended for a short radius because of the exceptionally wider joint width required between the brick nearer the top of the face of the arch rings. Laying the voussoirs in a rowlock position is recommended for an arch that has a short radius. Two or more rings of rowlocks are recommended (see Figure 22-14).

An arch must resist **compression,** the weight or forces above it. Compressive forces try to push an arch's voussoirs closer together. An arch's structural strength and resistance to compression rely on full mortar joints. Without full mortar joints, the forces can displace the voussoirs. It is recommended that open brick cores or open cells of hollow masonry units be filled solid with mortar.

## Keystones

An arch may have a uniquely sized voussoir at its center called the **keystone.** A keystone has more depth and is formed from other masonry materials such as stone or precast concrete, which makes it an attractive architectural detail (see Figure 22-15). As a rule of thumb, Brick Industry Association Technical Notes 31 recommends "that the keystone should not extend above adjacent arch brick by more than one third the arch depth."

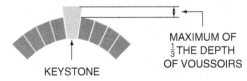

KEYSTONE

MAXIMUM OF $\frac{1}{3}$ THE DEPTH OF VOUSSOIRS

**Figure 22-15 The center voussoir is known as the keystone. Being a different color and larger makes it more prominent.**

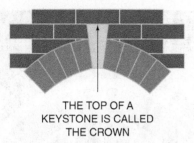

THE TOP OF A
KEYSTONE IS CALLED
THE CROWN

**Figure 22-16** The keystone crown is level with the brickwork.

According to Brick Industry Association Technical Notes 31, the keystone helps to prevent cracks from developing in mortar joints near the center of the arch. The mortar joint nearest the center of the span is the joint most likely to first crack. The Brick Industry Association concludes that a wider keystone positions the mortar joints farther from the center of the span and reduces the chances for cracked mortar joints.

Having the keystone crown or top no more than ⅓ the depth of the adjacent voussoirs and aligned level with the brick coursing for the wall facade is preferred since it simplifies aligning the keystone with course spacing (see Figure 22-16).

## Abutments

Walls or piers called **abutments** support an arch. The abutments must resist the lateral forces or horizontal thrust created by the arch. Exerting a wedging effect, lateral forces can result in inadequately designed abutments spreading apart, causing the supported arch to collapse.

The abutments must be designed to support the weight and thrust imposed by the arch. Mortared brick masonry walls resist lateral and compressive forces, transferring the weight of the arch and wall above it to each side of the opening (see Figure 22-17).

The amount of lateral force that is transferred to the abutments depends partly on the shape of the arch. Semi-circular arches transfer less lateral force than segmental arches having a low rise. The **rise** of an arch is a measure of its height from the **spring line**, the imaginary line at

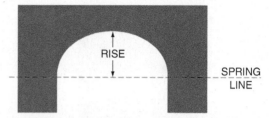

**Figure 22-18** The rise of an arch is measured from the center of the spring line to the bottom of the voussoirs at its center.

which an arch begins, to the bottom of the voussoirs at its center (see Figure 22-18).

In addition to factors such as arch depth, rise, and span, design elements such as expansion joints, type of mortar, and flashing determine the structural resistance to collapsing for an arch. Such factors require brick arches and their supporting abutments to be designed by qualified structural engineers. The structural integrity of an arch can be diminished by improperly placed expansion joints. Brick Industry Association Technical Notes 31 recommends that a "structural analysis of the arch should consider the location of expansion joints."

Each end of an arch begins at its skewback. Part of an abutment, a **skewback** is that part of an abutment supporting the voussoir at the end of an arch. For segmental arches, the skewbacks are sloping surfaces (see Figure 22-19).

The skewbacks are flat, horizontal surfaces for semicircular, multicentered, Gothic, and Tudor arches (see Figure 22-20).

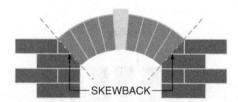

**Figure 22-19** Brick cut at an angle form the two skewbacks supporting the arch.

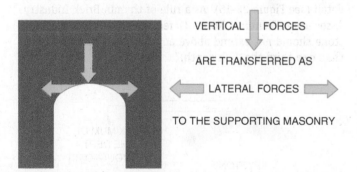

VERTICAL FORCES
ARE TRANSFERRED AS
LATERAL FORCES
TO THE SUPPORTING MASONRY

**Figure 22-17** The arch permits the weight above the opening to be transferred to the abutments.

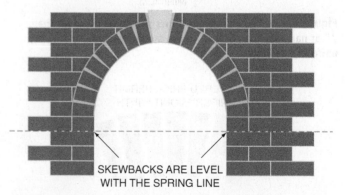

SKEWBACKS ARE LEVEL
WITH THE SPRING LINE

**Figure 22-20** The skewbacks are aligned level with the spring line and the brick coursing in the abutments.

# Constructing a Semicircular Arch

A semicircular arch is an arch appearing as half of a circle, formed by one radius line and 180 degrees of rotation. The considerations for constructing adequately supported arches require addressing several issues. A temporary form must be made to support the brickwork above the arched opening until the mortar cures, permitting it to be self-supporting. Adequately sized abutments must be constructed in accordance to engineered designs to support the weight transferred from above the arched opening. The inclusion of expansion joints should be at the discretion of a structural engineer. Both the pattern bond for the wall above the opening and the voussoirs creating the arch design must be laid out to ensure uniform joints and proper spacing of brick. Flashing and weeps may be required in the brickwork above the arch. Considerations for tooling joints between voussoirs and the arch form after the temporary form is removed must be addressed. Addressing each of these concerns helps ensure success when constructing arches of any type.

## Building an Arch Form

A semicircular arch is constructed above a temporary form or permanent structural steel arch lintel. A temporary arch form known as wood centering can be constructed from ¾" plywood and wood spacers (see Figure 22-21).

Two identical pieces of ¾" plywood are cut to the curvature specified for the arch. Two semicircles whose diameter is equivalent to the arch span are cut from ¾" plywood using a jigsaw. A radius line equivalent to half the arch span is used to mark the semicircle form (see Figure 22-22).

**Figure 22-21** A plywood form supports voussoirs laid in the rowlock position.

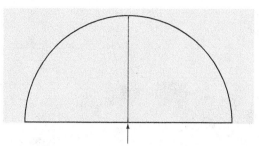

A RADIUS LINE REPRESENTING ONE-HALF THE ARCH SPAN IS USED TO DRAW A LINE ONTO ¾-INCH PLYWOOD.

**Figure 22-22** The red line represents the radius of the circle for which the form is designed.

**CAUTION**

Only those trained by a competent person should operate a jigsaw. Wear eye protection when using a jigsaw. Comply with the tool manufacturer's safety precautions.

To provide proper spacing between the two pieces of plywood, use 2 × 4 or 2 × 6 framing lumber cut 2" less than the width of the soffit, which is the horizontal underside of an arch, perpendicular to the face of a wall. The assembled form measures ½" less than the soffit width of the arch. This permits a ¼" extension of the brick beyond the face of the form on each side, providing clearance between the form and the mason's line. The two semicircular pieces of plywood must be aligned square with one another. Screws secure the framing lumber spacers to the plywood.

Arches having arch rings in the rowlock position and with a soffit width of a single brick require no additional support between the two pieces of plywood because the edges of the plywood support the ends of each brick. However, arches with more than the length of one brick forming the width of the soffit and arch rings laid in a soldier position necessitate applying a continuous solid covering atop the arch form that is capable of supporting the weight of the voussoirs. Fastening sheet metal or thin plywood to the semicircular form provides full support for the voussoirs. Figure 22-23 illustrates two types of arch forms.

**CAUTION**

Wear gloves to protect the hands from abrasions and cuts, and wear eye protection when handling and cutting sheet metal.

**Figure 22-23** The open-top form to the left is intended to support an arch whose soffit width is equivalent to the length of a single brick. The covered-top form to the right permits supporting a soffit width of more than one brick length.

## Laying Out the Voussoirs

The voussoirs are dry bonded along the arch form. Using standard brick rather than tapered voussoirs creates tapered mortar head joints. Head joints nearest the intrados, the imaginary curved line at the bottom edge of an arch, should be a minimum of ¼". Avoid having joints smaller than ¼" unless very uniform brick are used. Joints should be no larger than ¾" at the extrados, the imaginary curved line at the top edge of an arch's voussoirs. Marking the brick spacing onto the plywood form as determined by dry bonding assures uniform joint widths (see Figure 22-24). A keystone should be in the center of the arch.

**FROM EXPERIENCE**

Turning the form on its side and laying it on a flat, level surface enables dry bonding the voussoirs for the curvature of the arch form. When standard rectangular-shape brick are intended to be laid in the rowlock position, each brick is stood on end on the flat level surface upon which the arch form is laying on its side. The brick are spaced around the circumference of the semicircle to determine equal spacing between them. Marks are made on the face of the plywood form indicating spacing for the brick to be laid as voussoirs.

## Setting the Form

The arch form is supported using 2" framing lumber placed against each abutment and diagonally braced. The form is positioned with the radius line aligned level with the spring line. Raising the radius line of the arch form 1 to 2" above the spring line emphasizes the circular appearance for

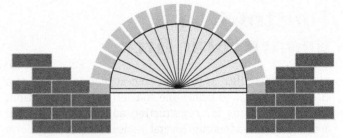

**Figure 22-24** Marking lines on the form provides equal spacing when laying the brick voussoirs.

**Figure 22-25** Wedges and braced props assure accurate leveling and positioning of the arch form.

a semicircular arch. Wooden wedges between the form and supports permit adjusting the form perfectly level (see Figure 22-25).

## Constructing the Abutments

The abutments are built above the height of the voussoirs. This procedure permits using a mason's line or straight edge to align the voussoirs with the rest of the wall. Aligning the skewbacks level with the brick coursing, no slanting angle needs cutting (see Figure 22-26).

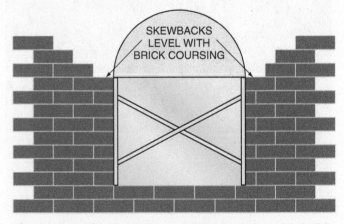

SKEWBACKS
LEVEL WITH
BRICK COURSING

**Figure 22-26** Skewbacks for a semicircular arch are level with the brick coursing.

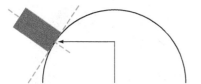

THE STRAIGHT LINE OF THE BRICK
TOUCHES THE CURVE AT ITS CENTER.
THE INTERSECTION OF THE TWO
BROKEN LINES (ARROW) IS THE
TANGENT POINT FOR THE ARCH AND
THE BRICK.

**Figure 22-27** **The brick first appears to rest flat against the arch form, when actually only the middle of the brick is touching as it is aligned tangent to the form.**

AVOID HAVING ONE EDGE OF THE
BRICK AGAINST THE FORM WHILE THE
OPPOSITE EDGE IS ABOVE THE FORM.

**Figure 22-28** **A poorly positioned voussoir results in an arch having an irregular curvature.**

## Aligning the Voussoirs

Beginning at the skewbacks, voussoirs are positioned against the arch form. For smaller radii, it may appear that only the center of each voussoir touches the form (see Figure 22-27). This condition indicates accurate alignment of the voussoir with the form, a result of having the voussoir tangent to the arch form.

Voussoirs not aligned tangent to the arch form, as illustrated in Figure 22-28, result in an arch having an irregular, undesirable curvature.

For voussoirs laid in the soldier position, using a radius line attached to the center of the arch form aligns each voussoir's edge at its proper angle (see Figure 22-29).

The center voussoir or keystone completes the arch. Mortar joints must be fully mortared to enable the arch to resist the forces and weight above it.

The temporary arch form prohibits tooling the mortar joints of the underside of the arch. Placing wooden dowel rods or ropes on the form and between each brick eliminates mortar at the bottom of the joints (see Figure 22-30). This procedure eliminates the need to cut out hardened, untooled mortar between voussoirs along the soffit once the form is removed. Once the form is detached, the wooden dowels or ropes are removed. The joints are filled with mortar and tooled. Also, this procedure helps prevent the undesirable results of trapped mortar protrusions staining brick and misaligning the brick tangent to the arch form.

**Figure 22-29** **Using a mason's line attached at the center of the circle as a radius line helps position each voussoir at its proper angle. In this illustration, the spring line is located below the center of the circle, highlighting the circular appearance of the arch.**

**Figure 22-30** **Ropes eliminate mortar at the bottom of the joints.**

## Laying Brick Outside of the Arch Ring

As voussoirs are laid, the brick wall facade is laid to the outside of the arch ring, aligned with the mason's line one course at a time as when building other walls. The leads, which may be the abutments, are laid before laying the voussoirs, permitting accurate alignment of both the voussoirs and the adjoining wall. Referred to as creepers, each brick adjacent to the arch voussoirs must be cut at an angle. The angle for cutting each creeper changes along the arch ring. The angle at which each creeper is to be cut can be determined by measuring the length required for each at its top and bottom edge of the face. A line connecting these two measurements forms the angle for the creeper. Another procedure is to use a t-bevel to determine the angle for a creeper.

*For step-by-step instructions on cutting arch creepers, see the Procedures section on page 330.*

## Removing the Arch Form

Brick Industry Association Technical Notes 31 recommends leaving a temporary arch form for a minimum of seven days after constructing the arch. Technical Notes 31 also advises that "longer curing periods may be required when the arch is constructed in cold weather conditions and when required for structural reasons." Disassembling the diagonal bracing and wooden supports, the form is removed below the arch. The dowel rods or ropes placed between the joints can now be removed. After the joints are cleaned of any debris or mortar and the voussoirs are dampened with water, the joints left void by the dowel rods or ropes are filled with mortar and tooled.

**FROM EXPERIENCE**

Using a grout bag or grout gun to fill the mortar joints requires less time and effort than using a trowel or a tuckpointer to fill the joints. Only those trained by a competent person should operate a grout gun. Inline ground fault circuit protection must be provided to protect operators of electric-powered grout guns from electric shock and electrocution.

## Structural Steel Arch Lintels

Placing dowel rods or sash ropes at the bottoms of joints is not necessary for an arch permanently supported by a structural steel arch lintel. The finished exposure of the soffit becomes the steel lintel rather than the brick, eliminating the need for tooling the bottoms of the joints (see Figure 22-31). Brick Industry Association Technical Notes 31 states that "if the arch is supported by a lintel, any arch depth may be used."

## Installing Flashing and Weeps

Flashing and weeps are needed above steel arch lintels and above all arches where water intrusion may be a potential problem. Flashing should be installed in the first bed joint above the keystone. Brick Industry Association Technical Notes 31 states that for arch spans less than about 3', "flashing should extend a minimum of 4" (100 mm) past the wall opening at either end and should be turned up to form end dams. This is often termed tray flashing. Weep holes should be provided at both ends of the flashing and should be placed at a maximum spacing of 24" (600 mm) on centers along the arch span, or 16" (400 mm) if rope wicks are used" (see Figure 22-32).

Step flashing can be installed in the bed joints above longer arches. Except for the top center flashing, end dams are formed only at the end of the stepped flashing nearest the arch center (see Figure 22-33).

Because flashing acts as a bond break, it affects the structural strength of the arch. For this reason, a structural engineer should give the specifications and design of an arch needing flashing.

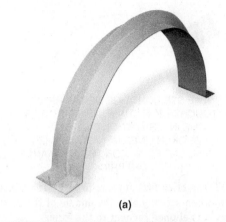
(a)

(b)

**Figure 22-31** The steel arch lintel (a), becomes a permanent structural member, serving as the exposed soffit (b). *(Courtesy of Powers Steel and Wire)*

FOLDING END OF FLASHING
FORMS A DAM

**Figure 22-32** Flashing installed in the first bed joint above the keystone prevents water migration.

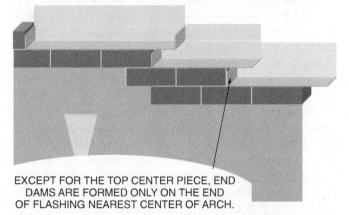

EXCEPT FOR THE TOP CENTER PIECE, END DAMS ARE FORMED ONLY ON THE END OF FLASHING NEAREST CENTER OF ARCH.

**Figure 22-33** Step flashing collects water, and weeps divert it out of the wall.

# Laying Out Other Arch Shapes

Segmental, multicenter, Gothic, and Tudor arches are constructed similar to a semicircular arch. A temporary form is positioned level with the springing line unless using a permanent structural steel arch lintel.

## Segmental Arches

The radius for a segmental arch is uniform but the arch ring is less than a semicircle (see Figure 22-34).

## Multicentered Arches

Using two seperate radii permits constructing a multicentered arch, also called an elliptical arch. The arch is laid out using the radius of a larger circle for the middle and that of a smaller circle at each end of the arch ring (see Figure 22-35).

## Gothic Arches

Having a rise equal to or greater than its span, a Gothic arch comes to a point at its center. Two overlapping circles of equal length radii form a Gothic arch (see Figure 22-36).

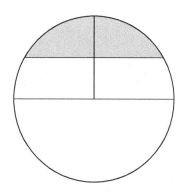

**Figure 22-34** The span of a segmental arch is less than the diameter of the circle forming its arc.

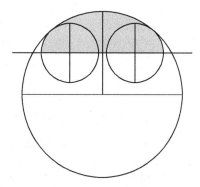

**Figure 22-35** Represented by the shaded area, a multicentered or elliptical arch is constructed from two different radii.

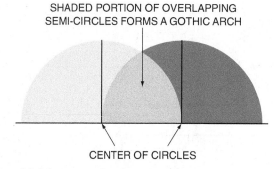

SHADED PORTION OF OVERLAPPING
SEMI-CIRCLES FORMS A GOTHIC ARCH

CENTER OF CIRCLES

**Figure 22-36** Two overlapping identical circles, indicated by the center shaded portion, form a Gothic arch.

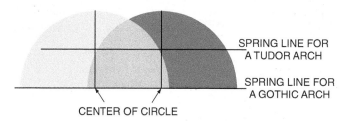

SPRING LINE FOR A TUDOR ARCH

SPRING LINE FOR A GOTHIC ARCH

CENTER OF CIRCLE

**Figure 22-37** Drawing two overlapping, identical circles forms both a Tudor and a Gothic arch. The span for the Tudor arch is greater than its rise.

## Tudor Arches

Like a Gothic arch, a Tudor arch comes to a point at its center. Unlike a Gothic, a Tudor arch has a greater span than rise (see Figure 22-37).

The shape for a Gothic or a Tudor arch can be altered by changing any of the following:

- The diameter of the two identical circles
- The location of the spring line
- The distance between the centers of the overlapping circles

## Jack Arches

Unlike any other arch, the intrados of a jack arch appears to be horizontal, nearly aligned level with brick coursing. Its rise above the spring line, known as **camber**, is very little or nonexistent. Collectively forming a wedge, the voussoirs support weight above an opening (see Figure 22-38).

Using a radius line attached to the center's opening below the spring line, each voussoir is positioned at its proper angle (see Figure 22-39).

The distance between the spring line and the centerpoint to which the string is attached determines the skewback angle and the angle for positioning each voussoir.

## Circular Arches

A circular arch or bull's eye arch is a 360-degree round opening bordered with voussoirs. A circular arch requires cutting and laying the brick below the horizontal centerline before laying the voussoirs (see Figure 22-40).

Figure 22-38 Resisting forces of compression, the voussoirs act collectively as a wedge supporting the weight above the jack arch.

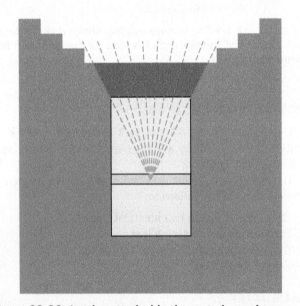

Figure 22-39 A string attached in the center's opening properly aligns each voussoir.

Figure 22-40 Brick below the circular arch are cut before laying the voussoirs.

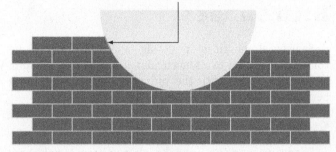

PLYWOOD TEMPLATE IS USED TO MARK BRICK FOR PROPER CUTTING

Figure 22-41 A plywood template assures accurate cutting for each brick.

Figure 22-42 Starting at the center, voussoirs are laid for the bottom half of the circle.

A pattern made from ¼" plywood can be used for marking the line of cut onto the brick (see Figure 22-41). This procedure requires first bedding the brick in mortar, marking it according to the pattern, removing it for cutting, and finally bedding the brick permanently in fresh mortar.

Beginning at the bottom, the voussoirs below the horizontal center line are laid (see Figure 22-42).

An arch form is required before laying the upper half of the circle arch. The upper half of a circular arch can be constructed by building an arch form in the same manner as for a semicircular arch and having the spring line the same as the centerline. The same procedures used for constructing a semicircular arch complete the upper half of the circular arch.

An attractive garden wall design is constructed using the techniques for constructing the lower half of the circular arch (see Figure 22-43).

## Horseshoe Arches

A horseshoe arch is a circular-shaped arch greater than the 180-degree design of a semicircular arch but less than 360-degree design, forming a circular arch (see Figure 22-44).

Techniques and procedures used for both the semicircular arch and the full circular arch are used to construct a horseshoe arch.

Figure 22-43  Construction techniques used for a circular arch are used to build this garden wall.

Figure 22-44  The horseshoe arch forms more of a circular pattern than a semicircular arch but less than a complete circle.

## Triangular Arches

Two straight inclined sides (see Figure 22-45) form a triangular arch.

## Venetian Arches

A Venetian arch is a semicircular arch flanked by a narrower horizontal span of brickwork on each side.

Figure 22-45  Similar to a Gothic arch, a triangular arch comes to a point at its center.

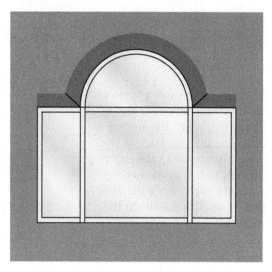

Figure 22-46  A Venetian arch supports brickwork above this Palladian window.

A Venetian arch can be found above a Palladian window. A design attributed to the sixteenth century Italian architect Andrea Palladio, the opening exhibits a semicircular top flanked on both sides by narrower, square-top openings (see Figure 22-46).

Many brick manufacturers design and produce special shapes for all types of arches, both bonded and unbonded, and with or without a center keystone. Using architectural drawings, brick manufacturers can make any shape voussoir for an arch. This improves production by eliminating cutting the brick at the job site.

## Summary

- Having engineered structural steel and wood supports, brick arches are no longer needed as structural support. However, their graceful curvatures account for their continuing popularity as architectural design elements.
- Arches are used today to help define the intended architectural stylings for buildings.
- Many considerations, including shapes, spans, depths, specially shaped units, abutments, and formwork, must be addressed when building masonry arches.
- An apprentice mason should have the supervision of an experienced mason when constructing brick masonry arches.

## Procedures

# Cutting Arch Creepers

**A** With the mason's line aligned with the course for which the creeper is required, adjust the angle of a t-bevel so that the blade of the t-bevel is tangent to the arch ring adjacent to the position of the required creeper, while the body of the t-bevel is parallel to the mason's line.

**A**

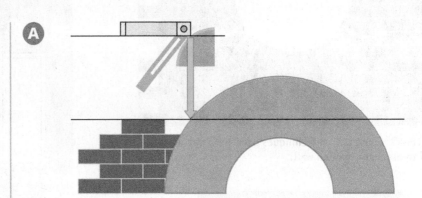

**B** Measure the length between the brick adjacent to the creeper and the arch ring. Subtract from this measurement the intended width of the two joints at either end of the creeper.

**B** MEASURE THE LENGTH BETWEEN BRICK ADJACENT TO CREEPER AND ARCH RING AT HEIGHT OF THE MASON'S LINE.

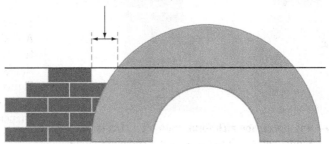

**C** Transfer the angle for which the t-bevel is adjusted to the face of the intended creeper, aligning the predetermined angle with a line indicating the length for the creeper at the top of its face.

**C**

**D** Cut the creeper and bed in mortar, aligning it to the mason's line.

**D**

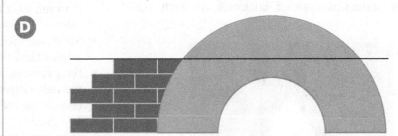

### CAUTION

**Wear eye, face, hearing, and respiratory protection when using masonry saws. Only those trained by a competent person should operate masonry saws.**

## Review Questions

*Select the most appropriate answer.*

**1** The span of a minor arch is

a. 2' or less

b. 4' or less

c. 6' or less

d. 8' or less

**2** Arch depth for semicircular and segmental arches should be

a. no less than 4"

b. no less than 1" per foot of span or 4", whichever is greater

c. constructed using tapered brick

d. both a and b

**3** Each masonry unit forming an arch ring is called a

a. creeper

b. keystone

c. skewback

d. voussoir

**4** The Brick Industry Association recommends that tapered head joints resulting from using standard brick around an arch ring be

a. No smaller than ⅛" or larger than ½"

b. No smaller than ¼" or larger than ½"

c. No smaller than ⅛" or larger than ¾"

d. No smaller than ¼" or larger than 1"

**5** Unless tapered brick are used for arches having a short radius, it is recommended to use

a. a single brick arch ring of brick laid in the rowlock position

b. two or more arch rings of brick laid in the rowlock position

c. an arch ring created with brick laid in the header position

d. an arch ring created with brick laid in the soldier position

**6** The masonry unit at the center of an arch ring is called a

a. cap

b. creeper

c. keystone

d. skewback

**7** An arch is supported at each end by that part of the abutment called the

a. cap

b. creeper

c. keystone

d. skewback

**8** The imaginary horizontal line at which an arch begins is called the

a. abutment

b. extrados

c. rise

d. spring line

**9** The height from the imaginary horizontal line at which an arch begins to the bottom edge of the highest point at its center is called the

a. depth

b. rise

c. run

d. span

**10** Each brick that requires cutting adjacent to and outside of the arch ring is called a

a. creeper
b. header
c. soffit
d. voussoir

**11** The minimum number of days recommended to have a temporary arch form remain is

a. 3
b. 5
c. 7
d. 14

**12** The horizontal underside of an arch, perpendicular to the face of a wall below an arched opening, is called the

a. facia
b. parapet
c. sill
d. soffit

**13** A type of arch whose bottom edge appears horizontal or flat is called a

a. jack arch
b. Gothic arch
c. Tudor arch
d. Venetian arch

**14** The type of forces above an arch for which it must be designed to resist is

a. compression
b. expansion
c. lateral
d. live load

# Chapter 23

# Cleaning Brick and Concrete Masonry

The pride in masonry workmanship is first evidenced in the care taken as walls are being constructed. Mortar stained and dirty walls distract from the appearances of even those walls exhibiting the best masonry skills and wall designs. Several sources contribute to soiled or stained masonry walls. Masonry units become soiled before installing. Mortar droppings splash surfaces, workers accidentally spatter mortar, winds impart mortar dust on wall facades, and rainwater splashes bare soil against walls at ground level. Preventing these occurrences minimizes cleaning time and expenses and contributes to a favorable reputation, for which a masonry contractor desires to be recognized. Still, some cleaning is most always necessary. Using approved cleaning agents and following recommended procedures returns masonry surfaces to their intended appearances without damaging the masonry units or mortar.

## OBJECTIVES

After completing this chapter, the student should be able to:

- Identify sources for construction dirt and mortar soiling masonry.
- List measures for preventing dirt-stained and mortar-stained masonry.
- Describe the different brick cleaning methods.
- Describe methods for cleaning concrete masonry units.

# Glossary of Terms

**bleeding** the result of excessive brick moisture content or mortar having too much water content, causing mortar stains on the face of walls as wet mortar gravitates from joints onto the face of walls.

**efflorescence** a deposit of soluble salts appearing on masonry walls.

**muriatic acid** also called hydrochloric acid, a water-based solution of hydrogen chloride gas recommended as a cleaning agent for removing mortar stains from only a few types of brick.

**propriety compounds** chemicals containing organic and inorganic acids, wetting agents, and inhibitors recommended for removing a variety of stains from brick, especially those susceptible to metallic oxidation staining.

**trisodium phosphate** a strong cleaner once found in household laundry and dishwashing detergents that is used to clean brick.

**white scum** white or gray stains on the face of brick masonry, typically related to the cleaning of brickwork with unbuffered hydrochloric (muriatic) acid solutions or inadequate prewetting or rinsing of the brickwork during cleaning.

# Preventing Soiled and Stained Walls Under Construction

Precautionary measures taken during construction minimize cleaning procedures for newly constructed brick and CMU walls. The lines of defense protecting materials and walls from soiling include the following:

- Protecting material stockpiles
- Exercising care when bedding brick and block in mortar
- Training workers for handling mortar
- Taking precautionary measures with scaffold planks
- Protecting walls at ground level
- Covering walls under construction

## Protecting Material Stockpiles

Masonry materials should be protected from mortar and ground soil. Elevating materials onto wooden pallets prevents stains caused by ground soils. Soils that have higher iron content cause difficulty in removing stains. Orange ground soil stains alter the units' colors before installation. Rain and other water sources dissolve ground minerals such as salts. Masonry units absorbing salt-laden ground water contribute to the white, chalky-appearing efflorescence stains.

Plastic coverings prevent material stockpiles from becoming soiled. Coverings also prevent rain saturation of masonry units. Laying masonry units too wet causes mortar bleeding. Mortar **bleeding** is the result of excessive brick moisture content or mortar having too much water content, causing mortar stains on the face of walls as wet mortar gravitates from joints onto the face of a wall. Keeping stockpiles at a safe distance from construction activities that create airborne debris prevents the soiling of brick and block. Figure 23-1 depicts properly stored materials.

**Figure 23-2** Elevated scaffold planking prevents materials from contacting the ground.

Elevating brick onto scaffold planking protects them from ground dirt and water-soluble ground salts (see Figure 23-2). However, the planking must remain relatively free of mortar droppings and mortar dust to prevent soiling the stockpiled materials.

## Exercising Care When Bedding Brick and Block in Mortar

Masons must exercise care with mortar. Improperly applied mortar joints result in mortar-stained brick and block. Also, removing mortar protrusions as units are bedded requires only those trowel techniques proven not to create mortar stains. Mortar protrusions that are permitted to fall from the face of the brick rather than retained on the trowel blade strike scaffold planking, scaffold bracing, and eventually the ground, splashing walls with mortar (see Figure 23-3).

**Figure 23-1** Brick are elevated above the ground and covered.

**Figure 23-3** Evidence of mortar droppings hitting scaffold bracing is seen here.

## Training Workers for Handling Mortar

Another cause of soiled walls is mortar carelessly placed onto mortarboards or into mortar pans. Having mortar splashed on walls and materials stockpiled beside the working area can be prevented if workers use care when shoveling mortar onto mortarboards or in mortar pans. Adding water to temper mortar is another source of mortar spatter stains onto walls. Adding small amounts of water at a time as mortar is tempered prevents splashing both the wall and materials stockpiled beside the mortarboard or pan.

## Taking Precautionary Measures with Scaffold Planks

Scaffold planking on which masons stand accumulates mortar droppings and other debris. When workers are not on the scaffolds, such as at the end of the work day, removing the scaffold plank nearest the wall prevents splashing rains from staining walls with mortar (see Figure 23-4).

**CAUTION**

Workers are prohibited from using scaffolds where scaffold planks are removed. Comply with OSHA regulations and have scaffolds fully planked before working from them.

## Protecting Walls at Ground Level

Water-soluble ground minerals are another source of wall stains. Splashing rains dissolve and deposit ground minerals onto wall surfaces. The resulting stains can be very difficult

Figure 23-5 **Placing straw on the ground prevents the wall from being splashed with soil during rains and causing mineral stains on the brick.**

to remove. Covering the ground with straw adjacent to walls protects them from rain-splashed ground minerals (see Figure 23-5).

## Covering Walls Under Construction

Periods of rain can wash out mortar joints before the mortar hardens. Not only does covering new work prevent mortar joint erosions caused by rainstorms, but it also eliminates mortar stains on walls as a result of washed-out joints (see Figure 23-6). Even for walls where the mortar is sufficiently cured, rainwater can saturate the tops of brick walls under construction, causing mortar to bleed and stain walls if construction resumes before the brick become dry. Newly constructed masonry walls should be covered with waterproof tarps at the end of each day and before rain begins.

Figure 23-4 **At the end of the day, moving scaffold planks nearest the wall prevents splashing rains from staining walls with mortar, as seen here.**

Figure 23-6 **A plastic tarp prevents the possibility of rainwater soaking the top of the unfinished wall.**

**Figure 23-7** Under certain conditions, using some cleaning agents to clean brick can result in efflorescence, soluble salts appearing as white deposits on the face of a wall.

# Classifications of Cleaning Agents

The three classifications of chemical cleaning agents are acids, propriety compounds, and detergents. Only those cleaning agents recommended by the brick producer and mortar manufacturer should be considered for use. Some cleaning agents can stain or discolor certain brick and colored mortars. Regardless of the cleaning method or cleaning agent selected, testing of the performance and results on a less noticeable section should be performed before widespread use. The probability for efflorescence, a deposit of soluble salts, to appear on masonry walls later increases when cleaning during freezing weather, in hot weather, or in extremely windy areas (see Figure 23-7). Efflorescence appears similar to white scum, white or gray silicate stains on the face of brick masonry. The Brick Industry Association Technical Notes 20 reports that white scum is typically caused by "failure to thoroughly saturate the brick masonry surface with water before and after application of chemical or detergent cleaning solutions."

Cleaning is not recommended when temperatures are below or expected to be below freezing. Cleaning in hot weather or when windy requires wetting the wall more frequently and cleaning smaller areas at a time to prevent rapid evaporation of the prewetting water and the cleaning solution.

## Muriatic Acid

Hydrochloric acid, also called muriatic acid, is a water-based solution of hydrogen chloride gas. Typically available in plastic 1-gallon containers at building supply centers, it is diluted in clean drinking water as recommended by the brick manufacturer. Typically it is initially diluted at a rate of 1 part acid to 20 parts water. More acid may be added if removing difficult mortar stains requires it. However, never use acid concentrations greater than recommended by brick producers or cement manufacturers. It should only be used to clean brickwork for which the brick producer and mortar manufacturer lists it as being suitable. Muriatic acid should never be used on masonry surfaces susceptible to acid staining. Muriatic acid may not be recommended by brick producers as the first choice for cleaning brick masonry walls. In addition, muriatic acid and its concentrated fumes should be considered extremely hazardous, potentially harmful to a person's skin, body organs, and the respiratory system. The environment, nearby building materials, and other objects can also be subject to damaging effects of acid exposure.

### ⚠ CAUTION

Only those trained by a qualified person should be exposed to, use, or dispose of muriatic acid. The fumes of muriatic acid can damage body tissue and the respiratory system. Wear a respirator that has an acid-approved filter meeting NIOSH standards. Because the wearing of a respirator may have harmful effects for the health of some people, one should have a medical evaluation to determine their ability to use a respirator. Wear approved acid-resistant clothing, boots, and gloves. Wear approved acid-resistant eye, face, and head protection. Post warning signs and keep those not protected as described here from accessing the work zone. Do not expose pets and animals to muriatic acid. Protect all surfaces, especially metal, from acid and its fumes. Use only in well-ventilated outdoor areas. Thoroughly wet with clean drinking water all plants and ground cover that may come into contact with the acid. Muriatic acid is considered to be a hazardous chemical. Never mix muriatic acid with anything other than clean drinking water. Seek medical care immediately if ingested and upon breathing concentrated fumes. Comply with manufacturers printed instructions, OSHA standards, environmental regulations, and all governing regulations for the use, storage, and disposal of acid.

Wearing personal protection equipment meeting the intent of OSHA standards, and with those in the area and everything else properly protected from exposure, the acid is carefully poured from the jug into a rubber, impact-resistant and acid-resistant bucket already containing sufficient water to dilute the acid as recommended for using as a cleaning solution. Having the acid immediately diluted on contact with the clean water minimizes the risks for the harmful effects of concentrated acid fumes.

**FROM EXPERIENCE**

Use impact-resistant rubber pails for holding solutions of acid and water. Accidental impacts and sharp objects can crack plastic pails, spilling the acid uncontrollably onto surfaces below it. Never use metal pails or buckets. Chemical reactions between the metal and the acid can corrode the metal and leave permanent stains on masonry walls. Rinse metal scaffold components frequently to prevent the corrosive effects of the acid solution.

## Propriety Compounds

**Propriety compounds** are chemicals containing organic and inorganic acids, wetting agents, and inhibitors for brick susceptible to metallic oxidation staining. For a particular brick, brick companies recommend using specific propriety compounds for cleaning newly constructed walls. The acids loosen the cement, wetting agents slow drying time by helping the acids cling to the surface, and inhibitors reduce metallic staining resulting form chemical reactions between the acids and the elements of the brick's composition. Diverse manufactured propriety compounds are available for cleaning a variety of stains from brick and concrete surfaces.

**CAUTION**

**Only those trained by a qualified person should be exposed to, use, or dispose of propriety compounds. Wear a respirator that has an acid-approved filter meeting NIOSH standards. Because the wearing of a respirator may have harmful effects for the health of some people, one should have a medical evaluation to determine their ability to use a respirator. Wear approved acid-resistant clothing, boots, and gloves. Wear approved acid-resistant eye, face, and head protection. Post warning signs and keep those not protected as described here from accessing the work zone. Protect all surfaces, especially metal, from chemicals and their fumes. Use only in well-ventilated outdoor areas. Thoroughly wet with clean water all plants and ground cover that may come into contact with the acid. Comply with all governing regulations for the use, storage, and disposal of chemicals. Comply with manufacturers printed instructions, OSHA standards, environmental regulations, and all governing regulations for the use, storage, and disposal of propriety compounds. Seek medical care immediately if ingested and upon breathing concentrated fumes. Do not expose pets and animals to propriety compounds.**

## Household Detergents

A solution of household detergent and clean drinking water can be used to remove some mud and soil stains. A solution of **trisodium phosphate**, a strong cleaner once found in household laundry and dishwashing detergents, and clean water also removes some mud and soil stains. Both the producers of the cement and the producers of the brick should be consulted when selecting cleaning agents. Never use products other than those recommended by the brick and cement producers. Use in accordance with manufacturer's printed instructions, OSHA standards, environmental regulations, and all governing regulations.

**FROM EXPERIENCE**

Always pour chemical cleaning agents into buckets already containing ample water to dilute the chemical as it is poured from the container. This procedure prevents workers from being exposed to potentially harmful fumes and vapors rising from the concentrate in otherwise open pails.

**CAUTION**

**Material Safety Data Sheets for each product are to be readily available at all sites where a product is stored, used, or disposed. Persons handling, storing, using, being exposed to, or disposing of chemical cleaners must be properly trained, clothed, and protected. Observe OSHA standards, Environmental Protection Agency regulations, all other governing regulations, and manufacturer's printed recommendations.**

## Cleaning Brickwork

Brick should not be cleaned before the mortar cures for the minimum time recommended by the cement manufacturer. Allowing the mortar to harden for seven days is a recommendation given in the Brick Industry Association Technical Notes 20. However, avoiding prolonged times between construction and the cleaning of the masonry is advised, because stains become more difficult to remove with the passage of time. Likewise, the cement and brick producers should be consulted for recommended cleaning methods, cleaning agents, and cleaning procedures. Chemical cleaners may affect colored mortars. Brick manufacturers identify specific categories of brick and recommended cleaning procedures for each category based on the composition of the brick. Cleaning agents should be selected according to the brick category to prevent chemical staining and discoloration. Power pressure washing or sandblasting systems should only be used if recommended by the brick manufacturer and then only in accordance with their recommended procedures, since these cleaning systems can cause irreversible structural and appearance damage.

Regardless of the method or chemicals used for cleaning, the following recommendations should be observed:

- Comply with all OSHA and other governing safety and health regulations, environmental regulations, manufacturer's printed instructions, and all other governing regulations.
- Protect all glass, wood, and metal surfaces from cleaning solutions and abrasive cleaning mediums. Read product information literature, and comply with recommended procedures, uses, and limitations.
- Protect stone and precast concrete from cleaning solutions and abrasive cleaning mediums.
- Use nonmetallic scrapers to remove large pieces of mortar before water-soaking an area.
- Test the proposed cleaning method and cleaning agent on a less noticeable wall area before continuing to clean.
- Post warning signs at the perimeter of the work zone and keep unprotected and unauthorized persons from entering the work zone.

The three methods recognized for cleaning new brick are the following:

1. The bucket and brush method
2. Pressure washing systems
3. Abrasive blasting

Each system has its advantages and limitations. No one method should be used before consulting the brick manufacturer's recommendations for the specific brick needing cleaning.

## Bucket and Brush Method

The bucket and brush method is a common practice for cleaning brick (see Figure 23-8). Only impact-resistant rubber or impact-resistant plastic buckets should be used. Metal buckets corrode easily and reactions between cleaning agents and metals can permanently stain brick.

**FROM EXPERIENCE**

A 16"-long-handle, plastic bristle scrub brush keeps one's body further from chemicals and their potentially harmful fumes and vapors.

**Figure 23-8 A clean Fortex rubber pail and a long handle plastic fiber or palmyra fiber brush are popular for bucket and brush cleaning.** *(Courtesy of Bon Tool Company)*

**CAUTION**

**Only those trained by a competent person should use any given chemical cleaner. Skin, eye, and respiratory protection are required when using chemical cleaning agents. Fumes and vapors are harmful to the skin, eyes, throat, lungs, and other body parts and body tissues. Comply with OSHA regulations, environmental regulations, and other local governing regulations when using chemical cleaners.**

Protective clothing must be worn when cleaning. This includes protective headgear, face shield and eye protection, acid-resistant clothing, acid-resistant gloves, and acid-resistant boots (see Figure 23-9).

Walls should be thoroughly prewetted with a continuous water stream to saturate the brick. Prewetting the brick prevents them from absorbing the cleaning solutions and dissolved mortar, either of which can result in brick staining. Nearby plants and ground covers should also be soaked with clean water and some plants may need to be protected completely from the cleaning solutions. Large areas are cleaned in portions, so that no single area is larger than can be cleaned and thoroughly rinsed without the risk of evaporation of

Figure 23-9 Acid-resistant gear protects the worker from serious injuries while using cleaning agents.

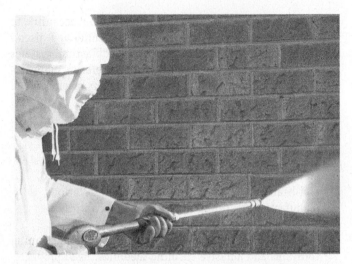

Figure 23-10 The wall is rinsed with clean water using a fan tip nozzle held approximately 8" from the face of the wall.

applied chemicals or prewetting water. Factors that govern the limiting of the size of an area to clean depend upon the absorption rate of the brick, the air temperature, the wind speed, and sunlight exposure.

Beginning at the top of a wall and working towards its bottom, cleaning solutions are applied with a scrub brush having acid-resistant bristles and permitted to remain on water-saturated, pre-wetted walls for time periods recommended by the manufacturer's printed instructions. Nonmetallic scrapers are used to loosen remaining mortar particles while the cleaning solution reacts with surface mortar stains. Reapplication of cleaning solutions may be recommended before rinsing. As the cleaned area is thoroughly rinsed with water, rinsing to the bottom of the wall below prevents lower areas from absorbing the loosened mortar and expended cleaning agent. Rinse nearby plants and ground cover also.

## Pressure Washing Method

The procedure referred to as pressure washing is misleading. Pressures above 50 psi are recommended for clean water rinsing only, in which cases, rinse water pressures between 200 and 300 psi should be the maximum. Brick Industry Association Technical Notes 20 recommends chemical cleaners be applied with low-pressure sprayers at pressures between 30 and 50 psi and using a 50-degree fan-shaped sprayer or by using the bucket and brush method. For rinsing walls with clean water, a pressure washer with a rated water flow of 3 to 8 gallons per minute, fitted with a 25 to 50-degree fan tip nozzle and held 6" to 8" from the wall is recommended (see Figure 23-10). As with the bucket and brush method, the area being cleaned and the wall area below it should be soaked with a continuous low-pressure water stream, saturating the brick to prevent chemical cleaners from being absorbed into the brick.

**CAUTION**

**Do not exceed water pressures above those recommended for the brick being cleaned. Avoid using less than 25-degree fan tips. Holding the water nozzle fan tip closer than recommended to the surface of brick can permanently damage the brick.**

Cleaning solutions are then applied with soft brushes first dipped into buckets containing the diluted cleaning solutions or with low-pressure tank sprayers operating at pressures between 30 and 50 psi. Cleaning solutions are permitted to remain on walls for time periods recommended by their manufacturers. Nonmetallic scrapers are used to loosen remaining mortar particles while the solution reacts with surface mortar stains (see Figure 23-11).

Figure 23-11 A heavy-duty plastic auto windshield ice scraper is used to remove mortar from the wall.

Reapplication of cleaning solutions may be recommended before rinsing.

**FROM EXPERIENCE**

A heavy-duty plastic auto windshield ice scraper works well to remove mortar from masonry walls.

Finally, the wall is thoroughly rinsed with medium pressure water streams, water pressures between 200–300 psi with a 25- to 50-degree fan-shaped tip, to remove loosened mortar and the expended cleaning solution, complying with the brick producer's recommendations for using pressure washers. High pressures, those greater than 700 psi, may alter a brick's appearance. The water wand should be moved in a horizontal, back-and-forth motion. Avoid "figure 8" motions with the hand-held water wand. Rinsing to the base of the wall prevents lower areas absorbing the loosened mortar and expended cleaning agent.

**CAUTION**

**Pressure Washer Safety**

- **Study the owner's manual and manufacturer's printed instructions for proper and safe use, especially topics related to health and safety.**
- **Become trained by a competent person before operating pressure washers.**
- **Wear approved body-, hand-, feet-, eye-, face-, and respiratory protection.**
- **Never direct high-pressure jets at persons, animals, electrical devices, or the machine itself.**
- **Post warning signs at the perimeter of the work zone, and keep unprotected and unauthorized persons from entering the work zone.**
- **Electric-powered pressure washers must have inline ground fault circuit protection to protect workers from electric shock and electrocution.**

**FROM EXPERIENCE**

In hot weather, work on the shaded side of the building as much as possible. This helps prevent rapid drying of cleaning solutions and water on the wall.

## Abrasive Blasting Method

Although generally not recommended, abrasive blasting is an alternative for cleaning those brick that may react with chemical agents and cause brick staining. Abrasive blasting is a method of cleaning brick masonry using compressed air and one of several types of abrasive materials, some harder than others. Examples include mined sand, crushed quartz, and crushed nutshells. The abrasive materials, stored in a blasting tank, are fed through an air hose and nozzle by an air compressor. The Brick Industry Association Technical Notes 20 recommends the material be delivered by systems capable of pressures between 60 and 100 psi with a minimum air flow capacity of 125 cubic feet per minute. Technical Notes 20 also recommends that the inside orifice or bore of the nozzle be $3/16$" to $5/16$" in diameter and the flow of abrasive materials to the nozzle be at a minimum rate of 300 pounds per hour. Water may be added to the air stream. Water reduces airborne dust and applies the abrasive materials more gently. Because of the potential of abrasives to permanently damage brick, abrasive cleaning should only be used when other methods cannot be used. The Brick Industry Association Technical Notes 20 does not recommend abrasive cleaning for brick having a sand finish or decorative surface coating. To operate abrasive equipment, a person must be well trained and experienced because of the high risk for abrasives to permanently damage brick and ruin the structural integrity of masonry walls. Abrasive blasting should be performed on dry walls only. The brick producer's recommendations should be observed because sand blasting damages many types of brick.

**CAUTION**

**Abrasive Blasting Safety**

- **Use in strict accordance with equipment manufacturer's printed instructions.**
- **Only persons trained and qualified should operate abrasive blasting equipment.**
- **Wear approved body-, eye-, face-, hearing-, and respiratory protection.**
- **Post warning signs at the perimeter of the work zone, and keep unprotected and unauthorized persons from entering the work zone.**
- **Electric-powered systems must have inline ground fault circuit protection to protect workers from electric shock and electrocution.**

Regardless of the cleaning method or cleaning agent selected, testing the performance and results on a less noticeable section should be performed before widespread use. Maintain a consistent air pressure, distance from, and angle to the wall façade throughout cleaning.

**Figure 23-12** An attached handle allows workers to hold the fluted silicon carbide rub brick. *(Courtesy of Bon Tool Company)*

## Cleaning CMU Walls

A concrete masonry unit (CMU) is a cement product. Because acid-based cleaners are intended to dissolve cements, they may have damaging effects on CMUs. Block may become etched from acid cleaning solutions. Fading and streaking may be apparent after applying acids. Only those cleaning agents recommended for CMUs should be considered. Dry masonry spatters can be removed with scrapers and silicon carbide, fluted-stone rub bricks (see Figure 23-12).

Mortar stains may be removed with the bucket and brush method. Prewetting the wall with water and scrubbing with a brush removes most dry mortar embedded between surface aggregates. Water alone removes most mortar. Pressure washing with water is another method for cleaning CMUs. However, do not exceed water pressures recommended by the block producer.

## Today's Trends

Several factors contribute to the emergence of specialized cleaning services for both new and older masonry construction. Complicated cleaning equipment and potentially harmful cleaning agents require workers to be well-trained. In some instances, regulatory agencies and governing regulations require workers to become certified to be eligible to use certain chemical cleaners. Environmental regulations require strict compliance when using many chemicals, sometimes requiring recovery of the expended chemicals. For these reasons, chemical cleaning services using sophisticated cleaning systems are more evident on larger construction projects than on smaller residential projects, where the bucket and brush method are still regularly employed.

## Removing Specific Stains

The following information is an excerpt from The Brick Industry Association, Technical Notes on Brick Construction #20 dated June 2006. It provides recommendations for removing specific stains from brick.

### CAUTION

Comply with all governing regulations for the use, storage, and disposal of specific chemicals. Use in accordance with manufacturer's printed instructions, OSHA standards, environmental regulations, and all governing regulations. Wear personal protection equipment required to meet OSHA standards for exposure to specific chemicals or processes and as recommended in manufacturer's printed instructions.

### Brick Dust

Dust produced from the cutting of brick sometimes adheres to the surface of brickwork. Compressed air, such as from a portable cylinder, has been found effective in removing this dust.

### Dirt and Mud

Dirt can be difficult to remove, particularly from a textured brick. In addition to propriety cleaners, scouring powder and a stiff bristle brush are effective if the texture is not too rough. For very rough textures, pressurized water cleaning can be effective.

### Egg Splatter

Brickwork vandalized with raw eggs has been successfully cleaned by prewetting the stain, applying a saturated solution of oxalic acid crystals dissolved in water, and rinsing with water. Mix the solution in a nonmetallic container, and apply with a brush. If the egg splatter is to be removed from brick that contain vanadium (typically light colored units),

**Figure 23-13** Soil stains along the bottom of this newly constructed wall may be difficult to remove. Placing straw on the ground adjacent to the wall prevents soil-stained walls.

**Figure 23-14 Manganese stains appear as tan or brown in color. Occasionally, they may appear as gray staining.** *(Courtesy of Brick Industry Association)*

**Figure 23-15 Algae and mold frequently appear on brick walls.**

a solution of 1.5 oz (10 g) washing soda (sodium carbonate) per gal (1 L) of water should be applied to the brickwork following the oxalic acid solution. Without this neutralizing solution, cleaning with oxalic acid may cause more severe staining.

## Manganese (Brown) Stain

Besides specially formulated propriety compounds, manganese stains have been effectively removed and their return prevented by carefully mixing a solution of acetic acid (80% or stronger), hydrogen peroxide (30 to 35%) and water in the following proportions by volume: 1 part acetic acid, 1 part hydrogen peroxide, and 6 parts water. After wetting the brickwork, brush or spray on the solution. Do not scrub. The reaction is usually very rapid and the stain quickly disappears. After the reaction is complete, rinse the wall thoroughly with water. Caution: Although this solution is very effective, it is a dangerous solution to mix and use. Acetic acid-hydrogen peroxide may also be available in a pre-mixed form known as peracetic acid. An alternate treatment sometimes suggested for new and mild manganese stains is oxalic acid crystals and water. Mix 1 lb of crystals (0.45 kg) to 1 gal (3.79 L) of water. The neutralizing wash mentioned in the "Egg Splatter" section should be considered when oxalic acid is applied to brown or light-colored brick.

## Oil and Tar Stains

Oil and tar stains may be effectively removed by commercially available oil and tar removers. For heavy tar stains, mix the agents with kerosene to remove the tar, and then mix with water to remove the kerosene. After application, the stains can be hosed off. When used in a steam-cleaning apparatus, cleaners have been known to remove tar without the use of kerosene. Where the area to be cleaned is small, or minimal cleanup is desired, a poultice using naphtha or trichloroethylene is most effective in removing oil stains. Dry ice or compressed carbon dioxide may be applied to make tar brittle. Then, light tapping with a small hammer and prying with a putty knife generally will be enough to remove thick tar splatters.

## Organic Growth

Occasionally, an exterior masonry surface remains in a constantly damp condition, thus encouraging moss, algae, lichen, or other organic growth. Applications of household bleach, ammonium sulfate or weed killer, in accordance with furnished directions, have been used successfully for the removal of such growths.

## Paint and Graffiti

Commercial and proprietary paint removers and organic solvents are most effective at softening or dissolving paint so that it can be removed with a scraper and a stiff bristle brush or rinsed away with water. For very old dried paint, organic solvents may not be effective, in which case the paint must be removed by sandblasting or scrubbing with a nonmetallic abrasive pad. Graffiti that has penetrated into masonry is best removed by a poultice, which is a paste or

**Figure 23-16 Chalking paints migrate with rainwater, causing stained brickwork.**

**Figure 23-17 Green or yellow vanadium stains develop on some brick.** *(Courtesy of Brick Industry Association)*

gel that can cling to the masonry to extend its working time on the stain.

## Smoke

Scrubbing with scouring powder (particularly one containing bleach) and a stiff bristle brush is often effective.

## Vanadium (Green) Stain

Applying a solution of potassium or sodium hydroxide, consisting of 0.5 lb (0.23 kg) hydroxide to 1 qt (0.95 L) water or 2 lb (0.91 kg) per gal (3.79 L) to brickwork is an alternative treatment for vanadium stains. The solution should be allowed to remain for two or three days and then be washed off. Use a hose to wash off any white residue remaining on the brickwork after this treatment. Sodium hypochlorite, the active ingredient in household bleaches, can also be used to remove mild vanadium stains. Spray or brush onto the stain, and then rinse off after the stain disappears. Oxalic acid is another chemical known to remove vanadium stains. A mixture of 3 to 6 oz (20 to 40 g) oxalic acid per gal (1 L) of water (preferably warm) should be applied to the brickwork, followed by the neutralizing wash described earlier in the "Egg Splatter" section. More severe staining may result if the oxalic acid solution is applied without the neutralizing wash.

## Welding Splatter

When metal is welded too close to brick stored on site or too close to completed brickwork, molten metal may splash onto the brick and melt into the surface. A mixture of 1 lb (0.45 kg) oxalic crystals and 0.5 lb (0.23 kg) of ammonium biflouride per gal (3.79 L) of water is particularly effective in removing welding splatters. This mixture should be used with caution, as it generates dangerous hydroflouric acid, which can also etch brick and glass. Scrape as much of the metal as possible from the brick. Apply the mixture in a poultice, and remove when it is dried. If the stain has not disappeared, use sandpaper to remove as much as possible and apply a fresh poultice. For stubborn stains, several applications may be necessary.

## Stains of Unknown Origin

Stains of unknown origin can be a real challenge. Laboratory tests of unknown stains may be necessary to determine their composition. Then the appropriate method may be implemented to clean the brickwork. The application of a cleaning agent without identifying the initial stain may result in stains that are more difficult to remove. The visual characteristics of a stain may be the first clues to its source. Identification of stains is discussed further in Technical Note 23, later in this chapter.

## Cleaning Historic Structures

Improper cleaning can cause irreparable damage to historic brickwork. Therefore, cleaning of structures with historic significance should be overseen by a restoration specialist. Before a historic structure is cleaned, consider the purpose of cleaning: to improve the appearance; to slow deterioration; or to provide a clean surface for evaluation or further treatments. With historic structures, it is imperative to use the least harmful cleaning method that will achieve the desired results. Cleaning methods and materials must be carefully matched to the substance to be cleaned, the type of soiling/staining to be removed, and the desired results.

The following recommendations for cleaning brick are reprinted, courtesy of BrickSouth East Headquarters of Charlotte, North Carolina.

### SPECIAL SYSTEMS FOR WET CLEANING THROUGH-THE-BODY LIGHT BRICK WHERE TYPE "S" MORTAR IS USED

Type "S" (and Type "M") mortar is very difficult to remove from the face of all brick, but is a special problem when through-the-body or light colored brick is used, due to the sensitivity of these brick to strong cleaning materials.

The following cleaning procedures are recommended according to age of masonry work:

A. After work is 10 days old:

1. Remove all large mortar particles with hand tools before applying cleaning solutions.
2. Mask and otherwise protect adjacent non-masonry materials.
3. Saturate wall with clean water.
4. Use cleaning brush to apply solution of Sure Klean Vanatrol, Diedrich 202 Vana-Stop (or equal) mixed 4 to 6 parts of water to 1 part of solution.
5. Allow solution to remain on wall for 3 to 5 minutes while brushing and scraping, reapply solution.
6. Thoroughly rinse and brush clean.

B. After work is 30 days old:

1. Use procedure described above in steps 1–5.
2. Use high water pressure equipment to rinse wall, using pressure not greater than 800 PSI with a 40-degree nozzle fan tip. Consult brick manufacturer before using high-pressure water system.

*(Test clean a sample area to determine effectiveness of cleaning compound and the total cleaning system and to check wall for possible damages caused by system. Approval of owner or owner's representative should be obtained before proceeding with operation.)*

## CLEANING GUIDE

### Red Brick—Textured

This category includes all textured red through-the-body brick.

Brick in this category may be cleaned by the bucket and brush method, high water pressure method, or by sandblasting.

### Red Brick—Heavy Sand Finish

This category includes all red through-the-body brick with various applied heavy sand finish faces.

Brick in this category may best be cleaned by the bucket and brush method, using plain water and scrub brush, or with lightly applied high pressure water system, with plain water being used, Sandblast cleaning is not recommended. If mortar stains are excessive, use of cleaning compounds may be required.

### White, Buff, Gray and Chocolate Brick

This category includes all textured and sand finish brick with through-the-body colors other than natural red.

Brick in this category may be cleaned by the bucket and brush method, or by lightly applied high pressure water system. Sandblast cleaning is also recommended, except in the cases where heavy sand finish is involved. In the two wet cleaning systems, no muriatic acid or compounds containing muriatic acid may be used. Only plain water and detergent, or Sure Klean Vanatrol, Diedrich 202V Vana-Stop, or equal may be used.

### Green Stains

Green staining is caused by the presence of vanadium salts. Color and solubility of these salts are dependent upon the acidity of the brick. Very often green stains are brought about by wrongful use of muriatic acid or compounds containing muriatic acid. When green stains appear, the brick manufacturer should be consulted before attempting to remove stain.

Green stains may be removed by using Sure Klean 800 Stain Remover, Sure Klean Ferrous Stain Remover, Diedrich 940 Iron and Manganese Stain Remover, or Diedrich 950 Acid Burn Remover (or equal).

### Brown Stains

Brown staining can be caused by presence of soluble manganese or iron oxides. Very often brown or manganese stains are brought on by wrongful use of muriatic acid or compounds containing muriatic acid.

If these stains are light, Brick Klenz may take them off with little difficulty.

Also, oxalic acid (1 pound mixed in a gallon of water) may do the job if stains are new and light in color.

Many brown stains can be removed with Sure Klean 800 Stain Remover, Sure Klean Ferrous Stain Remover, Sure Klean Restoration Cleaner, Diedrich 950 Acid Burn Remover, Diedrich 940 Iron and Manganese Stain Remover, or Diedrich 101G Brick Cleaner (or equal).

Each product should be tested for effectiveness and possible bleaching action on joints.

### White Scum—Insoluble

Insoluble white scum is generally caused by faulty cleaning, failure to adequately saturate wall before cleaning, and failure to flush wall after applying cleaning compound. As opposed to white efflorescence, this stain cannot be removed with detergents or regular cleaning compounds.

Currently known method of removal is to use Sure Klean White Scum Remover, Diedrich 930 White Scum Remover, or equal.

## SPECIALTY CLEANING

### White Efflorescence

White efflorescence is a water soluble salt that is brought to the surface of masonry by evaporation of either construction water or by evaporation of rain water that has penetrated the wall.

Water used in mortar, grout, etc. will sometimes cause this "New Building Bloom." As the wall dries out, and as successive rains wash the walls, the "Bloom" should disappear.

If the masonry has received its regular cleaning and white efflorescence appears or reappears, no further action should be taken until this wall has had an opportunity to dry out completely. Application of additional cleaning solutions may only aggravate the problem at this point. Also, application of clear waterproofing materials may lock in moisture and crystalline growth, causing more scumming and possible spalling of brick.

If efflorescence stains persist, it is likely that rainwater is penetrating the wall. An inspection of the stained areas should be made to determine if sizeable cracks or openings exist, permitting water penetration. Faulty flashing or a lack of flashing will contribute to staining.

Any large openings should be repaired. Where only very fine hairline cracks are assumed to be allowing water penetration, application of a penetrating water repellent may be the only solution to the problem short of a complete tuckpointing job.

Before applying waterproofing materials, all possible repairs should be made and all efflorescence removed. This may be removed by applying plain water and brushing the affected area. If water fails to remove stain, use dilute solution of commercial cleaning compounds such as Sure Klean 600 or Diedrich 202 New Masonry Detergent (or equal) for red brick and Sure Klean Vanatrol, Diedrich 202V Vana-Stop (or equal) for all others. Some heavy white stains, known as "lime runs" or "silicone deposits" may require special cleaning procedures for removal. Contact the Brick Association of the Carolinas for further details. Allow entire wall to dry out completely (over a period of little or no rainfall) before applying waterproofing solutions.

### Smoke Stains

Smoke stains can generally be removed by using one of the following cleaners: Brick Klenz or Sure Klean Smoke Remover. A follow-up cleaning with Sure Klean Restoration Cleaner, Diedrich 101G Brick Cleaner (or equal) may be required after using smoke removal products.

Follow the directions found on containers.

### Mud Stains

Mud stains are the most difficult of all to remove.

Currently known method of removal is as follows:

Apply Sure Klean Restoration Cleaner, Diedrich 101G Brick Cleaner (or equal) full strength, with stainless steel pressurized "orchard" sprayer. Allow to remain on wall 5 minutes. Flush off with high pressure water spray. Repeat if necessary. Sprayer nozzle should be held at 90 degree angle to wall, as should rinse water nozzle. Sure Klean Light Duty Concrete Cleaner or Diedrich 960 Heavy Duty Concrete Cleaner might be less likely to bleach joints than Sure Klean Restoration Cleaner or Diedrich 101G Brick Cleaner.

### Paint Stains

Paint stains are very difficult to remove from masonry. Probably sandblasting is the fastest way to remove paint, but this process is sometimes harmful to the masonry surface.

Commercial paint removers are effective in some cases.

Sure Klean Defacer Eraser, Sure Klean Heavy Duty Paint Stripper, Diedrich 505 Paint Stripper Diedrich 606 Multi-Layer Paint Remover, or equal are very good for paint removal. If these products do not completely remove all paint particles after following printed directions, apply Sure Klean Restoration Cleaner or Diedrich 101G Brick Cleaner (or equal) to the stained area. Allow to remain on wall several minutes, then "blast" the area with water hose. *Follow directions found on containers.*

Joints struck while excessively wet can become light in color. Joints struck when "thumbprint" hard should dry to a uniform color if mortar and sand properties remain consistent.

Normal variations in joint color will be eliminated after completion of one of the wet cleaning processes. Where wide color variations are found, a mild bleaching of all joints with increased concentration of cleaning solutions usually brings improvement. Caution should be taken in using this process with acid-sensitive brick and colored mortars.

Light joints may be darkened by painting the joints with pigments specially selected to produce the required shade.

**The following disclaimer accompanies the preceding information.**

"The information and recommendations made herein are based on our own research, and research and experience of others, and are believed to be accurate. However, no guarantee of their accuracy is made because we cannot cover every possible application of the products described, nor anticipate variations encountered in masonry surfaces, job conditions, and cleaning methods used."

## Summary

- First impressions of properly cleaned masonry walls are unquestionable. The true colors of brick emerge in dramatic fashion. The masons' pride in quality craftsmanship and attention to details are evident.
- Preventive measures keep jobs cleaner during construction and reduce cleaning efforts and their related expenses after construction.
- Before using any cleaning products and equipment, become thoroughly trained in the matters of personal health and safety issues, product applications and limitations, equipment safety and operations, and regulatory compliance issues.
- Use only those cleaning procedures and chemical cleaners as recommended by the producers of any and all of the materials used for wall construction, including the manufacturers of the brick, block, and mortar ingredients.

## Review Questions

*Select the most appropriate answer.*

**1** To minimize the cleaning of new walls after construction,

a. elevate building materials above the ground

b. cover material stock piles

c. cover ground adjacent to wall with straw

d. all of the above

**2** The minimum number of days mortar should be permitted to cure before cleaning masonry walls is

a. 1

b. 3

c. 7

d. 14

**3** The first choice for removing mortar stains from brick should be

a. high pressure water spraying systems

b. muriatic acid

c. propriety compounds recommended by the brick producer

d. sand blasting

**4** For cleaning masonry walls, muriatic acid is initially diluted in clean drinking water at a rate of

a. 1 part acid and 3 parts water

b. 1 part acid and 6 parts water

c. 1 part acid and 9 parts water

d. 1 part acid and 20 parts water

**5** When using the bucket and brush method for cleaning masonry walls, the cleaning solution should be put in

a. a steel pail

b. an impact-resistant rubber pail

c. any plastic pail

d. both a and b

**6** When using pressure washer systems, cleaning solutions should be applied at pressures

a. between 30 and 50 psi

b. between 100 and 200 psi

c. between 200 and 300 psi

d. above 300 psi

**7** The probability for efflorescence to develop may increase for walls cleaned with chemicals when

a. temperatures are expected to be at or below freezing

b. temperatures are high, during hot weather

c. high winds prevail

d. all of the above

**8** Used improperly, abrasive cleaning can

a. damage the face of brick, altering a wall's appearance

b. damage the face of brick, allowing excessive water penetration and damages associated with freeze-thaw weather cycles

c. have damaging results to a wall's structural integrity

d. all of the above

**9** Scrapers used to remove dry mortar from walls should be made of

a. plastic

b. wood

c. metal

d. both a and b

**10** Concrete masonry units should be cleaned using

a. water

b. a silicon carbide, fluted rub brick

c. muriatic acid

d. both a and b

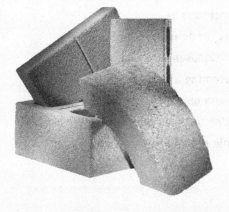

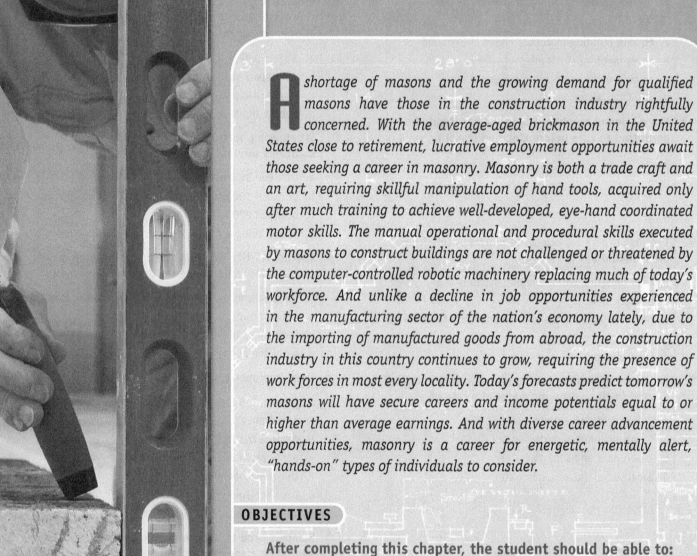

# Chapter 24 | Masonry as a Career

A shortage of masons and the growing demand for qualified masons have those in the construction industry rightfully concerned. With the average-aged brickmason in the United States close to retirement, lucrative employment opportunities await those seeking a career in masonry. Masonry is both a trade craft and an art, requiring skillful manipulation of hand tools, acquired only after much training to achieve well-developed, eye-hand coordinated motor skills. The manual operational and procedural skills executed by masons to construct buildings are not challenged or threatened by the computer-controlled robotic machinery replacing much of today's workforce. And unlike a decline in job opportunities experienced in the manufacturing sector of the nation's economy lately, due to the importing of manufactured goods from abroad, the construction industry in this country continues to grow, requiring the presence of work forces in most every locality. Today's forecasts predict tomorrow's masons will have secure careers and income potentials equal to or higher than average earnings. And with diverse career advancement opportunities, masonry is a career for energetic, mentally alert, "hands-on" types of individuals to consider.

## OBJECTIVES

**After completing this chapter, the student should be able to:**

- Describe the work performed by brickmasons.
- Describe the physical qualifications for doing masonry tasks.
- List possible career benefits for brickmasons.
- Explain the career pathways for becoming a brickmason.
- Describe the employment outlook and opportunities for brickmasons.
- List items of which a masonry contractor must be knowledgeable.
- Describe factors building a favorable reputation for a masonry contractor.

# Glossary of Terms

**apprentice** one learning a trade from one or more competent, experienced persons.

**architectural concrete** concrete that is permanently exposed to view and requiring special care in forming, placing, and finishing.

**architectural drawing** a drawing showing a completed project as it is to appear on a specific site.

**certification** assurance that the contractor is in compliance with applicable federal, state, and local regulations pertaining to the specific products or procedures.

**change order** documentation of an addition, deletion, or change of the contract.

**completed operations insurance** insurance providing coverage for damages that either products or services may cause after the work is completed.

**construction contract** a written agreement executed between the property owner of the proposed job or their representative and the contractor responsible for executing the job, giving the terms, rates, or prices for doing the job.

**construction manual log** a daily record of progress, including the number of workers on the job, the number of hours worked, the quantity of units laid, the wall or walls being constructed, and the daily weather.

**contract** a written agreement by which a contractor is compensated at an agreed upon rate or price for completing specific work.

**contract bond** an approved form of security guaranteeing complete execution of the contract and the payment of all debts pertaining to the construction of the project.

**contractor** the person or persons undertaking the terms of the contract.

**employee** one whose personal services are rendered to comply with instructions about when, where, and how work is done that is assigned by the employer.

**employee benefits** such amenities as medical insurance, paid holidays, paid vacation time, paid sick leave, and retirement income received by the employee at the expense of the employer.

**employer** the one or more individuals licensed as a business responsible for the hiring, training, safety, supervising, and paying of wages for the employee.

**estimate** a judgment of construction costs.

**estimator** one who judges or calculates either or both the materials and labor required for completing a given job.

**experience modification rating** also called EMR, the basis for adjusting worker's compensation premiums on the contractor's record of injury frequency and costs.

**job foreman** one responsible for supervising a group of workers.

**job superintendent** one responsible for the operations required for the completion of a job.

**journey-level worker** also called a journeyman, a competent, experienced worker possessing the skills required to perform without supervision the tasks for a particular trade.

**liability insurance** insurance providing protection for personal injuries and damage to the property of others arising from operations performed by the contractor.

**licensed contractor** a contractor whom government authority has granted permission to engage in contracting.

**markup** the dollar amount over and beyond material and labor costs needed for business operations.

**masonry contractor** one or more individuals whom government authority has granted permission to engage in masonry construction.

**masonry instructor** employed by industry, organized labor, local governments, or career and technical training centers, one teaching others masonry trade-specific technical information and procedures.

**Material Safety Data Sheets** also called MSDS, printed information available at the jobsite, including emergency and first aid procedures as well as storage and disposal information for every product used at the jobsite.

**profit** the amount of money remaining after all expenses have been met.

**project manager** one representing the building's owner and supervising all or parts of the construction details.

**scale drawing** a drawing showing all of the construction details smaller than actual size so that the project can be represented on paper.

**self-employed** one who works for another on a contract basis rather than as an employee.

**structural engineer** one meeting requirements for licensing to design a structure or parts of it to withstand all weights and forces that it may have to support.

**workers' compensation** insurance covering employee accident, injury, or death.

**working drawing** a scale drawing showing specific design details, including engineered specifications and required materials.

# Work Performed by Masons

A *mason,* also called a brickmason or bricklayer, is one who primarily installs mortared brick and concrete masonry units (CMUs). But in addition to brick and block, brickmasons or bricklayers use mortar to build walls consisting of architectural precast concrete units, structural facing tile, marble, and granite. Brickmasons doing residential construction, single-family or multi-family homes, are primarily involved with building block foundation walls, brick and CMU veneer walls, and flatwork such as mortared brick paving, porches, and steps (see Figure 24-1).

Brickmasons doing commercial construction, buildings such as stores, schools, churches, hospitals, and industrial sites, are as likely to be building composite cavity masonry walls as building anchored brick veneer walls. But in addition to brick and block, some homes and buildings have architectural cast stone components. These are made from precast concrete with a machine-honed finish, highly polished finish, or an exposed aggregate finish and in a variety of colors; they are used as wall panels, sill and band courses, coping and caps, window and door trim, and column facings (see Figure 24-2).

Many buildings have marble and granite wall panels that are installed by brickmasons. Where sanitary, easily cleaned walls are needed, bricklayers may be called upon to lay impervious-surfaced, structural hollow glazed tile (see Figure 24-3).

**Figure 24-1** A typical residential masonry construction project includes both brickwork and CMU walls.

**Figure 24-2** Enhancing the appearance of many homes and buildings are architectural precast concrete components that are installed by brickmasons. *(Courtesy of General Shale Brick)*

**Figure 24-3** Masons sometimes lay structural hollow glazed tile in areas requiring easily sanitized, impervious-surface walls. *(Courtesy of Elgin-Butler Brick Company)*

Although some bricklayers may have training to lay stone, one who builds with stone is called a stonemason, a trade requiring skills different from those of a brickmason or bricklayer. A brickmason may sometimes place concrete flatwork such as footings, walkways, patios, porches, and steps. However, a brickmason is not necessarily trained to place and finish architectural concrete, concrete that is permanently exposed to view and requires special care in forming, placing, and finishing. Likewise, training as a brickmason does not include those duties of forming, reinforcing, and placing structural concrete such as that used for roadways, bridges, or concrete walls. Masonry units dry stacked, such as those for retaining walls or bedded without mortar for paved surfaces, are more likely to be installed by landscapers or contractors specializing in these services alone.

## Physical Qualifications

Brickmasons work while standing, stooping, or kneeling. As walls are started, masons bend their back to reach down and install units. As walls are raised, a more erect body position is required. And eventually, they are likely to extend their arms above the shoulders to complete a wall. Such movements require masons to have good use of their body joints. Repeatedly lifting heavier masonry units, such as 30-pound blocks for periods of 8 hours, also requires good muscle development. Lifting units sometimes with only one hand and manipulating the trowel in the other hand necessitates strong finger gripping to secure the unit and flexible wrist motions for handling the trowel. Good eyesight and a sense of proportion are needed for observing the mason's line and aligning pattern bonds uniformly (see Figure 24-4).

**Figure 24-4 With one leg in a ditch and the other on an embankment, this brickmason works tirelessly, typically for 8-hour days, enduring summer's heat accompanied by the discomforts of humid air.**

Good physical stamina enables the mason to endure 8-hour days of summer's heat and winter's cold. For climbing ladders and working from scaffolds, masons need to overcome any incapacitating fears of heights.

## Career Benefits

A brickmason is either self-employed or works for an employer, the one or more individuals licensed as a business responsible for the hiring, training, safety, supervising, and paying of wages for the employee. The employer is usually a masonry contractor, one or more individuals whom government authority has granted permission to engage in masonry construction. Most masons are considered employees, those whose personal services are rendered to comply with instructions about when, where, and how work is done as assigned by the employer. Employees typically work for an hourly wage rate rather than a fixed weekly, monthly, or annual salary. Loss of time due to weather or work shortages reduces income earnings. But the hourly wages for masons, which in many cases are above average as compared with other occupations, enable masons to have annual incomes that are comparable to most other occupations requiring similar levels of training. Some employers offer their employees employee benefits, including paid medical and dental insurance, paid vacation time, paid sick leave, and retirement income. The benefits that a masonry contractor is likely to be able to afford to offer its employees are most likely directly related to the volume of business and the margin of profit. Contractors operating smaller businesses are more likely not to be able to offer medical and dental insurance programs for their employees than are contractors that have a large number of employees. Paying employees wages normally earned when working but for days of vacation, holidays, or personal illness are benefits offered by many employers. However, the amount of such benefits usually depends on the length of time the employee has worked for the employer that is offering the benefits. Retirement income savings programs that the employer, or both the employer and employee, contribute to and which the employee receives after retirement age are offered by some contractors.

One considered as self-employed works for another under the terms of a contract rather than as an employee. A contract is a written agreement by which a contractor is compensated at an agreed upon rate or price for completing specific work. The person undertaking the terms of the contract is called the contractor. A contractor may or may not have employees. Self-employed brickmasons are responsible for the financial obligations of providing their own personal benefits, such as medical insurance costs and retirement income savings. Time lost for illnesses, vacations, or holidays are at their own expense. There are notable differences between being employed as a brickmason and being a self-employed masonry contractor. Those self-employed do not necessarily have to personally render services. Rather, they may supervise the work of hired employees. Self-employed persons can usually set their own hours of work. Self-employed persons may be completing work for more than one client at a time, minimizing loss of time because of weather or job holdups beyond their control. There are several responsibilities a person must accept when becoming self-employed. Operations must be completed within agreed upon time deadlines, often requiring the hiring, paying, and supervising of employees. A self-employed mason must have a business license and file federal and state income taxes as required by federal and state governments. If hiring employees, a self-employed masonry contractor must file payroll taxes and perhaps pay workers' compensation insurance premiums. Although not required, self-employed contractors must consider offering hired employees wages and benefits that are competitive with those of optional employment opportunities from which the employees may choose.

# Employment Outlooks and Opportunities

The U.S. Department of Labor's Bureau of Labor Statistics predicts a shortage of skilled masons in the near future. The construction industry is one of the United State's fastest growing sectors for employment. Machines and robotics continue to replace workers performing manufacturing operations in many industries, but the products of robotic operations are limited to those permitted by assembly-line productions, computer-programmed robotic machines performing a specific manufacturing process while stationed at a moving conveyor line. The assembly-line production technology permitting mass production used for manufacturing processes is not adapted to building uniquely designed masonry walls on site. It may be that future generations will witness robotics replacing manual skilled labor, but most agree that the mason's personal skills are to be the standard for the twenty-first century.

## Career Pathways

Jobsite training, learning masonry techniques and skills from competent, experienced masons at construction sites, is the major training most receive to become a mason. Someone who learns a trade from one or more competent, experienced person is called an **apprentice.** The average apprentice mason can expect to serve an apprenticeship period working 40-hour weeks for three to four years before being considered a **journey-level worker,** a competent, experienced worker possessing the skills required to perform unsupervised the tasks for a particular trade. Organized labor and trade labor unions offer apprenticeship training to qualified workers,

those showing the desire and aptitude to become masons. Classroom settings enable the apprentices to learn related masonry technical information, governing safety regulations, blueprint reading, communication skills, and trade-related mathematics. Individual states offer registered apprenticeship programs through their departments of labor and industry. On-the-job training with a sponsoring employer and classroom training, often offered at local career and technical centers, insure that specific criteria are met before an apprentice earns nationally recognized certification as a journey-level worker. Many secondary schools, those offering high school grade levels, and post-secondary schools, community colleges or career and technical training centers, offer courses in masonry, enhancing the readiness skills for those desiring masonry as a career. Although those completing masonry programs offered at public high schools or at career and technical centers are not certified as journey-level workers upon completion of such programs, the programs prepare students with the fundamentals of masonry, increasing their likelihood of employment as apprentice masons.

There are advancement opportunities for journey-level brickmasons. Those demonstrating the potential for leadership can become a **job foreman**, responsible for supervising a group of workers, or a **job superintendent**, responsible for the operations required for the completion of a job. Some may become a **project manager**, representing the building's owner and supervising all or parts of the construction details. Brickmasons can also become trained as an **estimator**, estimating either or both the materials and labor required for completing a given job. A career as a **masonry instructor**, employed by industry, organized labor, local governments, or career and technical training centers, gives a brickmason the opportunity to help others become brickmasons. It is an opportunity to share with others masonry trade-specific technical information and procedures by becoming a mentor and experiencing the satisfaction of watching them develop confidence in themselves, and competencies leading to journey-level workers so much in demand for the construction industry's future. As mentioned earlier in this chapter, a brickmason can become a masonry contractor, working alone or hiring other employees. This remainder of this chapter is devoted to the important issues one must address when considering becoming a masonry contractor. The purpose of this material is to reveal the complex issues you must consider before becoming a masonry contractor rather than training you to become a masonry contractor. It takes more than just being a good mason to be a successful masonry contractor. Success as a masonry contractor depends upon the following:

- Knowledge and experience
- Personal attributes
- Legal compliance
- Honesty and reliability

# Knowledge and Experience

A masonry contractor must be able to understand and interpret architectural and working drawings(see Figure 24-5). An **architectural drawing** shows the completed project, as it is to appear on a specific site. A **working drawing** shows specific details, including engineered specifications, required materials, and scale drawings. A **scale drawing** shows all of the construction details smaller than actual size so that the project can be represented on paper.

A masonry contractor should have a good educational background, including competence in mathematics, oral and written communications, and interpersonal skills. In addition, the masonry contractor should be competent to safely and properly use masonry tools and procedures. Furthermore, a contractor must be able to interpret blueprints. A successful masonry contractor must be able to determine the labor costs and sometimes the material costs needed to complete a job. Competitive pricing of work is necessary. The contractor submitting unreasonably higher prices cannot compete with masonry contractors charging less. Although lower pricing may attract work, the contractor may not be able to meet expenses such as labor, tools, equipment, and insurance. Repeatedly too high or too low pricing brings failure to the contracting business.

Although materials and other expenses such as labor may be the same for one job as another, the cost of completing the work can vary. Factors influencing the total cost for doing work include the following:

- Travel time and distance
- Site accessibility
- Weather conditions
- State of the economy

Labor costs and vehicle expenses increase as travel time and distance become greater. Among other factors, production is related to site accessibility, temperature, and humidity.

The contractor must be aware of current and projected local, state, and national economic conditions that can impact the construction industry. A poor economy may result in less work and reduced earnings for the masonry contractor. To be successful, a business must be prepared to withstand losses of revenue due to periodical reduced income.

# Personal Attributes

Those having a successful business exhibit several desirable traits. Successful contractors are determined. They keep trying and do not let obstacles overcome them. Be it an out of level footing or undesirable weather, successful masonry contractors get beyond setbacks. Being creative generates good ideas that resolve problems. Successful businesses are lead by goal-oriented individuals. Successful contractors set goals and initiate plans to achieve those goals. Successful contractors are energetic people. They work long hours at the jobsite and spend much time in the evenings talking to others about present or future jobs. Successful contractors are natural leaders. They have the ability to inspire others to take pride in their work and to operate as a team.

Image promotes business success. Masonry construction can be considered as "dirty work," having clothing soiled with masonry dusts, mud, and body perspiration. A masonry contractor should consider having a professional image, clean clothed and well-groomed, when meeting prospective clients or appearing in public (see Figure 24-6).

**Figure 24-6** While demonstrating his techniques and procedures for building a brick lead to students enrolled in a masonry training program at a career center, this masonry contractor is "dressed for success," exhibiting self-respect and poise.

**Figure 24-5** A masonry contractor and the project's developer discuss the working drawings.

Successful contractors can be compared to successful coaches. Both must make decisions without delay, keep trying when faced with obstacles, make realistic but challenging goals, recognize both the strengths and limitations of each individual, and inspire all to work as a team.

# Operating a Masonry Contracting Business

A well-managed business complies with federal, state, and local laws and regulations. Successful businesses depend on good administrative decisions. The many related activities and expenses of operating a business include the following:

- Securing licenses and required certifications
- Providing insurance
- Purchasing and maintaining tools and equipment
- Continuing education and training
- Preparing estimates and contracts
- Documenting and recording

## Licensing and Certifications

A **licensed contractor** is a contractor whom government authority has granted permission to engage in masonry contracting. Larger masonry contractors are licensed by both local and state governments. Licensing by a locality may be all that is required for a minimal amount of contracting. The cost of becoming licensed varies depending on location and the amount of work performed. A contract bond is often required. A **contract bond** is an approved form of security guaranteeing complete execution of the contract and the payment of all debts pertaining to the construction of the project.

**Certification** assures that the contractor complies with applicable federal, state, and local regulations. As examples, government regulations may require that at least one employee on each job site be certified in first aid and CPR training. Certification may be required before an employee is permitted to operate certain machinery or use certain cleaning chemicals.

## Insurance

Several types of insurance are available to a masonry contractor. The type of insurance may be required or optional, depending on the size of the masonry business as well as the demands of the property owner.

A general **liability insurance** policy provides protection for personal injuries and damage to the property of others arising from operations performed by the contractor. The cost for liability insurance varies depending on the amount and extent of coverage as well as previous accident or injury claims and settlements. Every contractor should consider purchasing liability insurance. Workers' compensation

is an employer obligation established by law. **Workers' compensation** is insurance covering employee accident, injury, or death. The insurance rate is based on **experience modification rating** or EMR. An EMR adjusts workers' compensation premiums on the contractor's record of injury frequency and costs. Depending on the number of employees, a masonry contractor may be required to have workers' compensation. **Completed operations insurance** provides coverage for damages that either products or services may cause after the work is completed.

### FROM EXPERIENCE

In addition to basic liability coverage, every masonry contractor should consider purchasing a completed operations insurance policy. Review a policy with the insurance company agent to be sure the types of coverage desired are provided under the terms of the policy.

## Tools and Equipment

Tools and equipment are other expenses of the masonry contractor. A small contractor may mix mortar in a wheelbarrow using a mortar hoe, but most masonry contractors have a gasoline- or electric-powered mixer. In addition to mortarboards or pans, most contractors purchase corner poles, a builder's level and tripod, and a masonry saw. Steel scaffolds, scaffold accessories, and scaffold planks are necessary for building walls above the reach of the mason. A ladder is necessary for accessing scaffolding safely and a hoist or mobile equipment is necessary for getting materials to higher levels. One or more vehicles and trailers are needed to transport tools and equipment.

### FROM EXPERIENCE

Equipment such as scaffold setups and hoists can be rented from local equipment rentals when needed, sometimes a less expensive option to purchasing large amounts of specific equipment that is seldom needed.

Safety equipment is needed for protecting employees. OSHA safety standards regulate the personal protection equipment required for specific jobs or operational procedures. Specialized personal fall-arrest equipment, including a full-body harness and a shock-absorbing lanyard, may be required in some situations. Telescoping wall braces are needed to support taller walls during construction. Inspections, maintenance, and repair of tools and equipment are required periodically.

**FROM EXPERIENCE**

The cost of purchasing and providing personal protection equipment for employees, accessories for safe scaffold setups, telescoping wall braces, and other safety equipment and accessories is less expensive than the penalties and fines that a contractor pays for noncompliance with governing safety regulations.

A masonry contractor must purchase or lease and maintain one or more vehicles. A contractor may have additional expenses including the following:

- The purchase or lease of office and storage spaces.
- Accountant's fees for payrolls or filing taxes.
- Advertising expenses.

Not every masonry contractor has all of these expenses. But the larger a contracting business becomes, the more likely it is for the needs and related expenses to increase.

## Continuing Education and Training

A contractor must comply with legal regulations and building code requirements. Trade custom and local practice cannot be relied upon as acceptable performance. For example, even if the majority of contractors were not to install wall flashings and weeps or install sufficient wall ties supporting anchored brick veneer, no contractor is excused from complying with building code regulations. A contractor can attend education sessions and seminars, certification programs, and continuing education classes at colleges and technical centers, as well as subscribe to masonry trade publications to keep abreast of issues facing the masonry industry.

As one example, to stay informed of current trends emphasizing and promoting the design and construction of energy efficient and environmentally friendly buildings, a masonry contractor should become familiar with LEED Green Building Certification. The Leadership in Energy and Environmental Design (LEED) Green Building Certification system, developed by the U.S. Green Building Council and first used in 1999, is a voluntary program intended to encourage the construction of buildings whose construction materials and lifetime operation promote improved environmental quality and occupant health and well-being by addressing environmental issues and innovative designs. Based on a point system of a maximum of 69 obtainable points, a building may qualify for one of four levels of LEED certification: Certified, Silver, Gold, and Platinum. State and local governments may offer incentives to owners of buildings having Green Building Certification. With increasing interest in LEED Green Building Certification, a masonry contractor can benefit from understanding its goals and the ways in which masonry products and materials can be used to contribute to LEED credits. Being able to intelligently talk about important,

timely issues with clients, builders, architects, and engineers improves the contractor's credibility.

Many trade associations support the masonry industry by giving technical assistance to architects, designers, and contractors. A contractor should become familiar with these associations and their available technical assistance. These associations are often recognized as the authority on masonry construction. Construction industry publications and trade magazines address issues that include technical updates, new and improved products, and current federal, state, and local regulations.

Training of both the contractor and employees may be necessary to meet government regulations. For example, federal and state laws may require documented training for first aid and CPR certification, equipment operation certification, and government-controlled chemical usages certification.

In today's business environment, quality workmanship, productivity, and employee safety are directly related to employee training. Training is intended to:

- Remedy existing production problems.
- Present new techniques.
- Improve employees' skills.
- Introduce new products.
- Ensure proper equipment operation, inspection, and routine maintenance.

A masonry contractor should select and prioritize training topics. Training sessions can be held periodically on the jobsite.

**FROM EXPERIENCE**

Many contractors have regularly scheduled "tailgate meetings," times when all employees gather to share job-related concerns and to receive information, typically gathered at the jobsite parking area. Informal settings such as this are conducive to having workers share their concerns.

## Estimates and Contracts

An **estimate** is a judgment of construction costs. It can include material, labor, and equipment costs. Unlike a contract, an estimate is not a firm proposal. A **construction contract** is a written agreement executed between the property owner or their representative and the contractor. The construction contract includes drawings and specifications related to construction. Both parties must agree upon the rate or price for the work and time of payment to the contractor for services rendered. A description of materials, the party responsible for purchasing and securing materials, the party responsible for securing any necessary building permits, the provider of electricity and water, and, where applicable, construction deadlines should all be specified in the contract. Contracts can include the procedures for modifying the original contract or any other items specifically stipulated as being included in the contract documents. Depending on the details of the contract, attorneys may be paid to prepare the contract documents. After the terms and conditions stated in the contract are agreed upon by the parties involved, then the contract is signed and becomes a legal document.

Additional work and extended time resulting from owner-requested changes, errors or omissions in plans, and field problems require executed change orders. A change order is documentation of an addition, deletion, or change of the contract. Change orders should require the owner's signature before proceeding with changes or extra work not authorized in the contract. Procedures should be established in the terms of the contract for requesting additional payments and time extensions for change orders.

A masonry contractor must create revenue to be successful. Revenue is created once the contractor is recognized as a valuable resource who helps customers solve their problems and complete their projects. Revenue includes markups and profit margins. A markup is the dollar amount over and beyond material and labor costs needed for business operations. A markup produces a profit. A profit is the amount of money remaining after all expenses have been met. Profits may be used to expand the growth of the business. For example, new equipment may be purchased to improve quality and production. Profits can be shared with employees to reward present and to encourage future quality workmanship and productivity. Profits may be saved and invested for future retirement income. Successful contractors price their work to reflect fair profit margins.

## Documentation and Records

A masonry contractor must supervise the masonry construction so that it meets contract specifications and building code requirements. The contractor is responsible for the performance of all employees. The contractor needs to keep records. Records are kept in a construction manual log. The construction manual log provides a daily record of progress, including the number of workers on the job, the number of hours worked, the quantity of units laid, the wall or walls being constructed, and the daily weather. The construction manual log records may be needed to defend job delays beyond the control of the masonry contractor. All concerns that may affect time schedules should be put in writing to the general contractor or owner. For example, if the general contractor or owner is not ordering materials in time for the masonry construction to begin as scheduled, the masonry contractor may want to send them a memo advising having materials available or changing the deadline. Instructions coming from the general contractor or owner should be required in writing also. For example, materials ordered by the masonry contractor but selected by the general contractor or owner should be described and specified

in writing. Even the mason working alone as a masonry contractor should keep daily records and written documentations for future reference.

For safety and environmental concerns, a contractor is required by law to have Material Safety Data Sheets available at the jobsite. The information they provide includes 1) trade and chemical name of the product; 2) physical data, including appearance and odor; 3) physiological effects and physical hazards; 4) special protection information; 5) emergency and first aid procedures; 6) storage and disposal information; and 7) fire and explosion hazard data. Every employee must have access to and be able to interpret Material Safety Data Sheets.

A masonry contractor should document and resolve any existing problems before construction begins. As an example, the contractor may arrive at a new jobsite only to find that the footings are not level. But the masonry contractor's work must exhibit acceptable workmanship meeting the requirements of local building codes, engineered specifications, and the dimensions given on the working drawings. The masonry contractor must be able to cut the block while maintaining its engineered strength and use acceptable width mortar joints without sacrificing the reliability of the wall.

Pictures should be taken to record and document existing conditions. For example, a footing whose surface is not level creates a problem for the masonry contractor. But the masonry contractor assumes responsibility for the wall that it supports once it is constructed. If the masonry contractor finds that existing conditions prevent constructing an acceptable wall that meets engineered specifications and local building codes, then the wall should not be constructed until the party responsible for the placement of the footing resolves the problem. Figure 24-7 illustrates a problem arising from building a brick entry column on a concrete footing either too close or too shallow in relationship to the root system of the adjacent tree.

**Figure 24-7 The brick column exhibits good workmanship. However, the supporting footing was too close to grade level, permitting the roots of a tree to grow below it and tilt the column.**

Pictures can also be used to document preexisting property damage such as broken windows, damaged asphalt or concrete driveways, and damaged shrubbery or trees.

## Reliable Service

Quality workmanship alone does not ensure a successful business. Quality workmanship is expected of everyone in business. In addition to quality workmanship, a successful masonry contractor develops a reputation for honesty and reliability. Without these attributes, public trust is lost. And without trust, jobs are nonexistent. Agreed upon time schedules must be honored. Mistakes occur, but they must be admitted to and corrected if contractors want to keep the public's trust that they have earned in the past.

### FROM EXPERIENCE

Be prompt to deal with a client's concerns. Do all that is possible to correct any client concerns. Thank the client for bringing justifiable concerns to your attention. Building trust with the public is time and money well spent.

And finally, to prevent costly errors, a masonry contractor should not assume the responsibilities of an engineer. Unless given a working drawing with engineered specifications, a masonry contractor should consult a professional engineer before beginning any structural project. The professional engineer, not the masonry contractor, is legally qualified to specify wall requirements. A **structural engineer** meets the licensing requirements to design a structure or parts of it to withstand all weights and forces that the structure may have to support. Although the masonry contractor may be a competent mason capable of building quality masonry walls, a licensed structural engineer should design and specify the construction details for structural walls.

The wall in Figure 24-8 exhibits quality masonry construction. However, it collapsed because it was not designed to withstand the forces exerted by the ground behind it. A masonry contractor should request working drawings giving engineered specifications before contracting to build any structural wall or retaining wall. Otherwise, the masonry contractor may be held legally responsible for any damages caused by a wall's failure, as well as the costs for reconstructing walls to meet design requirements.

**Figure 24-8** Heavy rains saturated the ground behind this wall and caused it to fall. It had not been engineered to withstand the forces of the expanding earth behind it.

## Summary

- The masonry industry offers a variety of career opportunities, including job foreman, job superintendent, project manager, estimator, and masonry contractor.
- There are many avenues for becoming a journey-level worker, including on-the job training, state registered training programs, industry sponsored training programs, and masonry programs available to high school and post high school level students at public tax-supported career and technical centers.
- Opportunities for jobs requiring highly manual skilled workers such as brickmasons are excellent and seem to be increasing.
- Becoming a masonry contractor requires much more than being a good mason. It involves preparing cost estimations and fulfilling contract requirements. It requires understanding and abiding by laws and building codes, complying with safety regulations and tax deadlines, and motivating employees to produce professional results.
- Success depends on more than workmanship. Quality workmanship is always expected. But more importantly, successful contractors must build a reputation based on reliability and honesty that encourages customers to recommend them to others.

## Review Questions

*Select the most appropriate answer.*

**1** A brickmason is responsible for installing mortared

a. brick and concrete masonry units

b. architectural precast concrete units

c. structural facing tile, marble, and granite

d. all of the above

**2** One who works for another is called an

a. attendant

b. employee

c. employer

d. both a and b

**3** The physical qualifications for a brickmason require

a. good physical stamina

b. good eye sight and a sense of proportion

c. good use of body joints

d. all of the above

**4** The reason that robotics has not replaced the manual labor performed by brickmasons is because

a. robotic machinery is too expensive for the construction industry to consider

b. robotic operations are restricted primarily to mass production assembly line operations

c. robotic machinery cannot be programmed to perform operations normally performed by manual operations of an individual

d. all of the above

**5** One can be recognized as a journey-level mason after

a. completing an apprenticeship training program offered by a state's department of labor and industry

b. completing an industry sponsored apprenticeship training program

c. completing a two-year masonry training program offered to high school grade students at a career and technical center

d. both a and b

**6** A factor influencing the total cost for doing work is

a. travel time and distance

b. site accessibility

c. weather conditions

d. all of the above

**7** A contractor whom government authority has granted permission to engage in masonry contracting is called a

a. building contractor

b. general contractor

c. legal contractor

d. licensed contractor

**8** Government regulations may require that an employee be certified in

a. first aid/CPR training

b. heavy machinery operation

c. usage and disposal of hazardous chemicals

d. all of the above

**9** A type of insurance that a contractor purchases for coverage of personal injuries and damage to the property of others arising from operations performed by the contractor is called

a. completed operations insurance

b. homeowners insurance

c. liability insurance

d. workers' compensation insurance

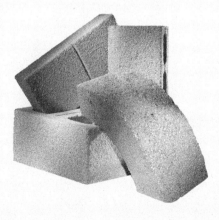

**10** A type of insurance that a contractor purchases for damages that either products or services may cause after the work is completed is called

a. completed operations insurance

b. homeowners insurance

c. liability insurance.

d. workers' compensation insurance

**11** A judgment of construction costs is called

a. a bill of materials

b. a contract

c. an estimate

d. a job bid

**12** A written agreement executed between the property owner of the proposed job or their representative and the contractor responsible for executing the job giving the terms, rates, or prices for doing the job is called a

a. a bill of materials

b. a contract

c. an estimate

d. a job bid

**13** A documentation of any additions, deletions, or changes of the contract is called a

a. change order

b. modification

c. revision

d. substitution

**14** A contractor should keep a daily record of progress called a

a. construction journal

b. construction manual log

c. job diary

d. site record

**15** The information included on Material Safety Data Sheets is

a. emergency and first aid procedures

b. material storage and disposal

c. fire and explosion hazard data

d. all of the above

**16** The person responsible for designing structural walls to withstand all weights and forces that they are to support is the

a. contractor

b. mason

c. project's owner

d. structural engineer

# Chapter 25 | Safety for Masons

Are you working in a safe environment? Don't bet your life on it! It's your health and your life, both of which you must give the utmost priority. While working, nothing should be more important than the prevention of accidents, injuries, and occupational illnesses. Workers who ignore safety regulations jeopardize their own safety and health as well as that of others. Following established safety procedures not only reduces the risk for injuries, but it eliminates safety violation citations, costly penalties, and employment termination.

## OBJECTIVES

After completing this chapter, the student should be able to:

- ⊗ Describe the proper work clothing, shoes, and personal protection equipment required for compliance with governing safety regulations when performing specific tasks.
- ⊗ Set up and maintain a safe work area in a masonry training lab.
- ⊗ Demonstrate the safe handling and storage of construction materials.
- ⊗ Explain why one must be trained by a competent person to use every tool, piece of equipment, or piece of machinery.
- ⊗ List procedures for reporting an accident or injury.
- ⊗ Recognize and report safety hazards.
- ⊗ Explain safety precautions for preventing electric shocks and fatal electrocution.
- ⊗ Identify supported scaffold components and explain safety regulation requirements related to each component.

# Glossary of Terms

**ANSI** a private organization acting as administrator and coordinator for voluntary standards.

**competent person** as defined by OSHA Construction Industry Regulations 1926.450(b) definitions, "one who is capable of identifying existing and predictable hazards in the surroundings or working conditions which are unsanitary, hazardous, or dangerous to employees, and who has authorization to take prompt corrective measures to eliminate them."

**coupling pin** a metal tubular pin inserted in tubular scaffold end frames permitting them to be stacked and locked for multitiered scaffold setups.

**Dense Industrial 65** the industry standard requirement for recognizing approved sawn scaffold planking.

**double insulated** electrical equipment with enclosures that are nonconductive to electricity.

**equipment grounding conductor** a wire going to ground connected to metal enclosures of electrical-powered tools, machinery, and equipment.

**ground fault circuit interrupters** also called GFCIs, a circuit breaker detecting as little as 5 milliamps of current difference between the amount of current going through one wire to a piece of electrical equipment and that returning by path of the other wire and within a fraction of 1 second, interrupting or stopping the current flow when the difference is greater than 5 milliamps.

**guardrails** rails installed along all open sides and ends of platforms serving as personal fall protection.

**Laminated Veneer Lumber** also called LVL, used for scaffolding planks, a manufacturing process of gluing together many thin layers of wood to produce engineered lumber.

**mid-rails** rails approximately midway from the top edge of the guardrail and the platform surface, intended to provide personal fall protection.

**mudsill** a 2" or thicker wood plank placed on the ground to prevent scaffold end frame supports from point loading, imposing a concentrated load on the supporting surface.

**NIOSH** an organization that conducts research and provides guidance, information, and services for subjects related to occupational safety and health.

**OSHA** a federal government agency that conducts inspections of and issues citations and proposed penalties for employers covered under the Occupational Safety and Health Act of 1970 for alleged violations of applicable safety and health standards

**personal protection equipment** also called PPE, required worker protection depending upon the hazards to which a worker is exposed, which may include eye protection, face protection, head protection, hearing protection, hand protection, foot protection, respiratory protection, protective clothing, and barriers.

**sawn planking** individual boards 1¾" to 2" thick and 10" to 11½" wide approved for scaffolds to support workers, materials, and equipment.

**scaffold** a temporary work platform designed to support workers, materials, and equipment.

**scaffold platform** a plywood or metal deck supported by an aluminum frame.

**supported scaffolds** platforms supported by legs, beams, or other approved rigid supports.

**suspension scaffolds** platforms suspended by ropes or other nonrigid means from overhead.

**toe boards** vertical barriers extending above a platform surface to prevent falling materials.

# Complying with Safety Standards

The construction industry depends on both government agencies and private organizations for the safety and health of those exposed to worksite environments. The Williams-Steiger Occupational Safety and Health Act of 1970 requires that employers comply with occupational safety and health standards, rules, and regulations issued under the act. The U.S. Department of Labor, Occupational Safety and Health Administration (OSHA) is a federal government agency that conducts inspections of and issues citations and proposed penalties for employers covered under the Occupational Safety and Health Act of 1970 for alleged violations of applicable safety and health standards. Employers can face monetary penalties and in some cases imprisonment for failure to comply with applicable standards. There are specific standards for the construction industry, covered in the 29 CFR 1926 OSHA CONSTRUCTION INDUSTRY REGULATIONS publication.

Because you are impacted by safety standards, you should be aware of the National Institute for Occupational Safety and Health (NIOSH). NIOSH is an organization that conducts research and provides guidance, information, and services for subjects related to occupational safety and health. For example, NIOSH conducts research leading to improved designs for whole-body, fall arresting harnesses. Some types of personal protection equipment must meet NIOSH standards.

The American National Standards Institute (ANSI) is a private organization acting as administrator and coordinator for voluntary standards. The design of safety eyewear and safety footwear meeting OSHA standards is also recognized as meeting specific ANSI requirements. For example, ANSI-approved safety footwear has stamps or labels that include reference codes explaining the safety rating of a particular shoe.

This chapter is addressed to those who are training to become masons and are enrolled in masonry programs at public high schools, technical training centers, or industry sponsored training sites. The recommendations are for the protection of the individual or groups of students working in masonry training labs, which are enclosed spaces having multiple training stations. Although some of the safety and health standards required by OSHA or the governing regulatory agency at a construction worksite may not be legal requirements in the masonry lab, it is important that students become accustomed to complying with those safety standards required of every employee at a construction worksite. In addition to pointing out many of the federal and state government standards required at a construction worksite, this chapter includes recommendations specifically for the masonry lab environment. These are recommendations intended for students' protection as they develop competencies for successful employment as apprentice masons.

**FROM EXPERIENCE**

All masonry students and persons exposed to construction jobsites should have knowledge of and access to the publication entitled "29CFR 1926 OSHA CONSTRUCTION INDUSTRY REGULATIONS," a reference to OSHA government regulations for the construction industry.

# Personal Protection Equipment

Personal protection equipment, also called PPE, is OSHA-required worker protection depending upon the hazards to which a worker is exposed and may include eye protection, face protection, head protection, hearing protection, hand protection, foot protection, respiratory protection, protective clothing, and barriers.

## Protective Clothing

Shirts having a neck collar and sleeves 4" or longer should be worn (see Figure 25-1). Shirts with a neck collar and sleeves protect your back, shoulders, neck, and chest area from harmful sun rays, overexposure to which causes skin cancer. Trousers should extend from the waistline to the ankles. Long trouser legs protect the legs from abrasions and sun rays. Loose fitting, baggy shirts or trousers should be avoided, since they are likely to be caught on other objects. Trouser legs extending below the ankles can easily be caught on objects and cause injuries.

**CAUTION**

**Shirts without neck collars do not offer any protection to the neck from the harmful effects of the sun's ultraviolet rays. Precancerous skin conditions and skin cancers can develop on the neck after significant sun exposure. Those exposed to the sun should wear shirts having neck collars and sleeves 4" or longer. Significant exposure to the sun can age the skin, damage its immune system, lead to precancerous skin conditions, and cause skin cancers. Malignant melanoma, the most deadly of all skin cancers, is best prevented by avoiding excessive sun exposure and sunburn. Applying protective sunscreen lotions having a high SPF (sunscreen protection factor) rating to exposed skin areas as directed by the supplier is recommended for those not allergic to the lotions. A lotion should offer protection from both UVA**

Figure 25-1 Shirts should have neck collars and sleeves 4" or longer. Trousers should extend from the waistline to the ankles.

Figure 25-2 Safety-toe footwear ankle high or higher, labeled by ANSI as having slip-resistant soles, is recommended for masonry trainees working on concrete floors.

light, penetrating the skin and causing deeper connective tissue damage, and UVB light, causing surface sunburn. Always wear protective clothing, because sunscreens do not completely protect you from the damaging rays of the sun.

## Foot Protection

Depending upon the hazards, safety-toe footwear may be required on construction worksites (see Figure 25-2). Footwear meeting the standards of the American National Standards Institute, ANSI Z41.1-1991, is required for meeting OSHA standards "...where there is danger of foot injuries due to falling or rolling objects, or objects piercing the sole . . ." (OSHA standard 1910.136a). It is recommended for students training in masonry labs to wear footwear meeting this OSHA standard. In addition, the following are footwear recommendations for students training in masonry labs.

1. Footwear meeting ANSI standards (the requirements of the American National Standards Institute) and listed as having slip-resistant soles should be worn in masonry labs because of the potential for slipping on wet concrete floors.
2. Hard, thick soles may reduce the risk of foot injuries caused by object penetration.
3. Shoes should extend above the ankles, with fully laced and snugly tied shoestrings to support the ankles and reduce the potential for debris entering the shoes.

### CAUTION

Sharp objects such as the pointed ends of nails and screws can penetrate shoe soles, piercing the feet and causing injuries. Never leave the sharp ends of nails or screws in building materials exposed. Promptly remove them from material or discard material inside a refuse container. Rusty nails and screws penetrating the skin can cause serious occupational illnesses, including tetanus, which can be fatal. Consult your family doctor for recommendations regarding a tetanus vaccination.

## Eye Protection

Eye protection meeting ANSI standards Z87.1-1968 are required to meet the intent of OSHA standards "...when machines or operations present potential eye or face injury from physical, chemical, or radiation agents" (OSHA standard 1926.102(a)1). To prevent potential eye injuries resulting from cement dust, mortar, brick or block fragments, and other airborne or falling debris, protective eyewear bearing the marking "Z 87" should be worn (see Figure 25-3). Those whose vision requires wearing corrective lenses must have protective lenses and side shields for their eyewear. Goggles may be worn over corrective lenses so long as the goggles do not alter the lens correction.

## Face Protection

In addition to eye protection, a face shield is required when activities present the potential for facial injuries, such as when performing grinding and sawing operations, using pressure washing systems, or being exposed to hazardous chemicals (see Figure 25-4).

## Hearing Protection

Noise levels must not exceed those listed in Section 1926.52, Table D-2, Permissible Noise Exposure, of the 1926 OSHA CONSTRUCTION INDUSTRY REGULATIONS. Personal protective equipment such as ear muffs, ear plugs, or ear pods are required when exposure to noise is not below the limits

**Figure 25-4** In addition to safety glasses, a face shield offers additional protection when performing procedures having the potential to cause facial injuries.

listed in the OSHA standards. Their effectiveness, rated in decibels for noise reduction, varies. Such protections do not totally eliminate noises, but they reduce noises below levels that have potentials for hearing loss. Students should have adequate hearing protection to wear in the masonry lab whenever noise levels require (see Figure 25-5). Those close to noise, such as that of any masonry saw, should be wearing hearing protection.

## Head Protection

On construction worksites, OSHA standards require the wearing of hard hats "...where there is a possible danger of head injury from impact, or from falling or flying objects, or from electrical shock and burns..." (OSHA standard 1926.100(a)). For example, hard hats can resist blows to the head from falling objects such as falling brick chips deflected by scaffold bracing. They can prevent injuries resulting from bumping into objects such as roof trusses at head level when working on scaffolds (see Figure 25-6). Hard hats

**Figure 25-3** Protective eyewear bearing the marking "Z 87" should be worn by those exposed to masonry dusts and the hazards of airborne debris.

Figure 25-5 Ear muffs protect the saw operator from high noise levels of the saw.

Figure 25-6 Hard hats should be worn by those working on or below scaffolds or ladders as well as any other time there is danger of head injury from impact, falling, or flying objects.

Figure 25-7 A particulate respirator provides protection from occupational diseases resulting from breathing airborne dust.

are to meet ANSI standards Z89.1-1969, having a 6-point web-strap suspension system to spread an impact over a larger area. Bump caps with a 4-point suspension system can be worn where hard hats are not required, protecting the masonry worker from head injuries when unthinkingly raising their head and bumping overhead objects.

## Respiratory Protection

Respiratory protection is required "...when such equipment is necessary to protect the health of the employee." (OSHA standard 1910.134 (a)(2)). Respiratory hazards to which masons are most frequently exposed are airborne cement, sand, and dusts created from dry-sawing masonry units. Those working in environments contaminated with airborne dusts associated with masonry are at risk for developing occupational-related respiratory diseases such as silicosis. Breathing air contaminated with cement dusts, products containing silica, or sand over a period of time can cause occupational diseases impairing the lungs.

### Dust Masks and Particulate-Filter Respirators

Dust masks prevent you from breathing nuisance dusts such as sawdust, gypsum, limestone, and calcium silicate. However, dust masks do not provide respiratory protection. NIOSH-approved respirators are required for protection from silica, toxic dusts, fumes, and mists. Students should wear particulate-filter respirator masks whenever there are dangers of breathing mineral dusts such as cement, products used for the mixing of "practice mortars" containing silica, airborne sand, or those created when using saws for dry-cutting operations (see Figure 25-7). Filters with minimum efficiency ratings of 99.97% are available, removing practically 100% of airborne dust that you breathe.

Consult your doctor before wearing a dust mask or a particulate-filter respirator to be certain that these devices do not pose potential threats to your health. For example, those with extra sensitive breathing passages or diagnosed with asthma should not be exposed to airborne masonry dusts or wear dust masks or respirators without first consulting their doctor.

### Ventilation

OSHA Construction Industry Regulations, Section 1926.57, require ventilation "whenever hazardous substances such as dusts, fumes, mists, vapors, or gases exist or are produced in the course of construction work, their concentrations shall not exceed the limits specified in 1926.55(a)." Ventilation systems must be provided whenever levels reach the limits specified. Appropriately sized and approved ventilation systems should be provided to remove airborne contaminates away from those exposed.

> **CAUTION**
>
> Wear a particulate-filter respirator, and operate an exhaust fan, duct, or hood of approved volume and velocity, designed and approved to collect and exhaust dust to the outside whenever exposed to visible levels of airborne dust in a masonry lab.

### Hand Protection

To protect the hands, general-purpose, industrial gloves are recommended when carrying masonry units, steel rebar, joint reinforcement, and lintels. Besides leather work gloves, latex-coated, cotton-lined, general-purpose industrial gloves that have high tear ratings and offer improved hand dexterity and grip are available. To eliminate the risk of allergic reactions, latex-free, synthetic rubber gloves are available. Anti-vib gloves, worn to absorb and dampen shock, are recommended when using impact tools, grout vibrators, and hammers and chisels. Chemical-resistant, liquid-proof gloves are required when exposed to brick cleaning chemicals. It is recommended that students wear gloves when disassembling masonry projects and when carrying masonry units (see Figure 25-8).

 **FROM EXPERIENCE**

The rough edges and surfaces of wood boards can become ensnared in the fabric of gloves. The careless practice of tossing boards when wearing gloves can cause injuries to hands, arms, and shoulders if a board ensnares the glove as it is thrown. The careless practice of pitching scaffold boards or planks from scaffolds should be avoided. You might be pulled off balance and fall from scaffolds if your gloves become caught on boards or planks tossed from scaffolds.

**Figure 25-8** Gloves protect the hands when handling masonry materials.

> **CAUTION**
>
> Remove jewelry when working. Finger rings can restrict blood circulation if the finger begins to swell following an injury. Objects falling on rings may compress them, limiting blood circulation and possibly severing fingers. Wristwatches, bracelets, necklaces, and other body jewelry should be removed to prevent them becoming entangled in construction materials and machinery components.

## Maintaining a Safe Work Area

In a masonry lab with training areas or stations, considerations for maintaining a safe work area include 1) setting up and maintaining a safe work station, 2) transporting

**Figure 25-9** Keep floors free from mortar droppings and debris.

materials and equipment to and from the work station safely, and 3) discarding waste and debris properly.

A clean lab is a safer workplace for everyone. Floor surfaces, typically concrete slabs, must be clear of mortar and scraps. Immediately remove dropped mortar from the floor and reuse. Put masonry scraps in a bucket (see Figure 25-9).

Wet concrete floors can cause you to slide and fall. Place a traffic caution cone at the spill, and wipe up spills immediately (see Figure 25-10). Sprinkling sand on damp floor areas increases foot traction.

Placing tempering water containers under a mortarboard stand or on a materials platform reduces the potential of water spills.

Storing shovels, hoes, scrapers and brooms on tool hangers is a recommended safety precaution taken in masonry labs (see Figure 25-11). It is unsafe to leave tools leaning against walls or other objects because they may fall to the floor, creating tripping hazards.

Place building materials no closer than 2' to the wall being constructed. Although material stockpiles in material storage areas are usually higher, stack materials no higher than 2', about knee high, at the work station where the lab project is being constructed (see Figure 25-12).

Transport wheelbarrows and materials carts within marked traffic lanes (see Figure 25-13). Limit foot traffic to traffic lanes also. Avoid transporting materials through areas identified as work stations to prevent damaging projects or injuring those at the work stations. Never overload wheelbarrows. Never fill a 6-cubic foot contractor's wheelbarrow more than half the height of its sides with mortar, sand, gravel, brick, or other building materials.

Place the shovel or hoe handle between the body and arm when wheeling a barrow (see Figure 25-14). Never leave these handles out in front of or off to the side of the wheelbarrow where they could strike others.

Put the brick tongs over the brick when transporting the cart or barrow on which the brick are loaded in order to secure the brick (see Figure 25-15).

Keep materials evenly distributed on the material carts or wheelbarrows. Stack and remove building materials side by

**Figure 25-10** Mop up spills immediately, and identify the potentially slick floor with a traffic caution cone.

**Figure 25-11** Store shovels, hoes, scrapers, and brooms in ways that prevent them from falling to the floor.

side, one layer at a time to prevent the cart or barrow from upsetting. Having materials on just one side causes the cart or barrow to upset (see Figure 25-16). Do not exceed the capacity weight limits posted on them.

**Figure 25-12** Place building materials no closer than 2' and no higher than 2' at the work station.

**Figure 25-13** The transport of wheelbarrows and materials carts is restricted to those areas of the masonry lab identified as traffic lanes.

When a loaded wheelbarrow is off balance and upsetting, let it fall on its side if no one is nearby who could be injured (see Figure 25-17). Trying to regain balance of an upsetting wheelbarrow can cause muscle and back injuries.

Have the wheelbarrow headed in the direction of intended travel before loading it (see Figure 25-18). Wheelbarrows can become off balanced when attempting to pivot their tires 180 degrees.

The plastic straps or steel bands securing cubes of new brick must be removed.

**CAUTION**

**Brick on uneven or soft terrain can fall, entrapping and injuring those nearby. Have a competent person who is authorized and certified to use appropriate machinery or equipment relocate the cubes that have the potential for falling.**

Following these procedures to remove straps or bands is suggested:

1. Make sure the cube of brick appears to be stable before cutting any packaging straps or bands. If the brick are on sloping ground and have the potential to fall after the strap is removed, take caution and stand on the upper side of the brick if the strap is to be removed rather than having a fork-lift move the cube to level ground. Cut first the band or strap securing the five separate straps of brick using a pair of tin snips (see Figure 25-19).

2. Place properly width-adjusted brick tongs across the top of the strap of brick. Cut each band or strap securing the five individual straps of brick with a pair of tin snips at the top of the strap or band (see Figure 25-20).

3. About 6" from the bottom of the strap or band, cut them free (see Figure 25-21). Make bends along their length, folding them over to fit into a refuse container.

4. The ends of the remaining metal bands can cause injury. Bend the ends of these bands over to prevent cuts (see Figure 25-22).

Figure 25-14 When transporting wheelbarrows, place long shovel or hoe handles between the body and arm so they cannot strike others.

Figure 25-15 Use brick tongs to help secure loads.

> **CAUTION**
>
> Never grab a metal band or plastic strap and pull on it in an attempt to remove it from under brick. You might lose your grip, causing the band or strap to slice your hands.

# Lifting and Carrying Materials by Hand

To prevent injuries associated with lifting, replace mortar on mortarboards or pans back into wheelbarrows using a shovel or trowel rather than trying to lift the entire board or pan to empty it. With arms fully extended from the body when attempting to carry a mortarboard or pan containing mortar alone, one can strain muscles or experience back injuries. If it

Figure 25-16 Do not have materials on only one side of a wheelbarrow or materials cart. As illustrated here, it upsets.

becomes necessary to lift a mortarboard or mortar pan loaded with mortar, always pair up with another (see Figure 25-23).

Lifting in unison, stoop and use your leg muscles when lifting heavy building materials rather than leaning over to

Figure 25-17 When no one is near, letting an off-balanced wheelbarrow fall on its side prevents potential back injuries when trying to regain balance of heavy loads.

Figure 25-18 Wheelbarrows should be loaded once positioned in the intended direction of travel.

Figure 25-19 Cut the strap or band securing the five separate straps of brick only after the brick are on level ground or when standing on the upper side of the brick cube if there is danger of any of the five straps of brick upsetting.

Figure 25-20 Having brick tongs placed across the brick, cut the metal band or plastic strap with tin snips near the middle of the top of the brick cube.

Figure 25-21 About 6" from the bottom of the cube and on each side, cut the band free and dispose of it in a refuse container.

Figure 25-22 Folding over the ends of the remainders of the bands reduces their potential for causing injuries. Wearing work gloves the entire time that bands are removed protects hands.

Figure 25-24 Bend at the knees, and use the leg muscles rather than the back to lift heavier objects. When lifting heavier items, you should wear heavy-duty leather work gloves providing a strong grip and extra protection between your hands and potential injuries.

place solid materials capable of supporting its weight below it to provide clearance for removing your hands and fingers.

### FROM EXPERIENCE

Two or more persons must lift in unison to prevent injuries. For example, lifting one end of a lintel without simultaneously lifting the other end tilts the opposite end down, pinching the hands and fingers of the person whose hands are under the opposite end.

### CAUTION

**Have good footing and traction when lifting and carrying heavier materials. Know the safe working weight limits for scaffold planks, ramps, and other surfaces, and do not exceed them when lifting and carrying heavy materials.**

Carry joint reinforcement wire or steel reinforcement rods by holding them in a position so as not to injure someone with the ends or make contact with electrical wires and other overhead obstructions (see Figure 25-25).

### CAUTION

**Contact with overhead electrical wires can be deadly. Electrically conductive wire joint reinforcement cannot be handled wherever it has the potential for coming closer than 10' to overhead high voltage electric wires.**

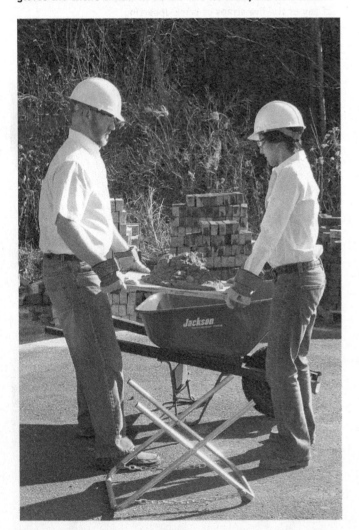

Figure 25-23 Pair up with another when lifting loaded mortar boards or pans, stooping if necessary to reach low.

lift, which is an unsafe procedure leading to back injuries (see Figure 25-24). Protect your hands and fingers with leather or other types of abrasion-resistant heavy-weight work gloves allowing a strong grip. Before setting down heavier objects,

**Figure 25-25** Carry joint reinforcement in such a manner as not to impale others with its ends or to come within 10' of high voltage overhead lines. Be aware of the location of both of its ends at all times. To be safe, all overhead lines should be considered potentially dangerous, high voltage lines.

# Maintaining a Clean and Safe Lab

Every student is responsible for leaving his or her work area clean and unobstructed at the end of each work day. Put all brick and block considered by the instructor as waste in the place the instructor designates. Temper unused mortar and return it to a mortar wheelbarrow, mortar box, or mixer as the instructor requests.

Move unused brick and block from the open floor area, placing them beside the assigned project, onto material carts or wheelbarrows, or in storage piles. Bringing no more materials to the work area than those expected to be used during a class period eliminates overcrowding work areas with materials. Sweep floors with brooms after all materials are properly stored. Place traffic caution cones at projects that are less noticeable. A well-cleaned masonry lab is illustrated in Figure 25-26.

**Figure 25-26** Floors are clean and traffic lanes unobstructed. Traffic caution cones are placed near projects that are less noticeable.

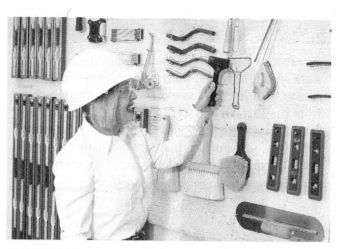

**Figure 25-27** A tool storage panel keeps tools readily accessible and easily inventoried.

Clean each tool and return it to its designated storage promptly after using (see Figure 25-27). Shovels, hoes, scrapers, and brooms should be returned to their assigned storage. Leaving long-handled tools leaning against walls is unacceptable, because they may fall to the floor and create tripping hazards.

Masonry construction has the potential for many safety hazards. A desire to work in a safe area is an attitude needed for a successful and safe employment experience. Students are evaluated on the efforts demonstrated for keeping a clean, well-organized, and safe masonry lab.

When it is time to stop working, immediately return tools and materials to their designated storage area. Working in teams, sweep the lab floor with brooms, and put debris in the place designated by the instructor.

# Using Hand Tools Safely

Do not use any tool until the instructor or other designated competent person demonstrates its proper use and care. Comply with OSHA PPE standards and the instructor's safety rules regulating the tool and the task for which it is used.

Inspect tools before using them. Give defective tools to the instructor rather than returning them to storage. Comply with all safety instructions provided by the manufacturer for each tool and with those safety instructions and procedural demonstrations given by the instructor.

## Hammers

Never strike the face of a hammer with another hammer. Hardened steel can chip, posing hazards to others as metal fragments become airborne. Strike metal chisel heads with only those hammers intended for striking hardened steel. Be aware of the presence of others when using the hammer. Turn away from others, and direct the hammer strikes toward the masonry lab floor whenever holding brick for breaking. Hold the brick above a bucket so that broken debris falls into it or is immediately placed in it.

**CAUTION**

In addition to safety eyewear, wear heavy-weight work gloves offering a strong grip when breaking masonry units with hammers. Keep fingers away from the surface impacted by the hammer, as illustrated in Figure 25-28. Placing a brick needing cutting atop a bucket of sand stabilizes and cushions it, typically permitting a more precise break.

## Chisels

Impacting chisels with hammers causes "mushrooming" of the metal where struck with hammers. Noticeable splinters of "mushroomed" metal should be removed using a file (see Figure 25-29).

Figure 25-29 A metal-cutting file removes metal fragments from the head of the chisel.

**CAUTION**

In addition to eye protection, wear a face shield and leather or other types of abrasion-resistant heavy-weight work gloves allowing a strong grip when using chisels to cut or chip material.

## Masons' Lines and Line Holders

Replace masons' lines that are showing wear with new lines. Worn lines, those showing evidence of separate strands weakening or broken, may break under tension and pose a hazard to nearby workers. Examine lines frequently each day, and discard worn lines (see Figure 25-30). Routinely replacing lines with new line is an inexpensive solution for preventing injuries caused by lines breaking under tension. Although it is not necessarily a safety hazard to remove damaged sections of line and tie knots to make longer lines, attempting to make lines longer by knotting is not recommended. Knotted lines

Figure 25-28 Wearing gloves to absorb impacts and cushion the hand, fingers grip the shank, preventing them from contacting the masonry unit once broken. A bucket of sand supports the brick.

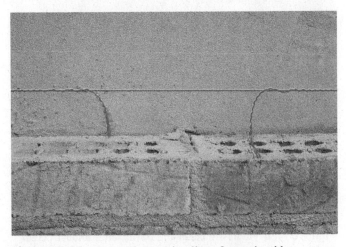

Figure 25-30 Frequently examine lines for noticeable wear or breaks, and replace damaged lines immediately.

can interfere with masonry units as they are being aligned to the line.

Use line blocks, line dogs, or line pins to attach masons' lines to walls. Never use nails to secure tensioned lines. Nails are easily dislodged from mortar joints, becoming airborne projectiles with the potential for serious injuries. Always use masons' line pins instead of nails to secure tensioned lines.

Be careful with twigs. Remove twigs from lines before tensioning them. If a line breaks while being tensioned, attached twigs become airborne, posing safety hazards to those nearby.

Remove tensioned lines from walls when unattended and always at the end of the work day. Removing masons' lines when not in use preserves the integrity of the lines and eliminates a source for potential injuries should a line break while tensioned.

## Levels

Placing the masons' levels in the open cells of 8" × 8" × 16" CMUs on the lab floor next to the mortarboard keeps them readily available at the work station (see Figure 25-31).

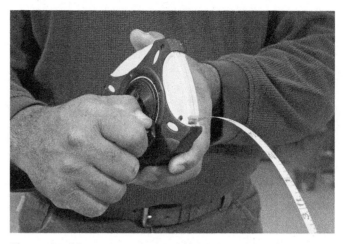

**Figure 25-32 Prevent slicing the fingers by being aware and keeping them away from the edges of tape-measuring blades.**

The weight of levels may cause smaller size block to tip over. The open cells of larger size block are too large to support levels upright. Levels laid on floors have the potential to become damaged and pose tripping hazards. Inspect levels daily, and give levels with broken cover glasses to the instructor for repair.

## Measuring Tapes

Keep fingers clear of measuring tape blades (see Figure 25-32). Measuring tapes can slice fingers in contact with the tape's edges as it is pulled from or retracted into the reel.

 **FROM EXPERIENCE**

Tools with sharp or pointed edges that are placed in pockets can cause injuries. Their edges can cause puncture wounds, especially when bending or stooping. Place only a 6' folding rule or curved jointer in hip pockets. Pencils should be clipped inside shirt pockets. All other tools should remain out of clothing pockets.

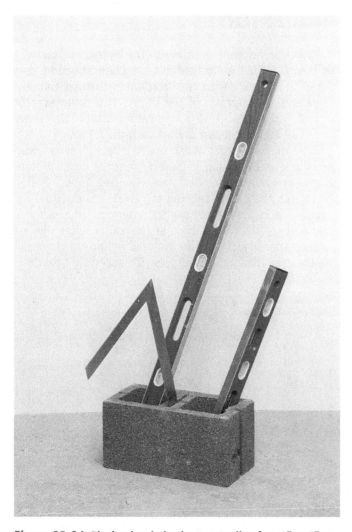

**Figure 25-31 Placing levels in the open cells of an 8" × 8" × 16" block keeps the levels accessible. Block with a center slot accommodate a framing square.**

# Electric Powered Tools, Machinery, and Equipment

Inspect electrical cords before using them. It is not permissible to use electrical equipment whose conductor insulation is damaged or missing. 120-volt tools and machines that

**Figure 25-33** Examine electrical plugs to be sure the equipment grounding conductor is intact.

**Figure 25-34** An in-line GFCI cord set is connected to the electric receptacle, providing ground fault protection to those exposed to electrical cords or equipment beyond the connection.

have metal enclosures must have three-prong plugs. The plug consists of two flat prongs that complete a 120-volt circuit and one round prong serving as an **equipment grounding conductor,** a wire going to ground connected to metal enclosures of electrical-powered tools, machinery, and equipment (see Figure 25-33).

The equipment grounding conductor is intended to help protect the operator from electrical shocks should a faulty wire contact the metal equipment housing. Never use electrical tools or machines whose round, equipment ground prong is missing. Even with an equipment grounding conductor, there is still a potential for one to experience electrical shocks and fatal electrocution. Faulty wires, plugs, or a grounding system can still cause equipment operators to receive dangerous electrical shocks. Equipment considered to be **double insulated,** that is having enclosures nonconductive to electricity, prevents operators from shocks originating from internal tool wiring failures. As with any electrical equipment, faulty or exposed wiring leaves potential for electrocution. For maximum personal protection, all 115-volt and higher powered tools on construction sites must be connected to power lines having approved **ground fault circuit interrupters** (GFCIs). A GFCI is a circuit breaker that detects as little as 5 milliamps of current difference between the amount of current going through one wire to a piece of electrical equipment and that returning by path of the other wire within a fraction of one second, and thereby interrupts or stops the current flow when the difference is greater than 5 milliamps. In-line GFCI cord sets—having quad or duplex outlet boxes, single or triple taps, and rated at either 120 volt/15 amp or 120 volt/20 amp—permit the maximum personal protection for equipment operators. A three-tap in-line 120-volt GFCI is shown in Figure 25-34. 220-volt GFCIs are available for machines such as stationary masonry saws, mixers, and other 220-240 volt machinery and equipment.

# Reporting Accidents and Injuries

Report an accident or injury to the instructor. Following an injury, 1) first aid procedures are administered by only those in authority with certification in first-aid training, 2) parents are informed of the injury, 3) if necessary, the injured party is taken to a medical facility for treatment, and 4) an accident report is filed with the school.

Have the instructor's assistance to properly use an emergency eye wash station if mortar, masonry dust, airborne debris, or chemicals contact the eyes (see Figure 25-35). Do not rub the eye, as rubbing can create more serious eye injuries. Seek professional medical attention after using an eye wash solution. Only those at medical facilities who are properly trained and have the required equipment can examine the eyes thoroughly and prescribe necessary treatment.

**Figure 25-35** An emergency eye wash station provides an appropriate sterile solution for flushing the eyes.

**FROM EXPERIENCE**

Sit down after an injury. An injury can cause you to pass out. Immediately sit down and call for the instructor or have another person get the instructor following an injury accompanied by pain. Some have been known to pass out after small injuries not considered to be medical emergencies. For example, a worker accidentally hit his finger with a hammer, bruising the finger but not cutting or fracturing it. The accompanying pain caused him to black out, temporarily losing consciousness and falling onto a concrete floor. An emergency situation, a fractured skull, preceded treatment for the bruised finger.

Do not administer first-aid to other students. Laws and governing school regulations typically require a person to not only be certified in first-aid training but also to have the custodial care or authority to treat students. Immediately tell the instructor or person supervising students when an injury or accident is witnessed. OSHA standards require that unless professional medical services are reasonably close in time and distance to a worksite, "... a person who has a valid certificate in first aid training... shall be available at the worksite to render first aid" (OSHA standard 1926.50(c)).

Accidents, injuries, and safety violations can be prevented if machinery, tools, equipment, and materials are used by students only after they are trained by a competent person. Even then, having little experience, the student should have permission and be supervised by the instructor or a competent person in charge. Filed records must show evidence that a student is knowledgeable of the safe and proper use for each specific hand tool, power tool, machinery, or equipment before using it. Printed records indicating that the student has demonstrated to the instructor proper and safe use of each tool before using it should also be filed. This is not unlike governing regulations for adult workers at worksites. Current and valid certifications, which are earned only after verified training and testing, are required before operating certain machinery, using equipment, or handling chemicals.

## Guidelines When Exposed to and/or Using Products and Chemicals

Never use products or chemicals without supervision, proper training, wearing apparel, and required personal protection equipment. Using chemicals such as those for cleaning brickwork requires knowledge of 1) required personal protection equipment, 2) the physiological effects and physical hazards of exposure to the chemical, 3) emergency and first-aid procedures relevant to exposure to the chemical, 4) environmental regulations controlling the use and disposal of the specific chemical, 5) the chemical manufacturer's recommended procedures for applying the chemical, 6) the brick and mortar manufacturers' recommendations for using a particular chemical, and 7) storage and disposal of the chemical.

## Material Safety Data Sheets

Understand the purpose for and information given in Material Safety Data Sheets (MSDS). Material Safety Data Sheets provide the following:

- The trade and chemical name for a product
- Physical data (appearance, odor, and so on) for a product
- Fire and explosion hazard data for a product
- Physiological effects and physical hazards for short-term and long-term exposure to a product
- Emergency and first-aid procedures related to a product
- Special protection information for using a product
- Storage and disposal information for a product

Know where MSDSs are kept. MSDSs should be accessible in the event of an accident or injury (see Figure 25-36). A MSDS must be on file at the worksite for every product present. For example, every type of masonry cement, masonry sand, chemical cleaners, acids, fuels, oils, lubricants, paints, stains, pigments, and aerosol products such as wasp and hornet sprays. When going for treatment of an injury or illness resulting from a product, the involved product's MSDS should accompany the injured or ill person so that medical personnel have information needed for treating the person.

**Figure 25-36 Copies of Material Safety Data Sheets for all relevant materials and products must be readily available both at the job site and in the masonry training lab.**

# Scaffolds

**CAUTION**

Scaffolds should be erected, dismantled, or accessed only by experienced and trained workers selected by a competent person. OSHA CONSTRUCTION INDUSTRY REGULATIONS 1926.450(b) defines a competent person as "one who is capable of identifying existing and predictable hazards in the surroundings or working conditions which are unsanitary, hazardous, or dangerous to employees, and who has authorization to take prompt corrective measures to eliminate them."

A scaffold is a temporary work platform designed to support workers, materials, and equipment. There are two classifications of scaffolds: supported scaffolds and suspension scaffolds. Supported scaffolds are supported by legs, beams, or other approved rigid supports. Suspension scaffolds are suspended by ropes or other nonrigid means from overhead. This chapter addresses only tubular welded steel frame scaffold, a type of supported scaffold commonly used by masons. There are two types of tubular welded steel frame scaffolds: 1) step-type end frames, and 2) open end frames. End frames are fabricated from round tubular steel having an outside diameter of either 1½", 1⅝", or 1¹¹⁄₁₆", and a 0.095" typical wall thickness. Because of heavy material loads, scaffold end frames used for masonry construction should be fabricated from 1⅝" or 1¹¹⁄₁₆" steel tubing. With only a few exceptions, OSHA construction industry regulation 1926.451(a)(1) requires scaffold and scaffold components to support "...its own weight and at least 4 times the maximum intended load applied or transmitted to it."

**CAUTION**

OSHA construction industry regulation 1926.451 (f)(6) standards require a minimum clearance of 10' between scaffolds and uninsulated overhead power lines less than 50 kv. To be safe, one should consider all overhead power lines as being uninsulated, maintaining a 10' minimum clearance. It is recommended to have the power company install protective coverings over electrical power lines in the vicinity of scaffolds to prevent electrocution should accidental contact with high voltage lines occur.

## Step-Type End Frames

The 5' wide by 5' high step-type end frame is a popular choice for house construction (see Figure 25-37). Moving the work platforms or wood planks from one horizontal step rail to another changes the working height.

Diagonal pivoted braces connected to end frames build a rigid scaffold (see Figure 25-38). The choice of braces permits 7' or 10' spacing between end frames. Because masonry materials are heavy, masons use scaffolds with a 7' spacing. Depending on the type of locking provided on the end frame, the ends of braces are available with either notches or holes.

**CAUTION**

OSHA standards require all braces to be in place and prohibit workers from climbing scaffold braces.

**Figure 25-37** This popular design for a 5' wide × 5' tall scaffold end frame permits adjusting the working height as a masonry wall is raised. *(Courtesy of Bon Tool Company)*

**Figure 25-38** A diagonal pivoting cross brace attaches between two end frames on both the front side and back side of the frames. *(Courtesy of Bon Tool Company)*

Although they are called step-type end frames, Section 1926.451 (e)(9) of the OSHA CONSTRUCTION INDUSTRY REGULATIONS gives standards with which all must comply when accessing, erecting, and dismantling scaffolds.

**CAUTION**

OSHA Construction Industry Regulations 1926.451 (e)(9)(iv) states that "cross braces on tubular welded frame scaffolds shall not be used as a means of access or egress."

## Open-End Frames

A 5' wide by 6' 4" or 6' 6" high open-end frame is another popular choice for masons' scaffolds (see Figure 25-39). Side brackets are attached to the end frames to support masons building the wall (see Figure 25-40).

Figure 25-40 Shown are the adjustable width side bracket (top), 20" side bracket (middle), and side and end bracket (bottom). *(Courtesy of Bon Tool Company)*

**CAUTION**

OSHA CONSTRUCTION INDUSTRY REGULATIONS 1926.452(c)(5)(iii) permit brackets supporting cantilevered loads to support personnel only unless a qualified engineer has designed the scaffold for other loads, and it is capable of resisting tipping forces.

# Scaffold Planking

Supported by the scaffold frames, sawn planks, engineered lumber, or fabricated work platforms carry the loads of workers, materials, and equipment.

## Sawn Planking

Sawn planking consists of individual boards 1¾" to 2" thick and 10" to 11½" wide. Sawn wood planks must meet the industry standard known as **Dense Industrial 65**. Select structural Douglas Fir, Southern Yellow pine, or their equivalents are recommended woods to use. Metal plank ties or drilled tie rods secured into both ends prevent sawn planks from splitting (see Figure 25-41).

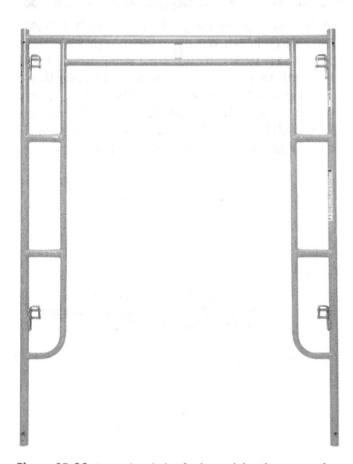

Figure 25-39 A popular choice for larger jobs, the open end frame scaffold is larger than the step-type end frame and has provisions for attaching side brackets from which masons work, reserving the entire width of the planked frames for laborers and materials. *(Courtesy of Bil-Jax)*

Figure 25-41 Plank ties, whose serrated edges are driven into the ends of scaffold planks, are typically used to prevent the ends of boards from splitting. *(Courtesy of Bon Tool Company)*

### CAUTION

Construction-grade material, such as 2" thick lumber used for floor joists, is not suitable for scaffolding planks because it is capable of supporting weight when set on edge rather than when laid flat. Construction-grade framing lumber may break when loads are imposed upon its wide flat surface.

### CAUTION

OSHA CONSTRUCTION INDUSTRY REGULATIONS 1926.451(f)(16) require planks to deflect no more than 1/60 of the span when loaded. To comply with this regulation, loads on scaffold planks spanning 7' end frame spacing should cause them to deflect not more than 1⅜" (see Figure 25-42).

## Engineered Lumber

Laminated Veneer Lumber, also known as LVL, is used for scaffolding planks. A manufacturing process of gluing together many thin layers of wood produces engineered lumber, the LVL-type scaffold planking. The LVL planking is usually 1½" or 1¾" thick and 9¼" or 11¾" wide. LVL-type scaffolding planks are heavier than sawn wood planks.

### CAUTION

With few exceptions, OSHA CONSTRUCTION INDUSTRY REGULATIONS 1926.451(b)(4) and 1926.451(b)(5)(i) require planking 10' or less in length to extend no less than 6" or more than 12" beyond the centerline of its support (see Figure 25-43). Where planks overlap to create longer platforms, "the overlap shall occur only over supports, and shall not be less than 12",

1⅜"
MAXIMUM
DEFLECTION

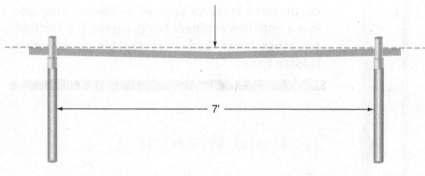

7'

Figure 25-42 Loads placed on planks spanning 7' must not cause them to deflect more than about 1⅜".

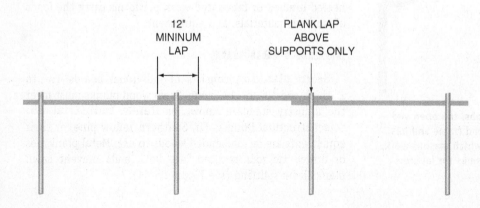

12"
MININUM
LAP

PLANK LAP
ABOVE
SUPPORTS ONLY

Figure 25-43 Where planks overlap to create longer platforms, the overlap must occur only over supports and be no less than 12" unless the planks are nailed together or otherwise restrained to prevent movement.

unless the platforms are nailed together or otherwise restrained to prevent movement" (Section 1926.451(b)(7).

## Scaffold Platforms

A scaffold platform consists of a plywood or metal deck supported by an aluminum frame. The ends of the platforms hook over the scaffold frames. Unlike planking, which has to be lapped at the ends, offset hooks permit arranging scaffold platforms in a smooth and evenly flat surface.

> **CAUTION**
>
> Regardless of the type of scaffold planking (sawn, LVL, or platform), OSHA CONSTRUCTION INDUSTRY REGULATIONS 1926.451(b)(1)(i) permit no more than 1" between adjacent planks or platforms and the space between platforms and frame uprights "except where the employer can demonstrate that a wider space is necessary (for example, to fit around uprights when side brackets are used to extend the width of the platform)."

## Mudsills

Mudsills are 2" or thicker wood planks placed on the ground to prevent scaffold end frame supports from point loading, imposing a concentrated load on the supporting surface. Mudsills prevent base plates or screw jacks from sinking in soils or breaking through surfaces below them.

> **CAUTION**
>
> Scaffold planks used as mudsills and stained with soil should not be used again as scaffold planks. OSHA CONSTRUCTION INDUSTRY REGULATIONS 1926.451(b)(9) prohibit using boards as scaffold planks whose surfaces are difficult to see.

## Base Plates and Leveling Screw Jacks

A base plate is a flat, ¼"-thick, 6" × 6" metal plate and welded stem inserted in the hollow, tubular bottom-end legs of scaffold end frames. Base plates keep debris out of the hollow end frame legs and distribute the scaffold's weight over a larger area than allowable by the round tubular steel legs. Leveling screw jacks are similar to base plates, except their screw adjustment permits leveling scaffold setups. A jack extension can be used with a base plate for leveling scaffolds. All three are illustrated in Figure 25-44.

> **CAUTION**
>
> To meet the intent of OSHA CONSTRUCTION INDUSTRY REGULATIONS 1926.451(c)(2), scaffold frame uprights "shall bear on base plates and mudsills or other adequate firm foundation."

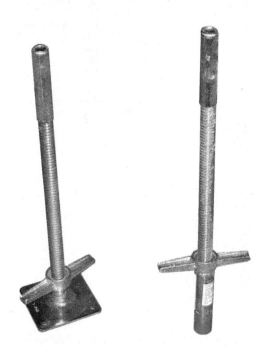

**Figure 25-44** A rigid base plate (left), screw jack with base plate (center), and jack extension (right) are illustrated here.

# Guardrails

**Guardrails**, rails installed along all open sides and ends of platforms serving as personal fall protection, and accessories, including guardrail posts, top rails, mid-rails, and toe board clips, are means for meeting the intent of OSHA CONSTRUCTION INDUSTRY REGULATIONS 1926.451(g)(1), requiring fall protection for those on scaffolds more than 10' above lower levels. A guardrail post slips over a steel coupling pin while an outer rigid support extends lower on the outside of the scaffold frame leg (see Figure 25-45).

Top rails attach to the tops of the guardrail posts (see Figure 25-46). They are to be between 38" and 45" above the platform (between 36" and 45" for scaffold manufactured or placed in service before January 1, 2000) and in most cases capable of supporting a minimum force of 200 pounds, Section 1926.451(g)(4)(ii) and Section 1926.451(g)(4)(vii).

**Mid-rails**, rails approximately midway from the top edge of guardrail and the platform surface, provide additional fall protection (see Figure 25-47). Mid-rails must be capable of supporting a minimum force of 150 pounds (Section 1926.451(g)(4)(ix).

**Figure 25-45** A guardrail post is fitted to a scaffold end frame and coupling pin, allowing both a top rail and mid-rail to be fastened to it.

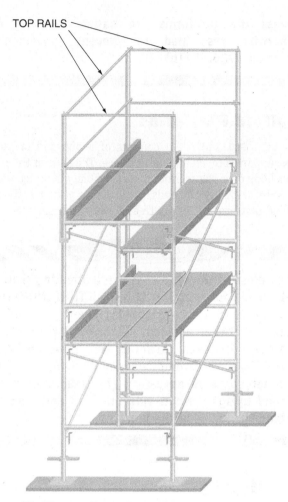

TOP RAILS

**Figure 25-46** Top rails are fastened to guardrail posts.

To prevent falling tools or materials from injuring those below scaffolds, OSHA standards require certain scaffold accessories unless the area below the scaffolds is barricaded and employees are not permitted to enter the area. **Toe boards**, vertical barriers extending above a platform surface to prevent falling materials, are one means to comply with the intent of OSHA standards. Toe boards are to be installed along the edge of platforms more than 10' above lower levels (see Figure 25-48). OSHA standards require toe boards within $\frac{1}{4}$" of the tops of platforms and to extend a minimum of $3\frac{1}{2}$" above the platform. Toe boards must be capable of supporting a minimum force of 50 pounds, Section 1926.451(h)(4).

> ### ◆◆◆ CAUTION ◆◆◆
>
> OSHA standards require a guardrail system with openings small enough to prevent falling objects where materials are above the height of toeboards, unless canopies or debris nets are used between the falling object hazards and workers below them (Section 1926.451(h)(2)(v).

**Figure 25-47 Properly installed mid-rails should be halfway between the platform and top rails.**

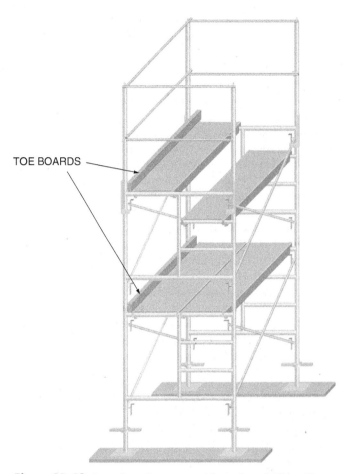

**Figure 25-48 A toe-board prevents objects from falling off the platform.**

# Coupling Pins

A coupling pin is a metal tubular pin inserted in tubular scaffold end frames so that they can be stacked and locked for multitiered scaffold setups (see Figure 25-49). By aligning holes in the pins with holes in the scaffold end frames, you can lock the sections together using self-locking pins or snap rings.

**Figure 25-49 An insert coupling pin permits stacking and locking end frames for multitiered scaffold setups.**

**CAUTION**

**OSHA standards require locking scaffold end frames together where uplift can occur (Section 1926.452(c)(4). Forces causing uplift include wind, climbing, hoist wheels and brackets attached to frames, and accidental contact of scaffolds with moving equipment or vehicles.**

# Scaffold Tipping Restraints

To meet the intent of OSHA CONSTRUCTION INDUSTRY REGULATIONS 1926.451(c)(1) , 5' wide supported scaffold are to be restrained from tipping at a height no more than 20' from the base with guys, ties, or braces. Restraint is to be repeated vertically thereafter every 26' of height or less. Such restraints must support both inner legs and outer legs, be at the ends of scaffolds, and be spaced no more than 30' along scaffold setups. OSHA standards require "scaffolds over 125' in height above their base plates shall be designed by a registered professional engineer and shall be constructed and loaded in accordance with such design" (Section 1926.452(c)(6).

# Fall Protection

OSHA CONSTRUCTION INDUSTRY REGULATIONS 1926.451 (g)(1) requires fall protection for workers on scaffold more than 10' above a lower level. When working on platforms supported by the scaffold frame, fall protection may be an approved guardrail system on all open sides as described earlier in this chapter or a personal fall arrest system (see Figure 25-50). When working or walking on surfaces other than scaffolds having an unprotected side or edge which is 6 feet or more above a lower level, workers must be protected from falling, Section 1926.501(b)(1). A personal fall arrest system consists of a lightweight body harness strapped to the legs and body. A shock absorbing lanyard with locking snap rings attached at one end to the body harness and to a secured attachment at the opposite end arrests or stops a fall, which reduces the force of a fall.

**Figure 25-50** A personal fall arrest system is designed to break a fall and support a worker in mid-air suspension. Immediate help is critical to the well-being of one suspended. Prolonged suspension in a body harness restricts blood circulation through the body and can be fatal.

## CAUTION

A person should be supported by a body harness following a fall no longer than required to get immediate help. Prolonged body suspension and support by the harness alone can have extremely dangerous effects on the blood, limiting its circulation through the body. Prolonged suspension of the body in a body harness can be fatal.

## CAUTION

Section 1926.451(g)(1)(vi) of the OSHA CONSTRUCTION INDUSTRY REGULATIONS requires each employee performing overhand bricklaying operations, reaching over a wall to lay brick, "from a supported scaffold shall be protected from falling from all open sides and ends of the scaffold (except at the side next to the wall being laid) by the use of a personal fall arrest system or guardrail system (with minimum 200 pound toprail capacity)."

# Summary

- Construction workers are exposed to a variety of hazards.
- Become trained by a qualified person before attempting any task and before using any tool, equipment, or machine.
- Whenever in doubt, consult a competent person before proceeding.
- Comply with all governing regulations intended to promote safety and health welfare.
- Learning and practicing proper safety procedures is most important for training and successful employment.
- Become properly trained by a qualified and competent person before erecting, dismantling, or accessing scaffold.

## Review Questions

*Select the most appropriate answer.*

**1** A federal government agency that conducts inspections of and issues citations and proposed penalties for employers covered under the Occupational Safety and Health Act of 1970 for alleged violations of applicable safety and health standards is called

a. ANSI
b. NIOSH
c. OSHA
d. all of the above

**2** Personal protection equipment includes protective

a. clothing
b. eyewear
c. respiratory equipment
d. all of the above

**3** To prevent skin cancers, those exposed to the sun should wear

a. shirts with neck collars and sleeves 4" or longer
b. wide-brimmed head protection that helps shade the neck, ears, and face
c. long trousers
d. all of the above

**4** One should wear particulate-filter respirator masks whenever there are dangers for breathing

a. mineral dusts such as cement, products used for the mixing of "practice mortars" containing silica
b. airborne sand
c. dusts created when using saws for dry-cutting operations
d. all of the above

**5** Hard hats should be worn where there is a possible danger of head injury from

a. impact
b. falling or flying objects
c. electrical shock and burns
d. all of the above

**6** Adequate hand protection should be worn

a. to absorb and dampen the vibrations of impact tools
b. for protection from chemicals
c. to protect hands when handling materials capable of damaging the skin
d. all of the above

**7** Keep floor areas of masonry labs safe by

a. immediately placing masonry scraps in buckets
b. keeping tools off of the floor
c. wiping up water spills
d. all of the above

**8** If it becomes necessary to lift a mortarboard or mortar pan loaded with mortar,

a. always pair up with another person
b. lift alone by bending over and keeping the legs straight
c. lift alone by bending at the knees and keeping the back straight
d. lift alone and wear hand protection

**9** Carry joint reinforcement wire or steel reinforcement rods by holding them in a position so as not to

a. injure someone with the ends
b. make contact with electrical wires and other overhead obstructions
c. be within 10' of overhead high voltage lines
d. all of the above

10  When transporting wheelbarrows in a masonry lab,

a. stay within those areas designated as traffic aisles or lanes

b. keep shovel and hoe handles between the body and arms so as not to strike others

c. have them headed in the direction of travel before loading them

d. all of the above

11  Students using hand tools are responsible for

a. visually inspecting the tool before using it

b. returning defective tools to the instructor

c. returning flawless tools to the designated place

d. all of the above

12  Before tensioning masons' lines,

a. examine them for breaks and excessive signs of wear

b. remove all twigs

c. wear approved eye protection

d. all of the above

13  120-volt tools and machines having metal enclosures

a. must have electrical cords with three-prong plugs, that is, two flat prongs completing a 120-volt circuit and one round prong serving as an equipment grounding conductor

b. must be connected to three-prong, grounded, 120-volt receptacles

c. do not require an equipment grounding conductor

d. both a and b

14  For maximum personal protection, all 120-volt powered tools and machines on construction sites can be

a. connected to only those receptacles having ground fault circuit interrupter protection

b. connected to properly sized extension cords having inline ground fault interrupter protection

c. connected to any 110-volt power source

d. both a and b

15  Should mortar contact your eyes,

a. have the instructor or one in charge assist with emergency eye wash procedures

b. seek medical attention at a medical facility

c. do not rub the eyes

d. all of the above

16  Printed copies of Material Safety Data Sheets

a. must be available at each jobsite

b. show emergency and first-aid procedures related to a specific product

c. show the hazard data for a specific product

d. all of the above

17  Before using chemicals and other hazardous materials, one should

a. be trained by someone who is experienced and competent

b. know the hazards and wear proper and adequate personal protection equipment

c. know the emergency and first-aid procedures

d. all of the above

18  Scaffolds should be erected, dismantled, or accessed

a. only by experienced and trained workers selected by a competent person

b. in accordance to OSHA regulations

c. by any employee at a jobsite

d. both a and b

19  In most cases, OSHA standards require supported scaffold to be no closer to overhead uninsulated high voltage power lines than

a. 5'

b. 10'

c. 15'

d. 20'

20  **Unless designed by a qualified engineer, side brackets attached to scaffold end frames are intended to support**

a. materials only

b. workers only

c. materials and workers

d. materials or workers, but not both at once

21  **Sawn scaffold planking**

a. consists of individual boards 1¾" to 2" thick and 10" to 11½" wide

b. must meet the industry standard known as Dense Industrial 65

c. should have metal plank ties or drilled tie rods secured into both ends to prevent them from splitting

d. all of the above

22  **The maximum deflection for scaffold planking for a 7' span between end frames is about**

a. ¾"

b. 1⅜"

c. 2"

d. 2½"

23  **Laminated veneer lumber scaffold planks are**

a. made by a manufacturing process of gluing together many thin layers of wood

b. usually 1½" or 1¾" thick

c. 9¼" and 11¾" wide

d. all of the above

24  **Scaffold planking should**

a. extend a minimum of 6" and no more than 12" beyond the scaffold end frames that support it

b. overlap no less than 12" adjoining planking

c. never be framing lumber

d. all of the above

25  **Scaffold end frames are supported by**

a. base plates

b. leveling screw jacks and base plates

c. mudsills

d. all of the above

26  **Protection above the scaffold planks preventing workers and materials from falling from the scaffolds includes**

a. guardrails

b. mid-rails

c. toe-boards

d. all of the above

27  **Fall protection is required for workers on scaffold that is above a lower level more than**

a. 4'

b. 6'

c. 8'

d. 10'

28  **Workers on scaffold can be protected from falls by**

a. approved railing on all open sides

b. personal fall arrest body harness systems

c. posted caution signs

d. both a and b

# Chapter 26

# Working Drawings and Specifications

Communication relies upon language, of which there are many different ones recognized throughout the world. Hand languages are interpreted using hand signs. Symbols create written text. Unless taught, people neither use or understand a given language. Similarly, those responsible for designing and overseeing the construction of a project must be taught to understand working drawings and job specifications, which make up the "language of construction." Lines, symbols, and scale drawings are used to express information required to erect a building specifically located and elevated on a site while adhering to the dimensions, materials, and appearance agreed upon by all parties involved in the project. The purpose of this chapter is not necessarily to prepare you to make or fully understand all working drawings and specifications, but to help you interpret the different lines, symbols, drawings, and specifications guiding those responsible for completing different parts of the construction.

## OBJECTIVES

**After completing this chapter, the student should be able to:**

- Identify the types of lines, symbols, and abbreviations used for drawings, and explain where they may be found on a drawing.
- Define and explain the types of working drawings that may be part of a construction document.
- Define, identify, and explain the purposes of elevation drawings, details, and sections.
- Explain the purpose of presentation drawings and how they differ from working drawings.
- Define that part of the construction documents called "specifications," and explain its purposes and contents.
- Become familiar with interpreting working drawings, noting those parts of each applicable to the masonry construction of the project.

# Glossary of Terms

**bench marks** points of known elevation used as a reference in determining other elevations.

**border lines** solid, bold lines indicating the outer perimeters for any details intended for the drawing.

**break lines** solid lines having diagonal lines extending beyond both sides of the lines, extending beyond a drawn object, indicating a break in the drawing with part of its scale size deleted.

**building material symbols** representations of common building materials. There are two types of building material symbols, elevations and sections.

**centerlines** lines appearing as a repetitious pattern of a long broken line followed by a shorter line or dot, indicating the center of objects such as doors, windows, and partition walls.

**change order** a written change to the original contract agreed upon by the owner, architect, engineer, and contractor. A change order identifies the specific change to the contract, changes in the monetary amount of the contract, and changes in the time allotted for completing the work in the contract.

**climate control symbols** representations of heating, air conditioning, and ventilation system components.

**construction details** also called **details,** drawings at a larger scale and showing in greater detail than the working drawings significant parts that require more information than is provided on the working drawings.

**construction documents** comprised of the working drawings and the specifications, which are the conditions of a contract upon which the owner and the contractor agree.

**dimension lines** solid lines having arrow heads, slashes, or dots at each end to indicate the terminations or end points between which the required measurement is given.

**drawing identification symbols** icons providing additional information for interpreting a set of drawings.

**electrical symbols** representations of electric fixtures, switches, receptacles, panel boxes, and other wiring systems and accessories.

**elevations** two-dimensional graphic scale representations of the front, sides, and back of a structure.

**extension lines** lines identifying the boundary of the dimension indicated by a dimension line.

**floor plans** scale drawings showing the overall length and width of the floor framing for a given floor level and the location of both exterior and interior walls and openings within the walls.

**foundation plans** scale drawings showing the foundation walls for the intended structure.

**framing plans** scale drawings showing floor framing and roof framing, which are the frameworks that serve as the skeleton of a structure to which coverings can be applied.

**general conditions** that part of the specifications defining the rights and the responsibilities of all parties involved in the work of the project.

**hidden lines** lines used to indicate objects hidden from view.

**leader lines** solid lines used where space does not permit showing dimensions between extension lines, ending with an arrow head identifying the notation at its opposite end.

**object lines** solid lines representing the outline shape of objects.

**plumbing symbols** representations of hot and cold water lines, waste lines, vent stacks, floor drains, gas lines, water meters, and a variety of plumbing fixtures and accessories.

**presentation drawings** perspective drawings, drawings appearing in a three-dimensional form as they would appear to the eye.

**project representative** one assigned to a job site to oversee the construction contract.

**scale drawing** a representation of an object drawn to a size other than its actual size, having all dimensions drawn to the same ratio as related to the actual dimensions.

**section** a drawing showing the vertical composition of a specific part of a working drawing.

**section lines** solid, parallel, slanting lines indicating a break in a vertical, cut-through part of a drawing.

**section reference line** also called a cutting-plane line, a line drawn for reference to a cross-section, a section perpendicular to the drawing shown elsewhere.

**special conditions** conditions of the contract other than general conditions or supplementary conditions setting forth requirements unique to the project.

**specifications** a written description of the conditions of the contract.

**supplementary conditions** that part of the specifications that supplements or modifies provisions of the general conditions.

**symbols** marks, letters, characters, figures, or combinations of these representing specific objects.

**topographical features** physical features of the land such as ground cover, wooded land, lakes, streams, and land contour.

**topographical symbols** representations of ground vegetation, surface water, roadways, utilities, wells, septic systems, fencing, property lines and corners, surface contours, and benchmarks.

**working drawings,** also called plans, graphic or pictorial representations of elements of a project showing their design, location, and dimensions.

# Lines

Lines are drawn differently to express specific meanings. This chapter addresses ten of the most frequently seen lines. These lines include object, dimension, extension, leader, centerline, break, section, section reference, hidden, and border. Once requiring highly skilled draftsmen using mechanical drawing equipment, today's lines and drawings are created much quicker using computer programs.

## Object Line

Object lines are solid lines representing the outline shape of objects. CMU foundation walls, for example, are represented by two object lines spaced in miniature scale representing the width of the CMUs required and the lengths of the walls (see Figure 26-1).

Another example are the object lines for drawings representing each side of a structure, including but not limited to such items as its windows, doors, chimneys, and roof lines (see Figure 26-2).

## Dimension Line

Dimension lines are solid lines that have arrow heads, slashes, or dots at each end to indicate the terminations or end points between which the required measurement is given (see Figure 26-3). Some examples of dimension lines indicate overall wall lengths, the measurement between the centerline of adjacent windows or doors, the measurement between the outside face of a wall and the centerline of a window or door, the width of wall openings, and the wall width itself.

Dimensions for narrow spaces may be given with the dimension lines drawn to the outer sides of the space (see Figure 26-4).

## Extension Line

Extension lines are solid lines extending from an object, perpendicular to and beyond the dimension lines (see Figure 26-5). Extension lines identify the boundary of the dimension indicated by the dimension line.

**Figure 26-1 A CMU wall is designated by object lines.**

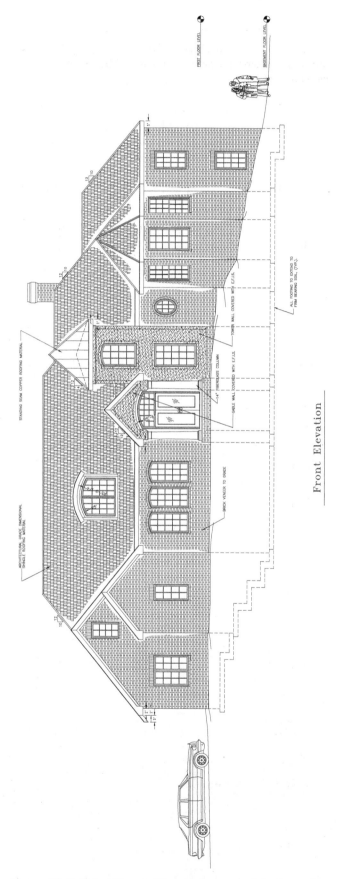

**Figure 26-2 Object lines outline the features of the front side of this home.** *(Courtesy of Design Vision)*

Figure 26-3 Dimension lines, whose ends are arrow heads in this drawing, show the points between which the dimension or measurement refers.

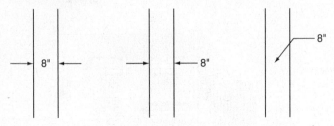

Figure 26-4 Three ways for dimensioning small spaces are illustrated here.

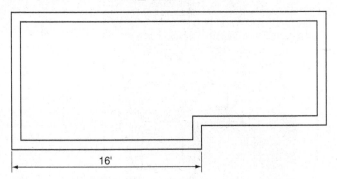

Figure 26-5 The extension lines help clarify the section to which the dimension refers.

## Leader Line

Where space does not permit showing dimensions between extension lines, solid leader lines are prescribed. Information at one end of a leader line is pertinent to the place referenced by the opposite end of the line, denoted by an arrow head (see Figure 26-6).

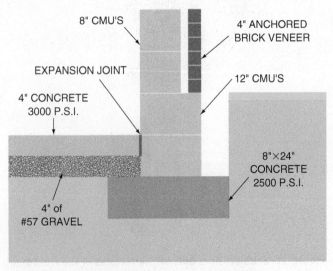

Figure 26-6 Leader lines indicate the object or space referenced by the accompanying notation.

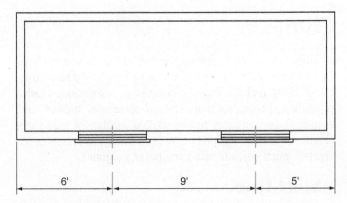

Figure 26-7 Centerlines, a long line followed by a short line with breaks along their inscriptions, denote the center point of an object or space.

## Centerline

Centerlines are lines appearing as a repetitious pattern of a long broken line followed by a shorter line or dot, indicating the center of objects such as doors, windows, and partition walls. Measurements are typically referenced from centerlines. For example, a measurement may be given along a dimension line to indicate the length between an extension line at the outside face of a wall and a centerline of a window (see Figure 26-7).

## Break Line

Break lines appear as solid lines with diagonal lines extending beyond both sides of the lines, extending beyond a drawn object, indicating a break in the drawing with part of its scale size deleted. Break lines are used to shorten a drawing, requiring less space for showing a drawing (see Figure 26-8). Break lines can only be used if no detail is deleted that would otherwise be shown.

## Section Line

Section lines are solid, parallel, slanting lines indicating a break in a vertical, cut-through part of a drawing (see Figure 26-9). Like break lines, section lines can only be used to shorten a vertical representation drawing when necessary details are not deleted from the drawing.

## Section Reference Lines

Section reference lines are lines appearing as a repetitious pattern of a long broken line followed by two shorter lines with an arrow head at each end (see Figure 26-10).

A section reference line, also called a cutting-plane line, is drawn for reference to a cross-section, which is a section perpendicular to the drawing shown elsewhere (see Figure 26-11).

The arrows indicate the direction in which the cross-section is viewed. Section reference lines are not to be confused with section lines used to save space by shortening a vertical drawing. Section reference lines are labeled differently, some

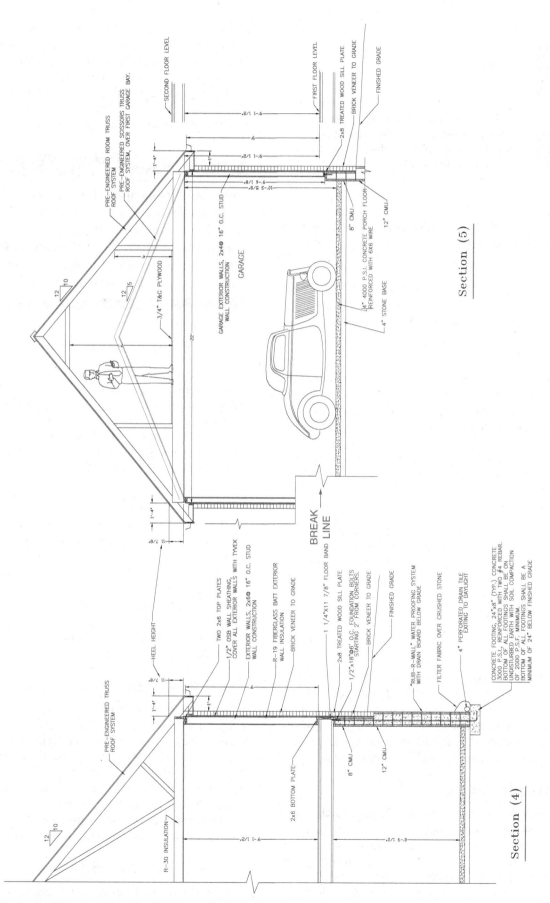

**Figure 26-8** Used only when necessary details are not deleted, the solid break line permits creating a drawing requiring less space. *(Courtesy of Design Vision)*

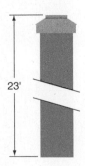

23'

Figure 26-9 Used only when necessary details are not deleted, the two diagonal break lines permit shortening the scale drawing created on paper.

providing more detailed information than others. A letter of the alphabet or a number identifies the section to which the reference applies (see Figure 26-12). This is called a drawing identification symbol.

But additional information, the sheet number where the drawing section is shown and the sheet number of the section reference line, may also be included (see Figure 26-13).

## Hidden Line

Hidden lines are drawn as a repetitious pattern of short, uniform-length lines. Hidden lines are used to indicate objects hidden from view. Since an object defined with hidden lines is not visible in the drawing, these lines are used only when necessary, perhaps to show the location of a hidden object and its relationship to details appearing on a drawing (see Figure 26-14).

## Border Line

Border lines are solid, bold lines indicating the outer perimeters for any details intended for the drawing. Rather than serving to give information to a drawing, border lines simply frame a drawing, leaving no doubt as to the intended information contained within them (see Figure 26-15).

## Symbols

Symbols are marks, letters, characters, figures, or combinations of these representing specific objects. Six categories of symbols are found on working drawings, 1) topographical symbols, 2) building material symbols, 3) electrical symbols, 4) plumbing symbols, 5) climate control symbols, and 6) drawing identification symbols.

## Topographical Symbols

Topographical symbols are representations of ground vegetation, surface water, roadways, utilities, wells, septic systems, fencing, property lines and corners, surface contours, and benchmarks. Getting familiar with topographical symbols helps masons become alert to the locations of buried utilities, wells, and septic systems. Symbols representing property lines and corners enable masons to be

aware of the proximity of proposed construction to property boundary lines. Benchmarks, points of known elevation used as a reference in determining other elevations, are identified with symbols. For example, a finished foundation height may be specified to be a certain height below or above a given stationary benchmark. Samples of topographical symbols are shown in Figure 26-16.

⚠ **CAUTION**

Although topographical plans may show buried utilities, they are not to be used as a reference to locations of buried utilities when beginning excavations and other ground diggings. Before digging, always have the utility companies' representatives locate and visually mark buried utilities.

## Building Material Symbols

Building material symbols are representations of common building materials. The two types of building material symbols are elevations and sections. Elevation symbols are pictorial representations of the face of a material. Section symbols represent the composition of a material. Both elevation symbols and section symbols are illustrated for several building materials in Figure 26-17.

## Electrical Symbols

Electrical symbols are representations of electric fixtures, switches, receptacles, panel boxes, and other wiring systems and accessories. Although an electrician may be responsible for their installations, a mason should know where electrical work boxes or panel boxes and rigid conduit are to be placed within masonry walls. This enables the mason to alert those responsible to perform their services as construction requires. Samples of electrical symbols are shown in Figure 26-18.

## Plumbing Symbols

Plumbing symbols are representations of hot and cold water lines, waste lines, vent stacks, floor drains, gas lines, water meters, and a variety of plumbing fixtures and accessories. As with electrical work, it is not the responsibility of the mason to install plumbing. But recognizing plumbing symbols enables masons to notify responsible parties to perform their services if plumbing is shown to be within open cells of CMUs, masonry chases, or between masonry wall wythes. Samples of plumbing symbols are shown in Figure 26-19.

## Climate Control Symbols

Climate control symbols are representations of heating, air conditioning, and ventilation system components.

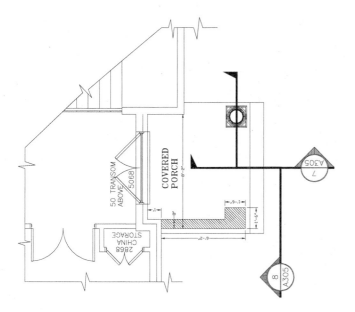

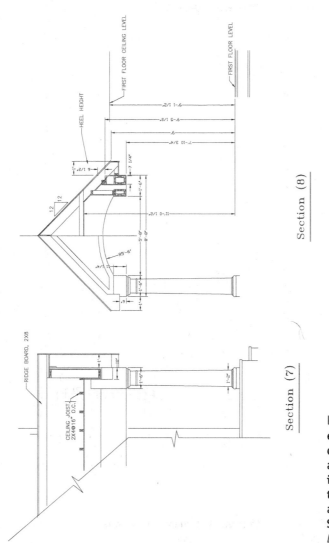

**Figure 26-10** The section reference line, ending with a circle, indicates a specific part of the drawing and the angle from which an accompanying section drawing of that part is to be viewed. Section reference lines on the drawing identified with the number 7 and 8 refer respectively to cross-sections **Section (7)** and **Section (8)**. *(Courtesy of Design Vision)*

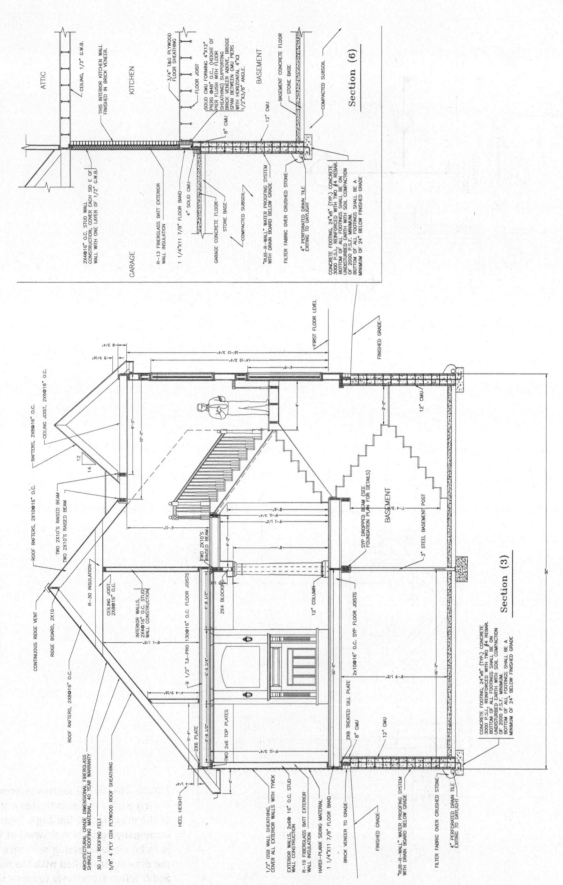

**Figure 26-11** Section (3) shows several details. Section (6) shows basement wall details for supporting a brick veneer, interior kitchen wall. *(Courtesy of Design Vision)*

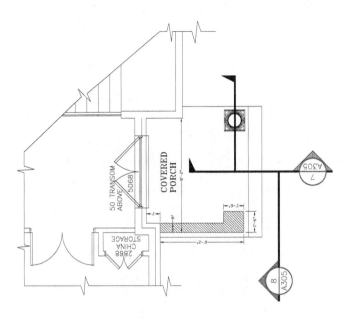

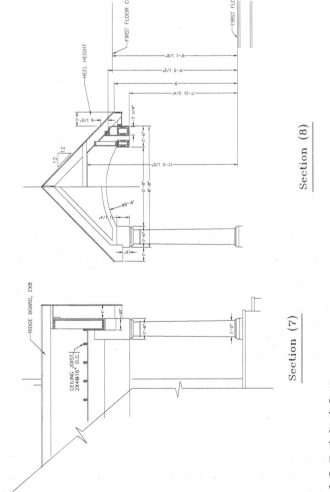

Section (8)

Section (7)

**Figure 26-12** Numbers 7 and 8 in the circles accompanying the section line indicate the section drawing, Section (7) and Section (8), to which you should refer, that is, the section identified with the same number. The arrowheads indicate the direction from which each section is to be viewed. *(Courtesy of Design Vision)*

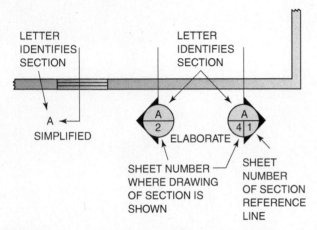

**Figure 26-13** Two additional ways for labeling section reference lines are shown here. A letter, "A" for this drawing, instead of a number is used to identify the section referred to.

Once again, recognizing plumbing symbols enables masons to notify responsible parties to have their work completed when necessary to continue masonry construction. Samples of climate control symbols are shown in Figure 26-20.

## Drawing Identification Symbols

**Drawing identification symbols** are icons providing additional information for interpreting a set of drawings. The more common symbols are those identifying section reference lines. Samples of identification symbols were shown earlier in Figure 26-12 and Figure 26-13.

## Abbreviations

Using only capital letters, abbreviations or acronyms are ways to denote types of materials, objects, and other pertinent information. Somewhat confusing, in some instances the same abbreviations are used to denote different materials. For example, the letter "I" is an abbreviation for either "I-beam" or "iron," and the letter "R" is an abbreviation for either "range" or "riser." Being aware of what the drawing is showing helps to determine the meaning of the abbreviation. As for the abbreviation "R," if it appears in a kitchen area, it can be assumed to identify a cooking range, but if it appears with a drawing of steps, it indicates the stair riser. A sample list of abbreviations is shown in Figure 26-21.

## Scale Drawings

A **scale drawing** is a representation of an object drawn to a size other than its actual size, having all dimensions drawn to the same ratio as related to the actual dimensions. Scale drawings can be smaller or larger than the actual size of the objects they represent. For example, small components

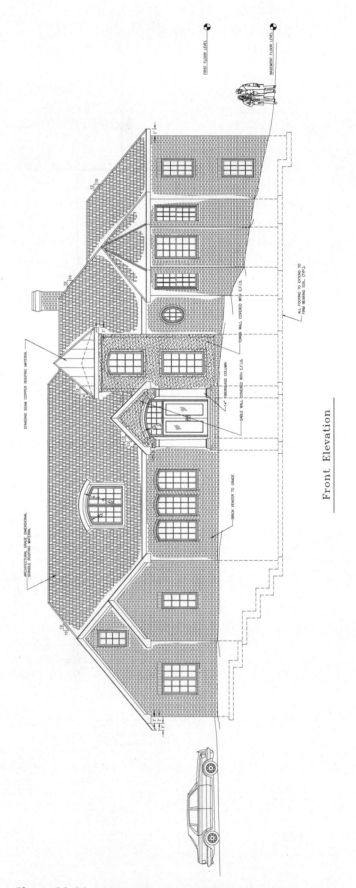

**Figure 26-14** Hidden lines, short uniform-length broken lines, indicate the depth below grade of the concrete footings for this home's foundation wall. *(Courtesy of Design Vision)*

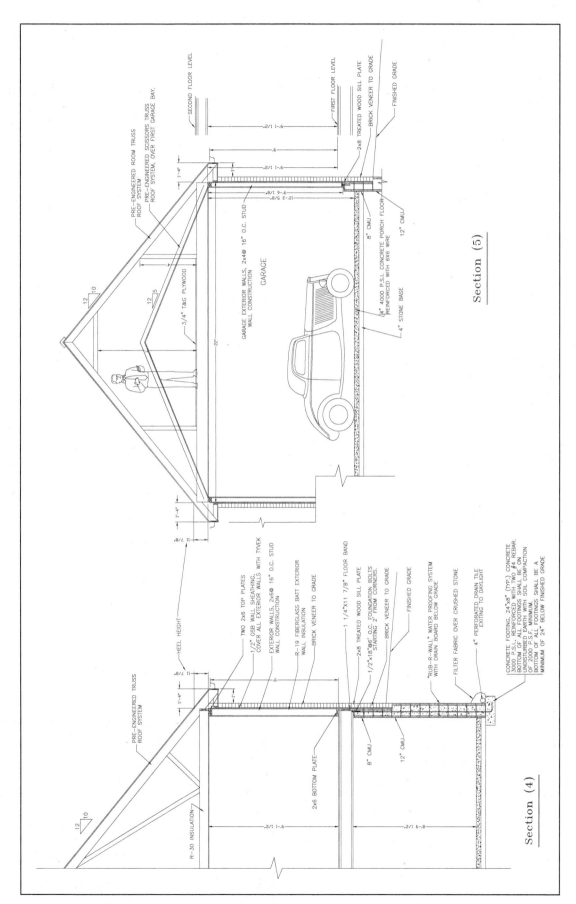

**Figure 26-15 Border lines frame the drawing, indicating the outer perimeters for which graphic information is provided within them.** *(Courtesy of Design Vision)*

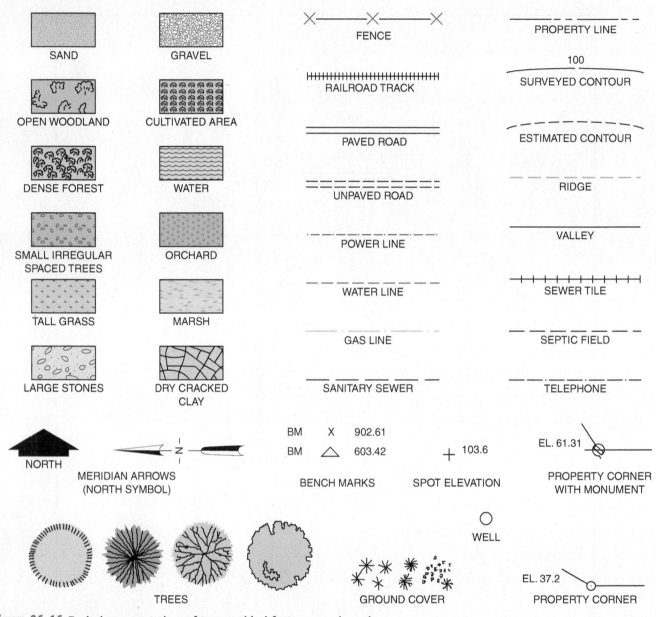

**Figure 26-16** Typical representations of topographical features are shown here.

of a wrist watch may be drawn larger than actual size, enabling those making the pieces to better view the details. On the other hand, construction drawings are drawn smaller than actual size so that the planned structure or section can be drawn or printed for viewing. Scales typically used for making construction drawings are the ¼" scale and the ⅛" scale. For ¼"-scale drawings, ¼" represents an actual length of 1', 48 times smaller than its actual size. Similarly, for ⅛"-scale drawings, ⅛" represents an actual length of 1', or 96 times smaller than its actual size. For a scale drawing intended to show the location of a structure in relationship to a building site property and its property lines, ¹⁄₁₆"-scale drawings are frequently used. A ¹⁄₁₆" scale permits drawing a 576' square plot of land on a 3'-square drawing area. Although the lines for scale drawings can be penciled on paper by those trained as draftsmen using mechanical

drawing equipment, computer programs expedite the process (see Figure 26-22).

**FROM EXPERIENCE**

Always refer to the dimensions given on the plans rather than attempting to determine the dimension by measuring it with a ruler. Where a dimension is not given or otherwise not visible, determine that dimension by subtracting other known dimensions from the overall dimension or consult the person responsible for the project. Trying to measure the dimension even when the scale is given can result in errors.

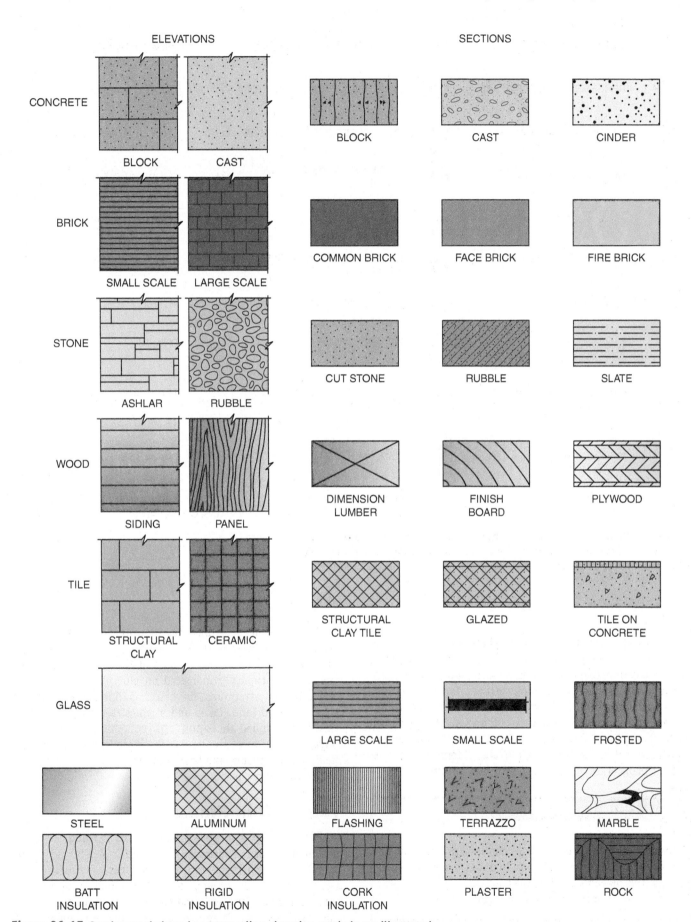

**Figure 26-17** Section symbols and corresponding elevation symbols are illustrated.

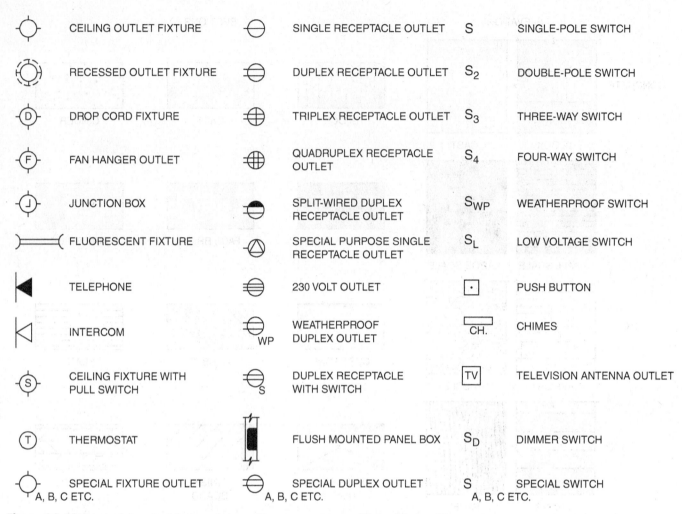

| | | |
|---|---|---|
| CEILING OUTLET FIXTURE | SINGLE RECEPTACLE OUTLET | S SINGLE-POLE SWITCH |
| RECESSED OUTLET FIXTURE | DUPLEX RECEPTACLE OUTLET | $S_2$ DOUBLE-POLE SWITCH |
| DROP CORD FIXTURE | TRIPLEX RECEPTACLE OUTLET | $S_3$ THREE-WAY SWITCH |
| FAN HANGER OUTLET | QUADRUPLEX RECEPTACLE OUTLET | $S_4$ FOUR-WAY SWITCH |
| JUNCTION BOX | SPLIT-WIRED DUPLEX RECEPTACLE OUTLET | $S_{WP}$ WEATHERPROOF SWITCH |
| FLUORESCENT FIXTURE | SPECIAL PURPOSE SINGLE RECEPTACLE OUTLET | $S_L$ LOW VOLTAGE SWITCH |
| TELEPHONE | 230 VOLT OUTLET | PUSH BUTTON |
| INTERCOM | WEATHERPROOF DUPLEX OUTLET | CH. CHIMES |
| CEILING FIXTURE WITH PULL SWITCH | DUPLEX RECEPTACLE WITH SWITCH | TV TELEVISION ANTENNA OUTLET |
| THERMOSTAT | FLUSH MOUNTED PANEL BOX | $S_D$ DIMMER SWITCH |
| SPECIAL FIXTURE OUTLET A, B, C ETC. | SPECIAL DUPLEX OUTLET A, B, C ETC. | S SPECIAL SWITCH A, B, C ETC. |

**Figure 26-18** Recognizing electrical symbols enables masons to coordinate efforts with electricians as masonry walls requiring accommodations for electrical components, materials, and hardware are built.

# Working Drawings

**Working drawings,** also called plans, are graphic or pictorial representations of elements of a project showing their design, location, and dimensions. Working drawings are scale drawings interpreted by the lines, symbols, and abbreviations addressed earlier in this chapter. Complete working drawings require enough details to permit building all that is shown without additional instructions. Plans typically included in construction documents are 1) plot plans, 2) foundation plans, 3) floor plans, 4) framing plans, and 5) landscaping plans.

**FROM EXPERIENCE**

Store plans by rolling the pages and inserting the rolled plans in a section of PVC pipe that has removable end caps. This prevents bending the plans and water damage, both of which makes the plans difficult to read.

## Plot Plans

Plot plans are scale drawings showing the location of the structure in relation to the site and the property corners and lines. As aerial or overhead views of the property, some of the features typically indicated on plot plans are existing and intended structures, driveways, parking areas, ground cover, wooded land, lakes and streams, fencing, and utilities. Also included are **topographical features,** which are physical features of the land such as ground cover, wooded land, lakes, streams, and land contour. A sample plot plan is shown in Figure 26-23.

## Foundation Plans

**Foundation plans** are scale drawings showing the foundation walls for the intended structure. Typically shown are overall wall length for each side, lengths of wall sections between adjoining corners, CMU or cast concrete supporting piers, concrete slabs, wall and concrete slab reinforcements, and wall thickness. The one responsible for the completion of

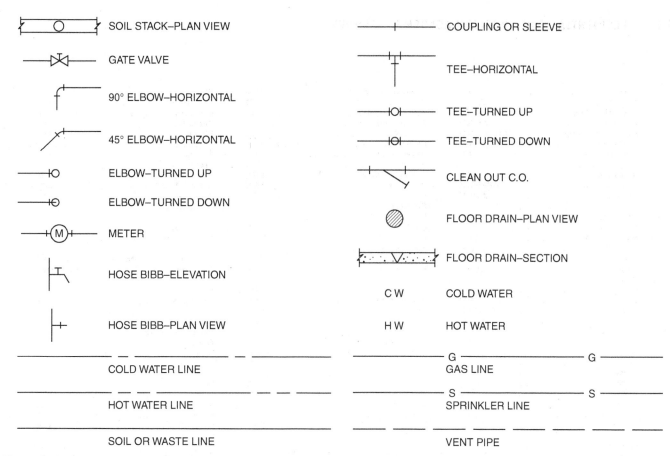

**Figure 26-19** Recognizing plumbing symbols enables masons to coordinate efforts with plumbers as masonry walls requiring accommodations for plumbing fixtures, equipment, and lines are built.

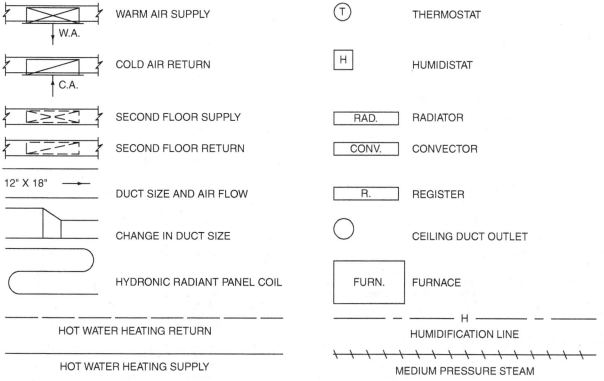

**Figure 26-20** Recognizing climate control symbols enables masons to coordinate efforts with heating and air conditioning technicians as masonry walls requiring accommodations for equipment and ductwork are built.

| | | |
|---|---|---|
| Access Panel . . . . . . . . . . . . . . . .AP | Dressed and Matched . . . . . . . .D & M | Plate . . . . . . . . . . . . . . . . . . .PL |
| Acoustic . . . . . . . . . . . . . . . .ACST | Dryer . . . . . . . . . . . . . . . . . . . . .D | Plate Glass . . . . . . . . . . . . . .PL GL |
| Acoustical Tile . . . . . . . . . . . . .AT | Electric Panel . . . . . . . . . . . . . .EP | Platform . . . . . . . . . . . . . . . .PLAT |
| Aggregate . . . . . . . . . . . . . . .AGGR | End to End . . . . . . . . . . . . .E to E | Plumbing. . . . . . . . . . . . . . . .PLBG |
| Air Conditioning . . . . . . .AIR COND | Excavate . . . . . . . . . . . . . . . .EXC | Plywood . . . . . . . . . . . . . . . .PLY |
| Aluminum . . . . . . . . . . . . . . . .AL | Expansion Joint. . . . . . . . . .EXP JT | Porch . . . . . . . . . . . . . . . . . . .P |
| Anchor Bolt . . . . . . . . . . . . . . .AB | Exterior . . . . . . . . . . . . . . . .EXT | Precast . . . . . . . . . . . . . . . .PRCST |
| Angle. . . . . . . . . . . . . . . . . . . | Finish . . . . . . . . . . . . . . . . .FIN | Prefabricated . . . . . . . . . .PREFAB |
| Apartment. . . . . . . . . . . . . . .APT | Finished Floor. . . . . . . . . .FIN FL | Pull Switch. . . . . . . . . . . . . . .PS |
| Approximate . . . . . . . . . .APPROX | Firebrick . . . . . . . . . . . . . . .FBRK | Quarry Tile Floor. . . . . . . . . .QTF |
| Architectural . . . . . . . . . . .ARCH | Fireplace . . . . . . . . . . . . . . . .FP | Radiator . . . . . . . . . . . . . . . .RAD |
| Area . . . . . . . . . . . . . . . . . . .A | Fireproof. . . . . . . . . . . . . . .FPRF | Random . . . . . . . . . . . . . . . .RDM |
| Area Drain . . . . . . . . . . . . . . .AD | Fixture . . . . . . . . . . . . . . . .FIX | Range. . . . . . . . . . . . . . . . . . .R |
| Asbestos . . . . . . . . . . . . . . .ASB | Flashing . . . . . . . . . . . . . . . .FL | Recessed . . . . . . . . . . . . . . .REC |
| Asbestos Board . . . . . . . . . . . .AB | Floor. . . . . . . . . . . . . . . . . . .FL | Refrigerator . . . . . . . . . . . . .REF |
| Asphalt. . . . . . . . . . . . . . . .ASPH | Floor Drain . . . . . . . . . . . . . .FD | Register . . . . . . . . . . . . . . . .REG |
| Asphalt Tile . . . . . . . . . . . . . .AT | Flooring . . . . . . . . . . . . . . . .FLG | Reinforce or Reinforcing. . . . .REINF |
| Basement. . . . . . . . . . . . . . .BSMT | Fluorescent . . . . . . . . . . . .FLUOR | Revision . . . . . . . . . . . . . . .REV |
| Bathroom . . . . . . . . . . . . . . . .B | Flush . . . . . . . . . . . . . . . . . .FL | Riser . . . . . . . . . . . . . . . . . . .R |
| Bathtub . . . . . . . . . . . . . . . .BT | Footing. . . . . . . . . . . . . . . .FTG | Roof . . . . . . . . . . . . . . . . . . .RF |
| Beam . . . . . . . . . . . . . . . . . .BM | Foundation . . . . . . . . . . . . .FND | Roof Drain. . . . . . . . . . . . . . .RD |
| Bearing Plate. . . . . . . . . . .BRG PL | Frame . . . . . . . . . . . . . . . . .FR | Room. . . . . . . . . . . . . . .RM or R |
| Bedroom. . . . . . . . . . . . . . . .BR | Full Size . . . . . . . . . . . . . . . .FS | Rough . . . . . . . . . . . . . . . . .RGH |
| Blocking . . . . . . . . . . . . . .BLKG | Furring . . . . . . . . . . . . . . . .FUR | Rough Opening . . . . . . . . . . .RO |
| Blueprint. . . . . . . . . . . . . . . .BP | Galvanized Iron . . . . . . . . . . . .GI | Rubber Tile . . . . . . . . . . . .R TILE |
| Boiler. . . . . . . . . . . . . . . . .BLR | Garage . . . . . . . . . . . . . . . .GAR | Scale . . . . . . . . . . . . . . . . . .SC |
| Book Shelves . . . . . . . . . .BK SH | Gas . . . . . . . . . . . . . . . . . . .G | Schedule . . . . . . . . . . . . . . .SCH |
| Brass . . . . . . . . . . . . . . . . .BRS | Glass . . . . . . . . . . . . . . . . . .G | Screen . . . . . . . . . . . . . . . .SCR |
| Brick . . . . . . . . . . . . . . . . .BRK | Glass Block . . . . . . . . . . . .GL BL | Scuttle . . . . . . . . . . . . . . . . .S |
| Bronze . . . . . . . . . . . . . . . .BRZ | Grille . . . . . . . . . . . . . . . . . .G | Section . . . . . . . . . . . . . . .SECT |
| Broom Closet . . . . . . . . . . . . .BC | Gypsum . . . . . . . . . . . . . . .GYP | Select. . . . . . . . . . . . . . . . .SEL |
| Building . . . . . . . . . . . . . .BLDG | Hardware. . . . . . . . . . . . . . .HDW | Service . . . . . . . . . . . . . . .SERV |
| Building Line . . . . . . . . . . . . .BL | Hollow Metal Door . . . . . . . .HMD | Sewer . . . . . . . . . . . . . . . .SEW |
| Cabinet. . . . . . . . . . . . . . . .CAB | Hose Bib . . . . . . . . . . . . . . . .HB | Sheathing . . . . . . . . . . . .SHTHG |
| Calking . . . . . . . . . . . . . . .CLKG | Hot Air. . . . . . . . . . . . . . . . .HA | Sheet . . . . . . . . . . . . . . . . .SH |
| Casing . . . . . . . . . . . . . . . .CSG | Hot Water . . . . . . . . . . . . . . .HW | Shelf and Rod . . . . . . . . .SH & RD |
| Cast Iron . . . . . . . . . . . . . . . .CI | Hot Water Heater . . . . . . . . .HWH | Shelving . . . . . . . . . . . .SHELV |
| Cast Stone . . . . . . . . . . . . . . .CS | I Beam . . . . . . . . . . . . . . . . . .I | Shower. . . . . . . . . . . . . . . . .SH |
| Catch Basin . . . . . . . . . . . . . .CB | Inside Diameter . . . . . . . . . . . .ID | Sill Cock . . . . . . . . . . . . . . . .SC |
| Cellar . . . . . . . . . . . . . . . . .CEL | Insulation . . . . . . . . . . . . . . .INS | Single Strength Glass . . . . . . . .SSG |
| Cement. . . . . . . . . . . . . . . .CEM | Interior. . . . . . . . . . . . . . . . .INT | Sink. . . . . . . . . . . . . . . .SK or S |
| Cement Asbestos Board . . . .CEM AB | Iron. . . . . . . . . . . . . . . . . . . .I | Soil Pipe . . . . . . . . . . . . . . . .SP |
| Cement Floor . . . . . . . . . .CEM FL | Jamb . . . . . . . . . . . . . . . . . .JB | Specification . . . . . . . . . . . .SPEC |
| Cement Mortar . . . . . .CEM MORT | Kitchen. . . . . . . . . . . . . . . . .K | Square Feet . . . . . . . . . . . .SQ FT |
| Center. . . . . . . . . . . . . . . .CTR | Landing . . . . . . . . . . . . . . .LDG | Stained . . . . . . . . . . . . . . . .STN |
| Center to Center . . . . . . . . .C to C | Lath . . . . . . . . . . . . . . . . .LTH | Stairs . . . . . . . . . . . . . . . . .ST |
| Center Line . . . . . . . . . . . .or CL | Laundry . . . . . . . . . . . . . . .LAU | Stairway . . . . . . . . . . . . . .STWY |
| Center Matched . . . . . . . . . . .CM | Laundry Tray . . . . . . . . . . . . .LT | Standard . . . . . . . . . . . . . . .STD |
| Ceramic . . . . . . . . . . . . . . .CER | Lavatory . . . . . . . . . . . . . . .LAV | Steel . . . . . . . . . . . . . .ST or STL |
| Channel . . . . . . . . . . . . . .CHAN | Leader . . . . . . . . . . . . . . . . . .L | Steel Sash . . . . . . . . . . . . . . .SS |
| Cinder Block. . . . . . . . . . .CIN BL | Length . . . . . . . . . .L, LG, or LNG | Storage . . . . . . . . . . . . . . .STG |
| Circuit Breaker . . . . . . . .CIR BKR | Library . . . . . . . . . . . . . . . .LIB | Switch . . . . . . . . . . . . .SW or S |
| Cleanout . . . . . . . . . . . . . . .CO | Light . . . . . . . . . . . . . . . . .LT | Telephone . . . . . . . . . . . . . .TEL |
| Cleanout Door. . . . . . . . . . .COD | Limestone . . . . . . . . . . . . . . .LS | Terra Cotta . . . . . . . . . . . . . .TC |
| Clear Glass. . . . . . . . . . . . .CL GL | Linen Closet . . . . . . . . . . .L CL | Terrazzo . . . . . . . . . . . . . . .TER |
| Closet. . . . . . . . . .C, CL, or CLO | Lining . . . . . . . . . . . . . . . . .LN | Thermostat . . . . . . . . . .THERMO |
| Cold Air . . . . . . . . . . . . . . . .CA | Living Room. . . . . . . . . . . . . .LR | Threshold . . . . . . . . . . . . . . .TH |
| Cold Water. . . . . . . . . . . . . .CW | Louver . . . . . . . . . . . . . . . . .LV | Toilet. . . . . . . . . . . . . . . . . .T |
| Collar Beam . . . . . . . . . . .COL B | Main . . . . . . . . . . . . . . . . .MN | Tongue and Groove . . . . . . .T & G |
| Concrete . . . . . . . . . . . . . .CONC | Marble . . . . . . . . . . . . . . . .MR | Tread . . . . . . . . . . . . . . .TR or T |
| Concrete Block . . . . . . . . .CONC B | Masonry Opening . . . . . . . . .MO | Typical. . . . . . . . . . . . . . . .TYP |
| Concrete Floor . . . . . . . .CONC FL | Material . . . . . . . . . . . . . . .MATL | Unfinished . . . . . . . . . . . . . .UNF |
| Conduit . . . . . . . . . . . . . . .CND | Maximum . . . . . . . . . . . . . .MAX | Unexcavated. . . . . . . . . . .UNEXC |
| Construction. . . . . . . . . . .CONST | Medicine Cabinet . . . . . . . . .MC | Utility Room . . . . . . . . . . . .URM |
| Contract . . . . . . . . . . . . . .CONT | Minimum. . . . . . . . . . . . . . .MIN | Vent . . . . . . . . . . . . . . . . . .V |
| Copper . . . . . . . . . . . . . . .COP | Miscellaneous . . . . . . . . . . .MISC | Vent Stack . . . . . . . . . . . . . .VS |
| Counter . . . . . . . . . . . . . . .CTR | Mixture. . . . . . . . . . . . . . . .MIX | Vinyl Tile . . . . . . . . . . . .V TILE |
| Cubic Feet . . . . . . . . . . . .CU FT | Modular . . . . . . . . . . . . . . .MOD | Warm Air. . . . . . . . . . . . . . .WA |
| Cut Out . . . . . . . . . . . . . . . .CO | Mortar . . . . . . . . . . . . . . . .MOR | Washing Machine . . . . . . . . .WM |
| Detail. . . . . . . . . . . . . . . . .DET | Moulding. . . . . . . . . . . . . .MLDG | Water. . . . . . . . . . . . . . . . . .W |
| Diagram . . . . . . . . . . . . . .DIAG | Nosing . . . . . . . . . . . . . . . .NOS | Water Closet . . . . . . . . . . . . .WC |
| Dimension . . . . . . . . . . . . . .DIM | Obscure Glass . . . . . . . .OBSC sL | Water Heater. . . . . . . . . . . . .WH |
| Dining Room . . . . . . . . . . . . .DR | On Center . . . . . . . . . . . . . .OC | Waterproof. . . . . . . . . . . . . .WP |
| Dishwasher. . . . . . . . . . . . . .DW | Opening . . . . . . . . . . . . . .OPNG | Weather Stripping. . . . . . . . . .WS |
| Ditto . . . . . . . . . . . . . . . . .DO | Outlet . . . . . . . . . . . . . . . .OUT | Weephole. . . . . . . . . . . . . . .WH |
| Double-Acting. . . . . . . . . . . .DA | Overall . . . . . . . . . . . . . . . .OA | White Pine . . . . . . . . . . . . . .WP |
| Double Strength Glass . . . . . . .DSG | Overhead. . . . . . . . . . . . . .OVHD | Wide Flange . . . . . . . . . . . . .WF |
| Down. . . . . . . . . . . . . . . . .DN | Pantry . . . . . . . . . . . . . . . .PAN | Wood . . . . . . . . . . . . . . . . .WD |
| Downspout . . . . . . . . . . . . . .DS | Partition . . . . . . . . . . . . . . .PTN | Wood Frame. . . . . . . . . . . . .WF |
| Drain . . . . . . . . . . . . . .D or DR | Plaster . . . . . . . . . . . . .PL or PLAS | Yellow Pine . . . . . . . . . . . . . .YP |
| Drawing . . . . . . . . . . . . . .DWG | Plastered Opening. . . . . . . . . .PO | |

**Figure 26-21 Frequently used abbreviations for working drawings are given in this chart.**

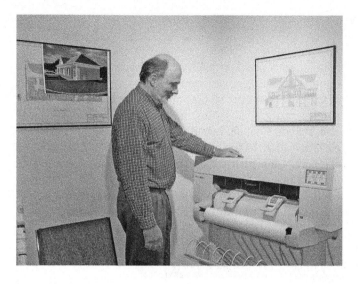

Figure 26-22 Today's designer relies upon computer technology, sophisticated computer software, and computer controlled equipment to create and print construction drawings. *(Courtesy of Design Vision)*

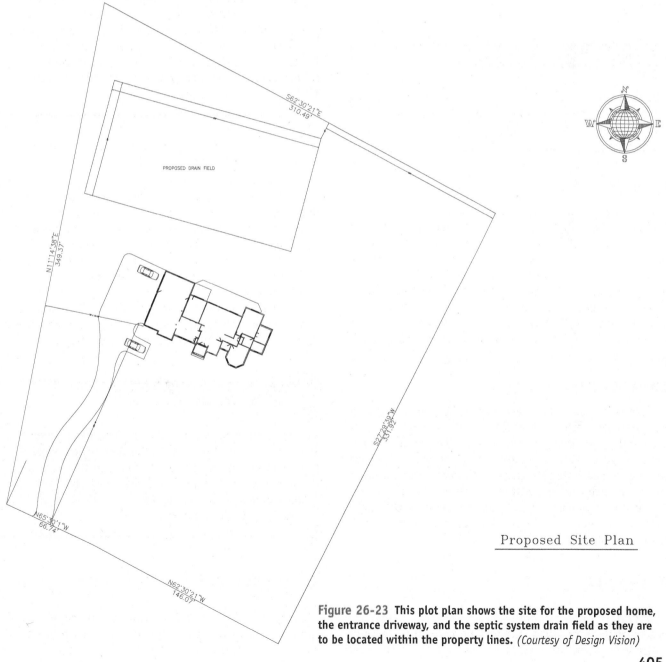

Proposed Site Plan

Figure 26-23 This plot plan shows the site for the proposed home, the entrance driveway, and the septic system drain field as they are to be located within the property lines. *(Courtesy of Design Vision)*

the foundation walls must be capable of interpreting foundation plans. Quite often, this person is the masonry contractor or the job foreman. The foundation plan in Figure 26-24 is typical of many of today's residential foundations, which often have offsets that are at angles other than square or 90°.

## Floor Plans

Floor plans are scale drawings showing the overall length and width of the floor framing for a given floor level and the location of both exterior and interior walls and openings within the walls. Masonry-related details such as fireplaces, hearth extensions, and interior masonry walls are indicated on floor plans. Each floor level requires a separate floor plan. The floor plan in Figure 26-25 is part of the same construction documents as the foundation plan in Figure 26-24.

**FROM EXPERIENCE**

Masons need to look at floor plans to determine the locations of interior masonry walls, masonry fireplaces, and fireplace surrounds and outer hearths.

## Framing Plans

Framing plans are scale drawings showing floor framing and roof framing, which are the frameworks that serve as the skeleton of a structure to which coverings can be applied. The framing plans show the direction in which the framing is oriented in relationship to the outside walls and the spacing and dimensions of the framing material. Additionally, roof plans indicate overhang, the projection of the roof beyond the wall below it. Figure 26-26 shows roof plans for the same home depicted in the preceding foundation and floor plans.

## Landscaping Plans

Landscaping plans are scale drawings showing both natural and manmade elements intended for both practical and aesthetic purposes. Examples of natural elements are ground covers, shrubs, trees, rock outcroppings, and water. Manmade elements are such items as paved walkways, water fountains and ponds, atriums, raised planters, and walls. A masonry contractor may be responsible for details such as brick walkways, garden walls, and other masonry architectural elements such as the tall outdoor clock shown in Figure 26-27.

## Elevations

Elevations are two-dimensional graphic scale representations of the front, sides, and back of a structure. Features typically represented in exterior elevations are footing and

foundation depths, grade lines, floor levels, exterior wall coverings material symbols, roof pitches and roofing materials symbols, chimneys, doors and windows, steps, stoops, porches, walkways, patios, and retaining walls. Elevations also include symbols referring to section locations. Interior elevations may be included in the construction documents to show built-ins such as kitchen cabinets and alcoves, large recesses in rooms typically separated by an arch. The elevations for the home represented in the preceding working drawings are shown in Figure 26-28.

**FROM EXPERIENCE**

Refer to the elevation drawings when constructing CMU foundation walls requiring a change of CMU size at grade level and permitting anchored brick veneer above grade. The elevation drawing for each side of the project indicates the finished ground level along its length. The intended finished ground level may not be the same as the existing ground level.

## Presentation Drawings

Presentation drawings are perspective drawings, drawings appearing in a three-dimensional form as they would appear to the eye. As the name implies, presentation drawings are shown by architects and designers to their clients so they can view a structure as it will appear to the eye when completed. Presentation drawings are similar to photos and are not considered as working drawings, which are drawings with details to permit building all that is shown without additional instructions. Figure 26-29 shows a presentation drawing of a single-story, brick veneer structure.

**FROM EXPERIENCE**

When beginning to build a foundation, it is good for the masonry contractor to show all employees the presentation drawing of the project. This enables them to visualize the finished building so that they have a better understanding and an appreciation of their role in making the project a reality.

## Construction Details

Construction details, also simply called details, are drawings at a larger scale and showing in greater detail the significant parts that require more information

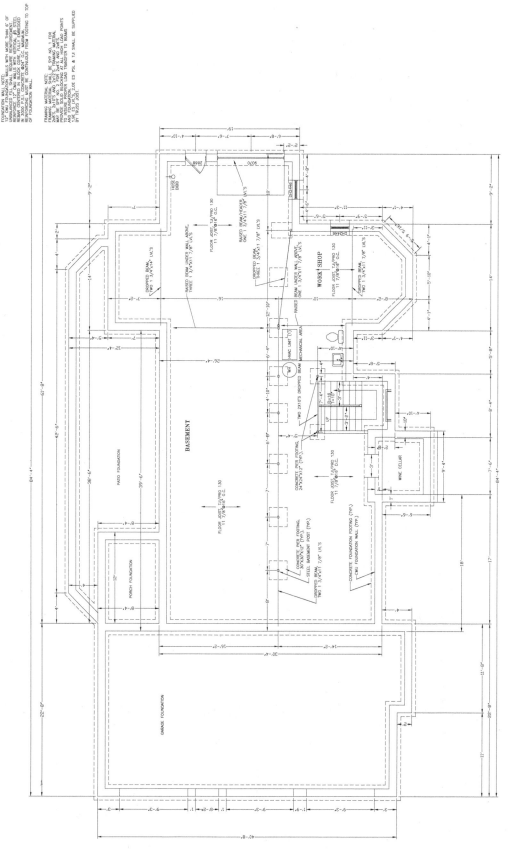

BASEMENT & FOUNDATION PLAN

**Figure 26-24** The foundation plan is key for having the remaining floors dimensioned as intended. The two broken lines indicate the width of concrete footings, and the two solid lines between the two broken lines indicate 12" CMU walls supported by the footings.
*(Courtesy of Design Vision)*

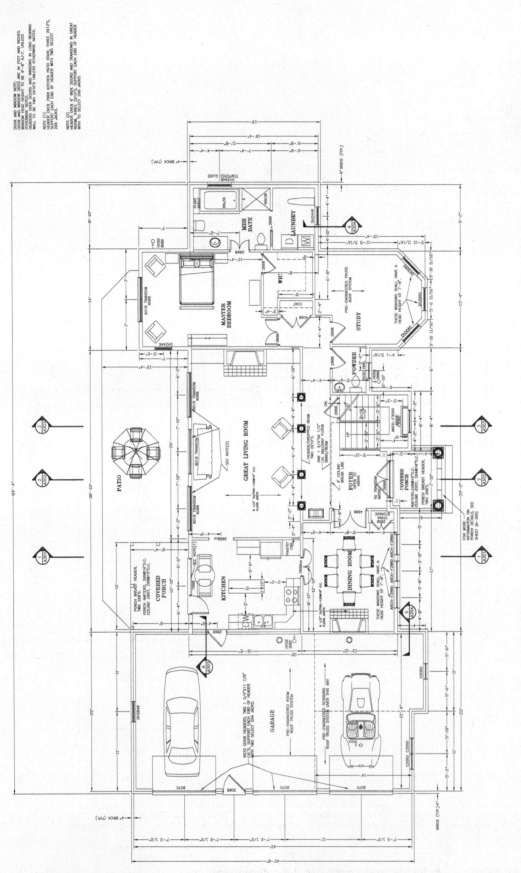

**Figure 26-25** This is the first floor plan for the same home represented by the foundation plan in Figure 26-24. *(Courtesy of Design Vision)*

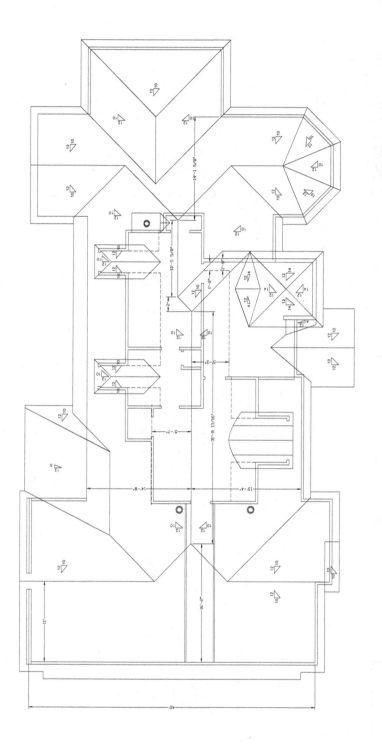

ROOF PLAN

**Figure 26-26 This drawing shows the roof ridges and valleys for the home in the preceding working drawings.** *(Courtesy of Design Vision)*

than is provided on the working drawings. Examples of construction details to which masons must be attuned are those showing fireplaces, steps, stoops, porches, and other masonry construction details. These details are typically shown in a **section,** a drawing showing the vertical composition of a specific part of a working drawing. Figure 26-30 is a section view showing the design for the brick interior kitchen wall shown on the floor plan in Figure 26-25. A photo of the wall as it appears in the finished kitchen is shown in Figure 26-31.

A section view of a masonry fireplace, shown in Figure 26-32, illustrates the type of information a mason may request or be provided when building a conventional masonry fireplace.

## Specifications

**Specifications** are a written description of the conditions of the contract. Along with working drawings, they comprise the **construction documents,** the conditions of

**Figure 26-27** The landscaping plans for this home indicated the location for a functional tower clock built of CMUs, brick, precast stone, glass block, and tile.

a contract upon which the owner and the contractor agree. Specifications address such topics as the quality of materials and workmanship required. The quantity, sizes, and capacities of specific equipment are listed. The information provided by working drawings should agree with the information contained in the written specifications. Usually, the information provided by the specifications is given priority over contradicting information found on the working drawings. But to avoid problems, information found contradictory to that of the working drawings should be clarified and agreed upon by the involved parties before construction begins. A **project representative**, one assigned to a job site to oversee the construction contract, should be approached when there are questions regarding the construction documents. The specifications include general conditions, supplementary conditions, and special conditions.

 **FROM EXPERIENCE**

Make no assumptions as to what is expected, and always consult responsible parties. Before parties agree to a contract, the following should be clarified: 1) the responsible party for pinning or marking the corners for foundation walls; 2) additional costs incurred if footings are not level and more time is required to align the foundation walls level; 3) the responsible party for work such as installing rebar and grout in walls, installing anchor bolts, installing foundation vents, and applying exterior foundation wall coatings; 4) additional costs for installing flashings and weeps; 5) the intended tooling to be used on mortar joints, and the one responsible for cleaning the masonry wall faces.

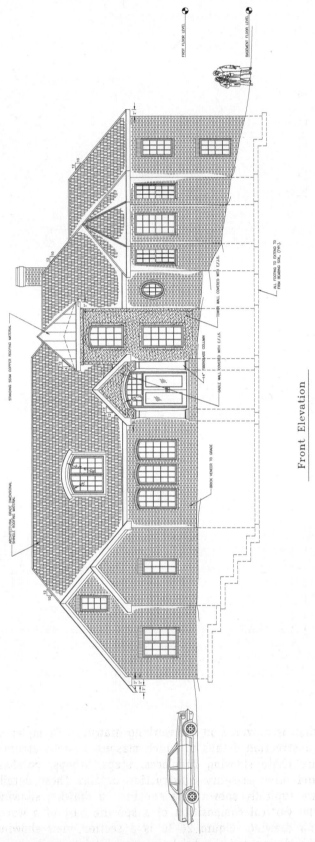

Front Elevation

**Figure 26-28** An elevation for each side of the home shown in foregoing drawings enables you to see the appearance of the home as it is intended to appear. Previous working drawings in this chapter refer to the same home. *(Courtesy of Design Vision)*

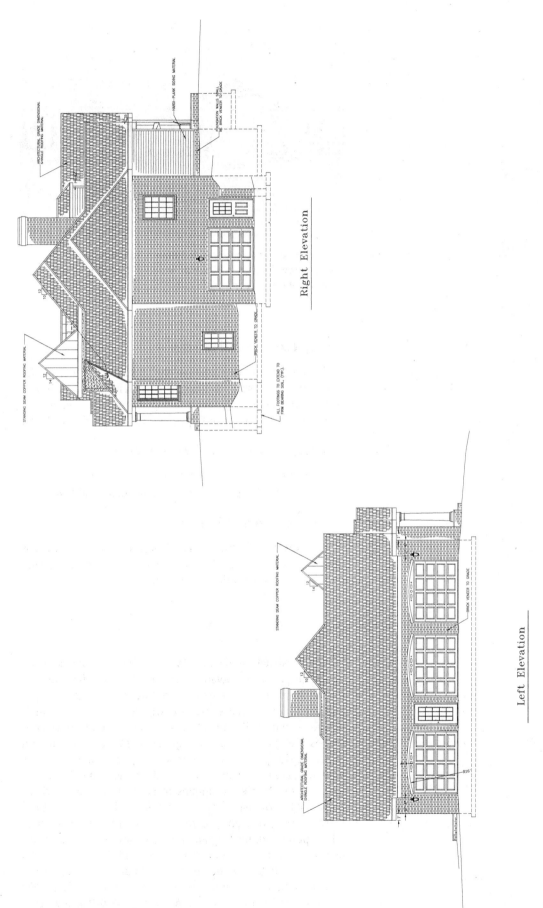

Right Elevation

Left Elevation

**Figure 26-28** (Continued)

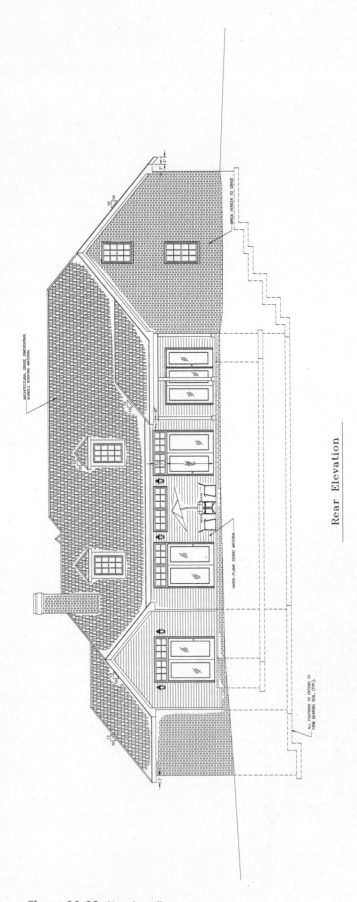

**Figure 26-28 (Continued)**

**Figure 26-29** The proposed building is illustrated in a presentation drawing, a pictorial image of how it is to appear once completed. *(Courtesy of Design Vision)*

## General Conditions

The **general conditions** in the specifications define the rights and the responsibilities of all parties involved in the work of the project. Subjects addressed include compliance with laws and regulations, payments, and time schedules.

## Supplementary Conditions

**Supplementary conditions** in the specifications supplement or modify provisions of the general conditions.

## Special Conditions

**Special conditions** are conditions of the contract other than general conditions or supplementary conditions setting forth requirements unique to the project.

# Change Orders

The construction documents are a binding agreement between all parties involved. There are three parties agreeing to the contract documents prepared by the architect and/or engineer. First is the owner for whom the construction is designed and built. Also, there is the general contractor, who is the main contractor responsible for the construction of the entire project. In addition, there are subcontractors who are responsible for only a part of the general contractor's responsibilities. Examples of subcontractors are those responsible for parts of the work such as excavation, masonry construction, framing and carpentry, roofing, electrical, plumbing, heating and air conditioning, interior wallboard, painting, cement finishing, and landscaping. No alterations, deletions or additions should be made to the construction documents unless a change order is agreed upon by the involved parties. A **change order** is a written

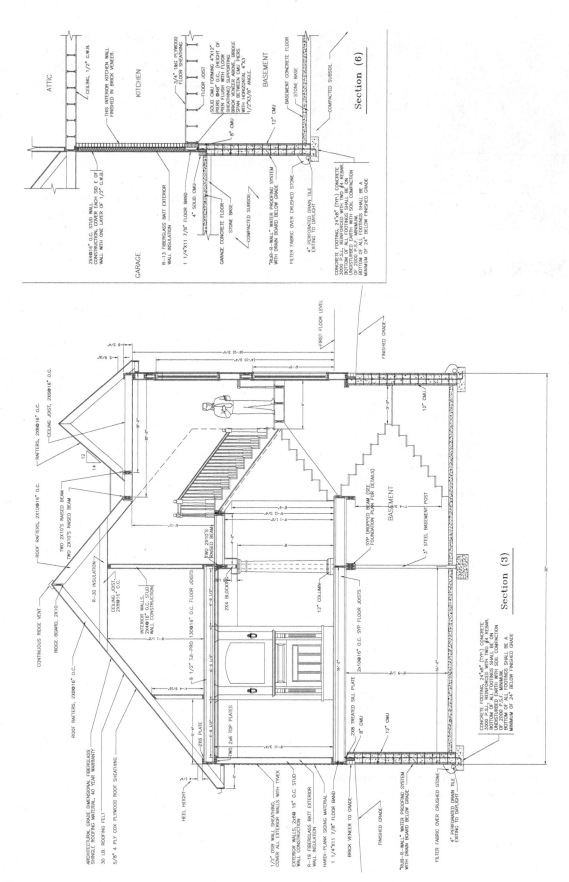

**Figure 26-30** Section (6) is a sectional drawing showing construction details for building an interior brick wall in the kitchen, enabling the mason to complete the construction as intended. *(Courtesy of Design Vision)*

**Figure 26-31** The finished anchored brick veneer wall, as it appears after completion, adds elegance to the kitchen. *(Courtesy of Design Vision)*

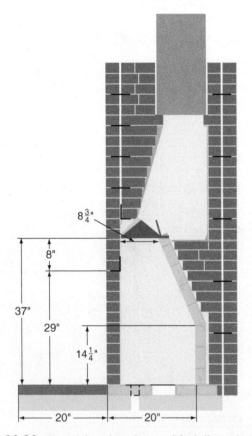

**Figure 26-32** The section view shows critical dimensions for the fireplace firebox and the outer hearth extension.

change to the original contract agreed upon by the owner, architect, engineer, and contractor. A change order details a change of work as specified in the original contract. Changes in the monetary amount of the contract and changes in the time allotted for completing the work in the contract are agreed upon in the change order.

# Summary

- Construction drawings are similar to languages, a set of characters used to express meanings. And as is true with languages, construction drawings can be interpreted only after you learn the character representations and their meanings.
- Although it is not the mason's responsibility to create the construction drawings, only those masons capable of interpreting them can proceed to build a project without assistance.
- Construction drawings give the information necessary for masonry contractors to determine the amount of time, materials, and equipment that is required to complete a job, thereby enabling them to make a bid proposition for a specific job.
- Masons rely on foundation plans to accurately build CMU foundation walls.
- Floor plans show masonry construction specified for each floor level, such as fireplaces, brick surround walls, and other interior masonry walls.
- Never make assumptions. The construction documents contain specifications for all details pertaining to the project. If needed information cannot be found or understood in the construction documents, consult a responsible party.

# Review Questions

*Select the most appropriate answer.*

**1** Solid lines representing the outline shape of objects are called

a. border lines
b. dimension lines
c. object lines
d. section lines

**2** Dimension lines terminate with

a. arrow heads
b. dots
c. slashes
d. all of the above

**3** Lines indicating the boundaries for dimension lines are called

a. border lines
b. extension lines
c. object lines
d. section lines

**4** Lines ending with an arrow head identifying the notation at their opposite end are called

a. break lines
b. leader lines
c. object lines
d. section lines

**5** Lines indicating the center of objects are called

a. break lines
b. centerlines
c. section lines
d. section reference lines

**6** Lines used to shorten a drawing, requiring less space for showing the drawing, are called

a. break lines
b. object lines
c. section lines
d. section reference lines

**7** Solid, parallel pairs of slanting lines indicating a break in a vertical, cut-through part of a drawing are called

a. break lines
b. object lines
c. section lines
d. section reference lines

**8** Lines indicating an accompanying cross-section are called

a. cutting plane lines
b. section reference lines
c. section lines
d. both a and b

**9** Solid, bold lines framing the outer perimeters for details intended for the drawing are called

a. border lines
b. dimension lines
c. object lines
d. section lines

**10** Examples of items represented by topographical symbols are

a. ground vegetation
b. property lines
c. roadways
d. all of the above

**11** Points of known elevation used as references in determining other elevations are called

a. bearing point
b. benchmarks
c. elevation points
d. plate marks

12  **A type of building material symbol is**

a. elevations

b. sections

c. segmentals

d. both a and b

13  **Types of building materials are indicated using**

a. capital-letter abbreviations

b. descriptive words

c. reference numbers

d. small-letter abbreviations

14  **The scale typically used for making construction drawings is the**

a. ⅛" scale

b. ¼" scale

c. ½" scale

d. both a and b

15  **Graphic or pictorial representations of a construction project are called**

a. plans

b. working drawings

c. sketches

d. both a and b

16  **Scale drawings showing the location of a proposed structure in relation to the site and the property lines are called**

a. construction surveys

b. boundary drawings

c. plot plans

d. topographical plans

17  **Foundation plans show**

a. concrete slabs

b. foundation walls

c. supporting piers

d. all of the above

18  **Floor plans show**

a. exterior walls        c. fireplaces

b. interior walls        d. all of the above

19  **Framing plans show the**

a. direction for which framing is oriented in relation to the walls

b. overhang or projection of the roof framing beyond the walls

c. spacing of the framing materials

d. all of the above

20  **Two-dimensional graphic representations of each side of structures are called**

a. elevations            c. site plans

b. floor plans           d. wall sections

21  **Drawings showing buildings as they should appear to the eye are called**

a. building drawings

b. construction drawings

c. isometric drawings

d. presentation drawings

22  **Drawings showing greater detail and drawn at a larger scale than working drawings are called**

a. construction details  c. exploded views

b. details               d. both a and b

23  **Written descriptions of the conditions of contracts are called**

a. agreements            c. orders

b. requirements          d. specifications

24  **The conditions of a contract upon which the owner and contractor agree are included in the**

a. building plans

b. construction documents

c. project proposal

d. site plans

25  **An individual assigned to oversee the construction of a project in order that the terms of the contract are met is called a**

a. job foreman

b. job inspector

c. project representative

d. job superintendent

# Appendix A

## Math Review

## Rules for Working with Simple Fractions

### Addition

1. Find a common denominator.
2. Add the numerators of the fractions.
3. Simplify the fraction.

### Subtraction

1. Find a common denominator.
2. Subtract the numerators of the fractions.
3. Simplify the fraction.

### Multiplying two simple fractions

1. Multiply the numerators.
2. Multiply the denominators.
3. Simplify the fraction.

### Multiplying a whole number and a fraction

1. Convert the whole number to a fraction, the whole number as the numerator, and assign the denominator the value of 1.
2. Multiply the numerators.
3. Multiply the denominators.
4. Simplify the fraction.

### Dividing simple fractions

1. Invert the second fraction (to the right of the division sign).
2. Convert the problem to multiplication by changing the division sign to a multiplication sign.
3. Multiply the numerators.
4. Multiply the denominators.
5. Simplify the fraction.

## Units of Conversion

### English Units of Conversion for Linear Measure

- 12" (in) = 1' (ft)
- 3' (ft) = 1 yard (yd)
- 1 yard (yd) = 36" (in)
- 1 mile (mi) = 5,280' (ft)
- 1 rod (rd) = 16.5' (ft)

### English Units of Conversion for Area Measure

- 1 square foot (sq ft) = 144 square inches (sq in)
- 1 square yard (sq yd) = 9 square feet (sq ft)
- 1 acre (A) = 43,560 square feet (sq ft)
- 1 square mile (sq mi) = 640 acres (A)

### English Units of Volume Measure

- 1 cubic foot (cu ft) = 1,728 cubic inches (cu in)
- 1 cubic yard = 27 cubic feet (cu ft)

### English Units of Capacity Measure

- 16 ounces (oz) = 1 pint (pt)
- 2 pints (pt) = 1 quart (qt)
- 4 quarts (qt) = 1 gallon (gal)

### English Capacity: Cubic Measure Equivalents

- 1 gallon (gal) = 231 cubic inches (cu in)
- 7½ gallons (gal) = 1 cubic foot (cu ft)

### English Units of Weight Measure

- 16 ounces (oz) = 1 pound (lb)
- 2,000 pounds (lb) = 1 net or short ton
- 2,240 pounds (lb) = 1 gross or long ton

## Metric Units of Linear Measure

- 1 millimeter (mm) = 0.001 meter (m)
- 1000 millimeters (mm) = 1 meter (m)
- 1 centimeter (cm) = 0.01 meter (m)
- 100 centimeters (cm) = 1 meter (m)
- 1 decimeter (dm) = 0.1 meter (m)
- 10 decimeters (dm) = 1 meter (m)
- 1 hectometer (hm) = 100 meters (m)
- 0.01 hectometer (hm) = 1 meter (m)
- 1 kilometer (km) = 1,000 meters (m)
- 0.001 kilometers (km) = 1 meter (m)

## Metric Capacity: Cubic Measure Equivalents

- 1 milliliter (mL) = 1 cubic centimeter ($cm^3$)
- 1 liter (l) = 1000 milliliters (mL)

## Metric Units of Weight Measure

- 1,000 milligrams (mg) = 1 gram (g)
- 1,000 grams (g) = 1 kilogram (kg)
- 1,000 kilograms (kg) = 1 metric ton (t)

# Formulas

## Formula for Perimeters of Quadrilaterals

- Quadrilateral: Perimeter (P) = sum of all sides

## Formula for Circumference of a Circle

- Circumference (C) = 3.14 × diameter (d)
- Circumference (C) = 2 × 3.14 × radius (r)

## Formulas for Areas (A) of Plane Figures

- Rectangle: Area (A) = length (l) × width (w)
- Parallelogram: Area (A) = base (b) × height (h)
- Triangle: Area (a) = ½ base (b) × height (h)
- Circle: Area (A) = 3.14 × radius² ($r^2$)
- Circle: Area (A) = 0.7854 × diameter² ($d^2$)

## Formulas for Volumes

- Cube: Volume (V) = length (l) × width (w) × height (h)
- Right circular cylinder:
  Volume (V) = area of base (AB) × altitude (h)

# Fractional Equivalents

Table A-1  Decimal and Metric Equivalents of Common Fractions of an Inch

| Fractional Inch | Decimal Inch | Millimeter |
|---|---|---|
| 1/16 | .0625 | 1.588 |
| 1/8 | .125 | 3.175 |
| 3/16 | .1875 | 4.762 |
| 1/4 | .25 | 6.35 |
| 5/16 | .3125 | 7.938 |
| 3/8 | .375 | 9.525 |
| 7/16 | .4375 | 11.112 |
| 1/2 | .5 | 12.7 |
| 9/16 | .5625 | 14.288 |
| 5/8 | .625 | 15.875 |
| 11/16 | .6875 | 17.462 |
| 3/4 | .75 | 19.05 |
| 13/16 | .8125 | 20.638 |
| 7/8 | .875 | 22.225 |
| 15/16 | .9375 | 23.812 |
| 1 | 1 | 25.4 |

| Equivalent Fractions |
|---|
| 1/8 = 2/16 |
| 1/4 = 2/8 = 4/16 |
| 1/2 = 4/8 = 8/16 |
| 3/4 = 6/8 = 12/16 |
| 1 = 8/8 = 4/4 = 16/16 |

Basic Trigonometric Formulas*

- sin θ = length of opposite side/length of hypotenuse
- cos θ = length of adjacent side/length of hypotenuse
- tan θ = length of opposite side/length of adjacent side
- Pythagorean Theorem $c^2 = a^2 + b^2$

*where θ is the measure of a certain angle in the triangle

# Appendix B

## Masonry Construction Materials

The following tables provide information for masonry construction materials.

- Table B-1 Rule of Thumb Figures
- Table B-2 Common Brick Sizes (Nominal Dimensions)
- Table B-3 Brick Classification
- Table B-4 Dimensional Tolerances
- Table B-5 Weathering Indices in the United States

- Table B-6 Mortar Materials Proportion Specification Requirements
- Table B-7 Property Specification Requirements
- Table B-8 ASTM C 476 Grout Proportions by Volume
- Table B-9 Concrete Block Grout Fill Quantities
- Table B-10 Steel Reinforcement Sizes
- Table B-11 Cleaning New Brick
- Table B-12 Pressure Washing Information

### Table B-1 Rule of Thumb Figures

**Block**

- 3 bags mortar per 100 block
- 1000 lbs of sand per 100 block
- 1.125 block per sq. ft. wall
- ASTM Specification C90

**Mortar**

- Typical walls (above grade) use Type N
- Below grade and block foundations use Type S
- Pavers use Type S*
- 1 part masonry cement mortar: 3 parts sand
- ASTM Specification C270

*Brick Industry Association Technical Notes 14A recommends using Type M mortar for exterior mortared brick paving on grade. Technical Notes 14A states that Type S mortar may be used when the pavement is not in contact with the earth or in interior applications.

| Brick | Modular (M/S) | Engineer Modular (E/M) | Queen Size (M/Q) | King Size (K/S) |
|---|---|---|---|---|
| Size<br>W × H × L<br>(Bed × Face × Length) | 3½" × 2¼" × 7⅝" | 3½" × 2¾" × 7⅝" | 3" × 2¾" × 7⅝" | 2⅝" × 2¾" × 9⅝" |
| Brick Per Sq. Ft. of Wall* | 6.75 | 5.76 | 5.76 | 4.6 |
| Bags of Mortar per 1000 brick | 7 | 8 | 6.5 | 7 |
| Wall Ties per 1000 brick | 55 | 65 | 65 | 80 |
| ASTM Specification | C216 | C216 | C216 | C652 |

*Add approximately 5% for waste.

**Sand**

- Order 1 ton of sand for every 7 bags of masonry cement mortar
- ASTM Specification C144

## Table B-1    *(Continued)*

### Pavers

- 4.5 Pavers per sq. ft.
- 2" sand setting bed; 1 ton covers 110 sq. ft.
- ASTM Specification C902 (Pedestrian and Light Traffic 1¼" units) and C1272 (Heavy Vehicular 2⅝" units)

### Sealers and Cleaners

*ProSoCo Products*

- Weatherseal Siloxane PD™ 1 gallon per 125 sq. ft. surface area
- Paver Enhancer™ 1 gallon per 125 sq. ft. surface area
- Sure Clean 600™ 1 gallon (undiluted) per 5000 brick or Vanatrol™

**Weeps**

- Order 1 weep for every 2 linear feet of base course, above and below all openings
- Order weeps for roof intersections on prefab chimneys

**Lintels**

- Minimum length = Masonry Opening plus 8" (Openings up to 6')
  Minimum length = Masonry Opening plus 12" (Openings between 6 to 9')
  Minimum length = Masonry Opening plus 24" (Openings greater than 9')
- Lintels are ordered in 6" length increments (order next size up)
- For vinyl windows, the out to out dimension is the same as the masonry opening

**Flashing**

- Around base course
- Above and below all window and door openings (double lintel length)

**Concrete**

- 4:2:1 mix 1 ton stone: 1/2 ton sand: 6 bags cement

(Courtesy of General Shale Brick.)

## Table B-2    Common Brick Sizes (Nominal Dimensions)

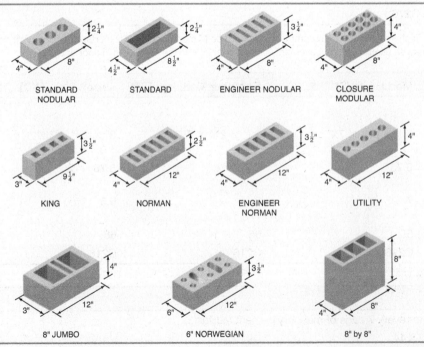

(Courtesy of The Brick Industry Association, *www.gobrick.com,* Technical Notes 9B—Manufacturing, Classification, and Selection of Brick, Selection, Part 3, Revised December 2003, Figure 1.)

Table B-3    **Brick Classification**

Depending on its use, brick can be classified by one of several specifications.

| Type of Brick Unit | ASTM[1] Designation | CSA[2] Designation |
|---|---|---|
| Building Brick | C 62 | — |
| Facing Brick | C 216 | A82.1 |
| Hollow Brick | C 652 | A82.8 |
| Paving Brick | C 902 | — |
| Ceramic Glazed Brick | C 126 | — |
| Think Brick Veneer Units | C 1088 | — |
| Sewer and Manhole Brick | C 32 | — |
| Chemical Resistant Brick | C 279 | — |
| Industrial Floor Brick | C 410 | — |

[1]American Society for Testing and Materials, 100 Barr Harbor Drive, West Conshoohocken, PA 19428-2959

[2]Canadian Standards Association, 178 Rexdale Boulevard, Etibicoke, Ontario, Canada, M9W 1R3

Since chemical-resistant brick and industrial floor brick are special applications, they will not be addressed in these *Technical Notes*.

## Uses

As the names imply, the uses of brick are similar to their respective ASTM designations.

**Building Brick:** Building brick are intended for use in both structural and nonstructural masonry where appearance is not a requirement. Building brick are typically used as a backing material.

**Facing Brick:** Facing brick are intended for use in both structural and nonstructural masonry where appearance is a requirement.

**Hollow Brick:** Hollow brick are identical to facing brick but have a larger core area. Most hollow brick are used in the same applications as facing brick. Hollow brick with very large cores are used in walls that are reinforced with steel and grouted solid. Larger cores or cells in hollow brick allow reinforcing steel and grout to be placed in these units, whereas it would be difficult to do so with building brick, facing brick, or some hollow brick.

**Paving Brick:** Paving brick are intended for use as a paving material to support pedestrian and light vehicular traffic.

**Ceramic Glazed Brick:** Ceramic glazed brick are units with a ceramic glaze fused to the body and used as facing brick. The body may be either facing brick or other solid masonry units.

**Thin Brick:** Thin brick veneer units are fired clay units with normal face dimensions but a reduced thickness. They are used in adhered veneer applications.

**Sewer and Manhole Brick:** Sewer and manhole brick are intended for use in drainage structures for the conveyance of sewage, industrial wastes, and storm water, and in related structures, such as manholes and catch basins.

(Courtesy of The Brick Industry Association, *www.gobrick.com,* Technical Notes 9A—Manufacturing, Classification, and Selection of Brick—Classification, Part 2, June 1989.)

### Table B-4 Dimensional Tolerances

Because of the variations in the raw materials and the manufacturing process, brick may vary in size. The permitted size variation is based on the brick type and the dimension being measured. These variations in size are listed in Table B-5. The variation is plus or minus from the specified dimension. Generally, "through-the-body" colored brick with a wide color range will also have a wide variation in dimensions. Brick without a wide color range may vary one way or another. Size variation becomes important when constructing an assembly with units aligned vertically or in wall sections with short horizontal dimensions.

| ASTM Standard | | Maximum Permissible Variation (in.) +/− | | | | | |
|---|---|---|---|---|---|---|---|
| | | <3 in. | Over 3 to 4 in. | Over 4 to 6 in. | Over 6 to 8 in. | Over 8 to 12 in. | Over 12 to 16 in. |
| C 62 | | 3/32 | 1/8 | 3/16 | 1/4 | 5/16 | 3/8 |
| C 216 | FBX | 1/16 | 3/32 | 1/8 | 5/32 | 7/32 | 9/32 |
| | HBS & HBB | 3/32 | 1/8 | 3/16 | 1/4 | 5/16 | 3/8 |
| | HBA[1] | — | — | — | — | — | — |
| C 652 | HBX | 1/16 | 3/32 | 1/8 | 5/32 | 7/32 | 9/32 |
| | HBS & HBB | 3/32 | 1/8 | 3/16 | 1/4 | 5/16 | 3/8 |
| | HBA[1] | — | — | — | — | — | — |
| C 902 | PX | 1/16 | 3/32 | — | 1/8[2] | — | — |
| | PS | 1/8 | 3/16 | — | 1/4[2] | — | — |
| | PA[3] | — | — | — | — | — | — |
| C 126[4] | | — | — | — | — | — | — |
| C 1088 | TBX | 1/16 | 3/32 | 3/32 | 5/32 | 7/32 | 9/32 |
| | TBS | 3/32 | 1/18 | 1/8 | 1/4 | 5/16 | 3/8 |
| | TBA[1] | — | — | — | — | — | — |
| C 32 | | | | +/− 1/8 in transverse direction  +/− 1/4 in length | | | |

[1]As specified by the purchase
[2]Over 5" to 8"
[3]No limit
[4]Special Requirements—see ASTM C 126

(Courtesy of The Brick Industry Association, *www.gobrick.com,* Technical Notes 9A—Manufacturing, Classification, and Selection of Brick—Classification, Part 2, June 1989.)

### Table B-5 Weathering Indices in the United States

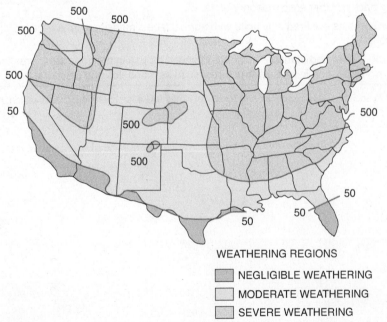

WEATHERING REGIONS

▢ NEGLIGIBLE WEATHERING
▢ MODERATE WEATHERING
▢ SEVERE WEATHERING

(Courtesy of The Brick Industry Association, *www.gobrick.com,* Technical Notes 9B—Manufacturing, Classification, and Selection of Brick, Selection, Part 3, Revised December 2003, Figure 2.)

**Table B-5**  *(Continued)*

### Recommended Minimum Grade for Face Exposures

| Exposure | Weathering Index | |
|---|---|---|
| | Less than 50 | 50 and greater |
| In vertical surfaces: | | |
|   In contact with earth | MW | SW |
|   Not in contact with earth | MW | SW |
| In other than vertical surfaces: | | |
|   In contact with earth | SW | SW |
|   Not in contact with earth | MW | SW |

(Courtesy of The Brick Industry Association, *www.gobrick.com,* Technical Notes 10B—Brick Sizes and Related Information, September 2001.)

**Table B-6**  Mortar Materials Proportion Specification Requirements

NOTE—Two air-entraining materials shall not be combined in mortar.

| Mortar | Type | Proportions by Volume (Cementitious Materials) | | | | | | | | Aggregate Ratio (Measured in Damp, Loose Conditions) |
|---|---|---|---|---|---|---|---|---|---|---|
| | | Portland, Blended Hydraulic or Hydraulic Cement | Mortar Cement | | | Masonry Cement | | | Hydrated Lime or Lime Putty | |
| | | | M | S | N | M | S | N | | |
| Cement-Lime | M | 1 | — | — | — | — | — | — | 1/4 | |
| | S | 1 | — | — | — | — | — | — | over 1/4 to 1/2 | |
| | N | 1 | — | — | — | — | — | — | over 1/2 to 1¼ | |
| | O | 1 | — | — | — | — | — | — | over 1¼ to 2½ | |
| Mortar Cement | M | 1 | — | — | 1 | — | — | — | — | |
| | M | — | 1 | — | — | — | — | — | — | |
| | S | 1/2 | — | — | 1 | — | — | — | — | |
| | S | — | — | 1 | — | — | — | — | — | Not less than 2¼ and not more than 3 times the sum of the separate volumes of cementitious materials |
| | N | — | — | — | 1 | — | — | — | — | |
| | O | — | — | — | 1 | — | — | — | — | |
| Masonry Cement | M | 1 | — | — | — | — | — | 1 | — | |
| | M | — | — | — | — | 1 | — | — | — | |
| | S | 1/2 | — | — | — | — | — | 1 | — | |
| | S | — | — | — | — | — | 1 | — | — | |
| | N | — | — | — | — | — | — | 1 | — | |
| | O | — | — | — | — | — | — | 1 | — | |

(Courtesy of The Brick Industry Association, *www.gobrick.com*, Technical Notes 8, June 2003.)

Table B-7    Property Specification Requirements[1]

| Mortar | Type | Average Compressive Strength at 28 Days, min. psi (Mpa) | Water Retention min. % | Air Content max. % | Aggregate Ratio (Measured in Damp, Loose Conditions) |
|---|---|---|---|---|---|
| Cement-Lime | M | 2500 (17.2) | 75 | 12 | |
| | S | 1800 (12.4) | 75 | 12 | |
| | N | 750 (5.2) | 75 | 14[2] | |
| | O | 350 (2.4) | 75 | 14[2] | |
| | | | 75 | | |
| Mortar Cement | M | 2500 (17.2) | 75 | 12 | |
| | S | 1800 (12.4) | 75 | 12 | Not less than 2¼ and not more than |
| | N | 750 (5.2) | 75 | 14[2] | 3½ times the sum of the separate |
| | O | 350 (2.4) | 75 | 14[2] | volumes of cementitious materials |
| | | | 75 | | |
| Masonry Cement | M | 2500 (17.2) | 75 | 18 | |
| | S | 1800 (12.4) | 75 | 18 | |
| | N | 750 (5.2) | 75 | 20[3] | |
| | O | 350 (2.4) | 75 | 20[3] | |

[1]Laboratory prepared mortar only.

[2]When structural reinforcement is incorporated in cement-lime or mortar cement mortar, the maximum air content shall be 12%.

[3]When structural reinforcement is incorporated in masonry cement mortar, the maximum air content shall be 18%.

(Courtesy of The Brick Industry Association, *www.gobrick.com,* Technical Notes 8, June 2003.)

Table B-8    ASTM C 476 Grout Proportions by Volume

| Grout Type | Portland Cement or Blended Cement | Hydrated Lime or Lime Putty | Fine Aggregate[1] | Coarse Aggregate[1] |
|---|---|---|---|---|
| Fine | 1 | 0 to 1/10 | 2¼ to 3 times the sum of the volumes of the cementitious materials | NONE |
| Coarse | 1 | 0 to 1/10 | 2¼ to 3 times the sum of the volumes of the cementitious materials | 1 to 2 times the sum of the volumes of the cementitious materials |

[1]Aggregate measured by volume in a damp, loose condition.

(Courtesy of The Brick Industry Association, *www.gobrick.com,* Technical Notes 3A, December 1992.)

Table B-9    Concrete Block Grout Fill Quantities

| Wall Construction | Cubic Measure of Void Volume Per 100 Square Feet of Wall Surface Area | |
| --- | --- | --- |
| Item | Cubic Feet (CF) | Cubic Yards (CY) |
| 12" Stretchers | 49.9 | 1.85 |
| 12" Corner Breakers | 42.9 | 1.59 |
| 1:1 Blend of 12" Stretchers and 12" Corner Breakers | 46.4 | 1.72 |
| 12" Retaining Wall Block | 56.0 | 2.07 |
| 10" Stretchers | 34.4 | 1.27 |
| 8" Stretchers | 29.7 | 1.10 |
| 8" Corner Breakers | 27.2 | 1.00 |
| 2:1 Blend of 8" Stretchers and 8" Corner Breakers | 28.5 | 1.06 |
| 8" Retaining Wall Block | 27.8 | 1.03 |
| 6" Stretchers | 20.7 | 0.77 |

| Bond Beams and Lintels | Cubic Measure of Void Volume Per 100 Lineal (Running) Feet of Beam or Lintel | |
| --- | --- | --- |
| Item | Cubic Feet (CF) | Cubic Yards (CY) |
| 12" Bond Beam Block with Center Web | 27.2 | 1.00 |
| 12" Bond Beam Block without Center Web | 32.9 | 1.22 |
| 8" Bond Beam Block with Center Web | 15.9 | 0.59 |
| 8" Bond Beam Block without Center Web | 21.2 | 0.79 |
| 8" U-Lintel Block | 16.7 | 0.62 |
| 6" Bond Beam Block | 21.1 | 0.78 |

Quantities represent calculated volumes. No allowance has been included for waste, settlement, workmanship, or other construction factors.
(Courtesy of General Shale Brick.)

Table B-10    Steel Reinforcement Sizes

| Reinforcement Type | Designation | Diameter, in. (mm) | Area, in.² (mm²) |
| --- | --- | --- | --- |
| | No. 3 | 0.375 (10) | 0.110 (71) |
| | No. 4 | 0.500 (13) | 0.196 (126) |
| | No. 5 | 0.625 (16) | 0.307 (198) |
| | No. 6 | 0.750 (19) | 0.442 (285) |
| Bars | No. 7 | 0.875 (22) | 0.601 (388) |
| | No. 8 | 1.000 (25) | 0.785 (506) |
| | No. 9 | 1.128 (29) | 1.000 (645) |
| | No. 10 | 1.270 (32) | 1.267 (817) |
| | No. 11 | 1.410 (36) | 1.561 (1007) |
| | W1.1 (11 gage) | 0.121 (3.1) | 0.011 (7) |
| | W1.7 (9 gage) | 0.148 (3.8) | 0.017 (11) |
| Wires | W2.1 (8 gage) | 0.162 (4.1) | 0.021 (14) |
| | W2.8 (3/16 in.) | 0.188 (4.8) | 0.028 (18) |
| | W4.9 (1/4 in.) | 0.250 (6.4) | 0.049 (32) |

(Courtesy of The Brick Industry Association, www.gobrick.com, Technical Notes 3B, May 1993.)

**Table B-11    Cleaning New Brick**

The Brick Association of the Carolinas offers the following recommendations for cleaning new brick.

**Special Systems for Wet Cleaning Through-the-Body Light Brick Where Type "S" Mortar Is Used**

Type "S" (and Type "M") mortar is very difficult to remove from the face of all brick, but it is a special problem when through-the-body or light colored brick is used due to the sensitivity of these brick to strong cleaning materials.

The following cleaning procedures are recommended according to the age of masonry work:

**A.** After work is 10 days old:

1. Remove all large mortar particles with hand tools before applying cleaning solutions.
2. Mask and otherwise protect adjacent nonmasonry materials.
3. Saturate wall with clean water.
4. Use cleaning brush to apply solution of Sure Klean Vanatrol or Diedrich 202 Vana-Stop (or equal) mixed 4 to 6 parts of water to 1 part of solution.
5. Allow solution to remain on wall for 3 to 5 minutes while brushing and scraping, and then reapply solution.
6. Thoroughly rinse and brush clean.

**B.** After work is 30 days old:

1. Use procedure described in the preceding steps 1–5.
2. Use high water pressure equipment to rinse wall, using pressure not greater than 800 psi with a 40-degree nozzle fan tip. Consult the brick manufacturer before using a high pressure water system.

*(Test clean a sample area to determine the effectiveness of the cleaning compound and the total cleaning system and to check the wall for possible damages caused by system. Approval of the owner or owner's representative should be obtained before proceeding with operation.)*

| Cleaning Guide |  |
| --- | --- |
| **Red Brick- Textured** | This category includes all textured red through-the-body brick. Brick in this category may be cleaned by the bucket and brush method, high water pressure method, or by sandblasting. |
| **Red Brick: Heavy Sand Finish** | This category includes all red through-the-body brick with various applied heavy sand finish faces. Brick in this category may best be cleaned by the bucket and brush method, using plain water and scrub brush, or with **lightly applied** high-pressure water system, with plain water being used. Sandblast cleaning is not recommended. If mortar stains are excessive, use of cleaning compounds may be required. |
| **White, Buff, Gray and Chocolate Brick** | This category includes all textured and sand finish brick with through-the-body colors other than natural red. Brick in this category may be cleaned by the bucket and brush method, or by lightly applied high-pressure water system. Sandblast cleaning is also recommended except in the cases where heavy sand finish is involved. In the two wet cleaning systems, no muriatic acid or compounds containing muriatic acid may be used. Only plain water and detergent, or Sure Klean Vanatrol or Diedrich 202V Vana-Stop (or equal) may be used. |
| **Specialty Cleaning** |  |
| **White Efflorescence** | White efflorescence is a water-soluble salt that is brought to the surface of masonry by evaporation of either construction water or by evaporation of rainwater that has penetrated the wall. Water used in mortar, grout, and so on will sometimes cause this "New Building Bloom." As the wall dries out, and as successive rains wash the walls, the "Bloom" should disappear. If the masonry has received its regular cleaning and white efflorescence appears or reappears, no further action should be taken until this wall has had an opportunity to dry out completely. Application of additional cleaning solutions may only aggravate the problem at this point. Also, application of clear waterproofing materials may lock in moisture and crystalline growth, causing more scumming and possible spalling of brick. If efflorescence stains persist, it is likely that rainwater is penetrating the wall. An inspection of the stained areas should be made to determine if sizable cracks or openings exist, permitting water penetration. Faulty flashing or a lack of flashing will contribute to staining. Any large openings should be repaired. Where only very fine hairline cracks are assumed to be allowing water penetration, application of a penetrating water repellent may be the only solution to the problem short of a complete tuckpointing job. Before |

Table B-11    (*Continued*)

| Cleaning Guide |
| --- |

applying waterproofing materials, all possible repairs should be made and all efflorescence removed. This may be removed by applying plain water and brushing the affected area. If water fails to remove the stain, use a dilute solution of commercial cleaning compounds such as Sure Klean 600 or Diedrich 202 New Masonry Detergent (or equal) for red brick, and Sure Klean Vanatrol or Diedrich 202V Vana-Stop (or equal) for all others. Some heavy white stains, known as "lime runs" or "silicone deposits" may require special cleaning procedures for removal. Contact the Brick Association of the Carolinas for further details. Allow entire wall to dry out completely (*over a period of little or no rainfall*) before applying waterproofing solutions.

**Green Stains**    Green staining is caused by the presence of vanadium salts. Color and solubility of these salts are dependent upon acidity of the brick. Very often green stains are brought about by the wrongful use of muriatic acid or compounds containing muriatic acid. When green stains appear, brick manufacturer should be consulted before attempting to remove the stain. Green stains may be removed by using Sure Klean 800 Stain Remover, Sure Klean Ferrous Stain Remover, Diedrich 940 Iron and Manganese Stain Remover, or Diedrich 950 Acid Burn Remover (or equal).

**Brown Stains**    Brown staining can be caused by the presence of soluble manganese or iron oxides. Very often brown or manganese stains are brought on by the wrongful use of muriatic acid or compounds containing muriatic acid. If these stains are light, Brick Klenz may take them off with little difficulty. Also, oxalic acid (one pound mixed in a gallon of water) may do the job if stains are new and light in color. Many brown stains can be removed with Sure Klean 800 Stain Remover, Sure Klean Ferrous Stain Remover, Sure Klean Restoration Cleaner, Diedrich 950 Acid Burn Remover, Diedrich 940 Iron and Manganese Stain Remover, or Diedrich 101G Brick Cleaner (or equal). Each product should be tested for effectiveness and possible bleaching action on joints.

**White Scum-Insoluble**    Insoluble white scum is generally caused by faulty cleaning, that is, failure to adequately saturate the wall before cleaning and failure to flush the wall after applying the cleaning compound. As opposed to white efflorescence, this stain cannot be removed with detergents or regular cleaning compounds. Currently known method of removal is to use Sure Klean White Scum Remover, or Diedrich 930 White Scum Remover (or equal).

**Smoke Stains**    Smoke stains can generally be removed by using one of the following cleaners: Brick Klenz or Sure Klean Smoke Remover. A follow-up cleaning with Sure Klean Restoration Cleaner, or Diedrich 101G Brick Cleaner (or equal) may be required after using smoke removal products. Follow the directions found on containers.

**Mud Stains**    Mud stains are the most difficult of all to remove. Currently known method of removal is as follows: Apply Sure Klean Restoration Cleaner and Diedrich 101G Brick Cleaner (or equal) full strength, with a stainless steel pressurized "orchard" sprayer. Allow to remain on the wall 5 minutes. Flush off with high-pressure water spray. Repeat if necessary. Sprayer nozzle should be held at a 90-degree angle to the wall, as should the rinse water nozzle. Sure Klean Light Duty Concrete Cleaner or Diedrich 960 Heavy Duty Concrete Cleaner might be less likely to bleach joints than Sure Klean Restoration Cleaner or Diedrich 101G Brick Cleaner.

**Paint Stains**    Paint stains are very difficult to remove from masonry. Probably sandblasting is the fastest way to remove paint, but this process is sometimes harmful to the masonry surface. Commercial paint removers are effective in some cases. Sure Klean Defacer Eraser, Sure Klean Heavy Duty Paint Stripper, Diedrich 505 Paint Stripper, or Diedrich 606 Multi-Layer Paint Remover (or equal) are very good for paint removal. If these products do not completely remove all paint particles after following printed directions, apply Sure Klean Restoration Cleaner or Diedrich 101G Brick Cleaner (or equal) to the stained area. Allow to remain on the wall several minutes, and then "blast" the area with a water hose. *Follow directions found on containers.*

**Cleaning Masonry Laid with Colored Mortar**    Colored mortar is highly sensitive to masonry cleaning solutions. While mineral oxide pigments are inert and are not affected by most cleaning materials, the materials will dissolve surrounding cement paste, allowing pigment to be washed away, exposing sand grains, and causing a change in mortar color and texture. Most manufacturers of colored mortar recommend cleaning with detergent and water only. Where mortar stains are heavy, a 1 to 6 solution of Sure Klean Vanatrol, Diedrich 202V Vana-Stop (or equal), and water may be used; but a curing period of 3 to 5 weeks is recommended before cleaning with anything other than detergent and water.

(*continued*)

**Table B-11    (*Continued*)**

| Cleaning Guide |
|---|

Sandblast cleaning is usually acceptable, as is high-pressure water cleaning with approved cleaning compounds. Protection of brick face must also be considered in selecting a cleaning system. (*As with all cleaning jobs, test clean a sample area to determine the effectiveness of the cleaning compound and the total cleaning system and to check the wall for possible damages caused by the system. Approval of owner or owner's representative should be obtained before proceeding with the operation.*)

**General Information**

**Light and Dark Joints**     Color change in mortar joints may be attributed to a change in quality of masonry cement, type of masonry cement, change in type or gradation of sand, and change in methods of cleaning. Also, the color of joints can be affected by variations in moisture content of individual brick surrounding the joint. Joints struck while excessively wet can become light in color. Joints struck when thumbprint hard should dry to a uniform color if mortar and sand properties remain consistent.

Normal variations in joint color will be eliminated after completion of one of the wet cleaning processes. Where wide color variations are found, a mild bleaching of all joints with an increased concentration of cleaning solutions usually brings improvement. Caution should be taken in using this process with acid-sensitive brick and colored mortars.

Light joints may be darkened by painting the joints with pigments specially selected to produce the required shade.

(Courtesy of Brick SouthEast Headquarters.)

**CAUTION**

The following disclaimer accompanies the preceding information.

"The information and recommendations made herein are based on our own research, and research and experience of others, and are believed to be accurate. However, no guarantee of their accuracy is made because we cannot cover every possible application of the products described, nor anticipate variations encountered in masonry surfaces, job conditions and cleaning methods used."

**Table B-12    Pressure Washing Information**

Because serious damage can be caused by improper pressure washing procedures, it is General Shale's policy not to recommend it for cleaning brick masonry. (General Shale, whose headquarters are in Johnson City, Tennessee, is one of the leading exterior building materials manufacturers in the United States.) This applies to all types of projects but is most important on residential jobs where experienced pressure washing personnel may not be available.

Many of today's brick are coated for color and texture effects, and these coatings may be as subject to abrasive removal as the mortar smears on the brick face. *Do not* use a pressure washer on this type of brick; brick manufacturers' generic terms for this type of brick include *Sand Faced Brick, Lime Faced Brick, Slurry Brick,* and *Molded Brick.*

If pressure wash cleaning methods are used, it is strongly suggested that a test be performed on a disposable panel or noncritical wall area to ensure the cleaning method is compatible with the brick.

Additional brick cleaning information and guidelines for pressure washing methods can be found in the BIA (Brick Industry Association) Technical Note Number 20. The following guidelines also provide some information that may be helpful.

**Table B-12**   *(Continued)*

*General Shale disclaims any and all responsibility for damages resulting from cleaning methods and materials.*

## General Guidelines and Specifications

- Use a moderate pressure 400–800 psi.
- Use a 25-degree to 35-degree fan tail nozzle.
- Use an average flow rate of 4 gallons per minute.
- Move wand in a horizontal motion.
- Maintain a consistent distance (6" to 8") of nozzle from wall surface.
- Ideal cleaning times:
  - After initial mortar set up to a maximum of 21 days after brick is laid (Type N).
  - After initial mortar set up to a maximum of 18 days after brick is laid (Type S).
- Apply cleaning solution with low pressure (40 psi max).

### Removal of Excess Mortar

- Use a wooden paddle or other objects that will not mar the brick.
- Do not use metal scrapers. This may leave shavings that could create stains.
- Do not use your pressure washer to remove large mortar particles. This can result in scarring of the face of the brick.

### Prewetting

- Prewetting prior to cleaning is one of the most important steps in the cleaning process.
- A saturated wall will not absorb contaminants or allow chemicals to dry on the surface.
- Improper prewetting is one of the major causes of masonry stains.

### Application of Cleaning Solution

- Do not apply cleaners with high pressure. This will drive the cleaning solution deep into the substrate, causing staining.
- Apply cleaning solutions with a low-pressure sprayer (max 40 psi).
- Allow the cleaning solution to dwell for 3 to 5 minutes before removal.

### Removal of Cleaning Solution

- Remember, use a 25-degree to 35-degree fan tail tip nozzle.
- Keep the wand 6" to 8" from substrate.
- Work in a horizontal motion (no figure eights).
- Keep wand at same distance to substrate.
- Rinse wall thoroughly!

### Steps in Pressure Washing

1. Perform test cleaning on a panel or noncritical wall area.
2. Provide protection for personnel and adjacent areas from pressure spray and chemical damage.
3. Remove excess mortar.
4. Prewet wall.
5. Apply cleaning solution.
6. Remove cleaning solution, and pressure rinse.
7. Thoroughly rinse wall.

### Pressure Washing Don't-Evers

- Don't use zero-degree nozzles.
- Don't use pressure washer to remove large particles of mortar.
- Don't wash your hands with a pressure washer.
- Don't use a figure eight motion.
- Don't write words on the wall with a pressure washer.
- Don't pressure wash sensitive brick.

(Courtesy of General Shale Brick.)

# Glossary

**6-6-10-10 wire reinforcement** 10-gauge wire welded at right angles on 6" centers.

**abutments** walls or piers supporting an arch.

**estribos** paredes o malecones que apoyan un arco.

**accelerators** admixtures that increase the masonry cement's rate of hydration.

**aceleradores** mezclas que aumentan la proporción del cemento de la albañilería de hidración.

**actual size** the true dimensions of an object.

**tamaño real** las verdaderas dimensiones de un objeto.

**adjustable anchor assemblies** assemblies used to anchor brick veneer to wood or steel framing.

**asambleas del ancla ajustables** las asambleas que fijan la chapa del ladrillo al aparato de madera o acero.

**adjustable assemblies** masonry wall reinforcement consisting of ladder-type or truss-type reinforcement embedded in the inner wythe with extended tabs or eye hooks to which rectangular adjustable sections are connected and embedded in the joints of the outer wythe.

**asambleas ajustables** refuerzo de pared de albañilería que consiste en escalera de mano o refuerzo del braguero-tipo empotrando el wythe interno con etiquetas o ganchos del ojo extendidas a que conectan las secciones ajustables rectangulares y se empotran en las junturas del wythe exterior.

**adjustable reinforcement assemblies** ladder and truss-type joint reinforcement assemblies used to anchor brick veneer to CMU walls.

**asambleas del refuerzo ajustables** la escalera de mano y braguero-tipo juntura de refuerzo, asambleas que fijan la chapa del ladrillo a las paredes de CMU.

**admixtures** materials added to masonry cements for specific purposes.

**mezclas** los materiales agregados al cemento de la albañilería para los propósitos específicos.

**air-circulating fireplace** a fireplace circulating room air through baffled, heat-exchange chambers behind the firebox walls.

**hogar aire-circulante** un hogar que circula el aire a través de las cámaras del intercambio de calor detrás de las paredes del firebox.

**air-entraining agents** additives increasing the air content in cement.

**agentes para empujar el aire** aditivos que aumentan el volumen aéreo en el cemento.

**air infiltration barrier** a sheathing membrane installed to the exterior of framing for preventing air infiltration.

**barrera de la infiltración aérea** una cubierta fina instalado en el exterior para evitar la infiltración aérea.

**air intake** a passageway for bringing outside air for combustion into the firebox.

**succión aérea** un pasadizo para atraer el aire de fuera para su combustión en el firebox.

**air space** the space between the backside of the brick veneer wall and the exterior side of the backer wall, permitting air circulation.

**espacio aéreo** el espacio entre la parte atrás de la pared de chapa de ladrillo y el lado exterior de la pared apoyante que permite la circulación del aire.

**American bond** a pattern consisting of a course of brick laid in the header position typically between every five or six courses of brick laid in the stretcher position.

**enlace americano** un diseño que consiste en una serie de ladrillos puestos típicamente en la posición de la cabeza arriba entre cada cinco o seis ladrillos puestos en la posición horizontal elástico.

**anchor bolt** a threaded bolt secured in and projecting from the top of a foundation wall to which the wood sill plate is fastened.

**cerrojo anclador** un cerrojo con filetes abrochado y proyectado de encima de la pared de fundación a la que está abrochada la placa de l umbral de Madera.

**anchor strap** a sheet-metal strap secured in and projecting from the top of a foundation wall to which a wood sill plate is fastened.

**cinturón anclador** un cinturón de metal en lamina abrochado y proyectado de encima de la pared de fundación a la que está abrochada la placa de l umbral de Madera.

**anchored brick veneer** a single-wythe brick wall supporting no weight other than its own, requiring vertical support and anchored to either a load-bearing or non-load-bearing wall, having an air space between the brick wall and the backup wall.

**chapa del ladrillo anclada** una pared de solo una capa de ladrillos wythe que no apoya otro peso que suyo, que necesita el apoyo vertical y está anclado a una pared que tenga un peso apoyado o no y que tiene un espacio aéreo entre la pared del ladrillo y la pared auxiliar.

**anchored veneer** a single-wythe masonry wall, requiring vertical support and anchored to either a load-bearing or non-load-bearing wall, with an air space between it and the backup wall.

**chapa fijada** un pared de albañilería con sólo una capa de ladrillos wythe, que necesita el apoyo vertical y está anclado a una pared que tenga un peso apoyado o no y que tiene un espacio aéreo entre la pared del ladrillo y la pared auxiliar.

**appliance chimney** a chimney designed to vent solid fuel appliances, those fired with coal and wood, and appliances fired with oil, natural gas, or liquid petroleum propane gas.

**chimenea del aparato** una chimenea diseñado para dar salida los aparatos de combustible sólidos, aquellos que usan carbón y madera, y los aparatos que usan el aceite, gas natural, o el gas de propano de petróleo líquido.

**apprentice** one learning a trade from one or more competent, experienced persons.

**aprendiz** uno que aprende un comercio de las personas más competentes con mucha experiencia.

**arching action** forces within a brick wall permitting the brick to partially support its own weight.

**acción de abovedar** las fuerzas dentro de una pared del ladrillo que permite al ladrillo apoyar su propio peso parcialmente.

**architectural concrete** concrete that is permanently exposed to view and requiring special care in forming, placing, and finishing.

**hormigón arquitectónico** hormigón expuesto a la vista que requiere cuidado especial para adquirir una forma, para colocarse y para acabarse.

**architectural CMUs** structural concrete masonry units whose one or more faces are produced in a multitude of texture and color combinations, suitable for both interior and exterior walls.

**CMUs arquitectónico** unidades de concreto estructurales de albañilería algunas cuyas fachadas están creadas como una multitud de texturas y combinaciones de colores, que son convenientes tanto para el interior como para el exterior.

**architectural drawing** a drawing showing a completed project as it is to appear on a specific site.

**dibujo arquitectónico** una exposición del dibujo de un proyecto entero tal y como aparecece en uuun sitio específico.

**areaway** a walled area below grade that permits ventilation, light, or egress.

**pasillo** una área amurallada de baja calidad que permite ventilación, luz, o salida.

**ash pit** a hollow space below the inner hearth for the deposit of ashes.

**hoyo de ceniza** un espacio vacío debajo del hogar interno para el depósito de cenizas.

**autoclaved CMUs** block cured in sealed steam kilns at steam pressures above 120 psi and a temperature of 350°F, which is harder and less effected by changes in weather than block cured in steam kilns at atmospheric pressures at lower temperatures.

**CMUs autoclavicados** el bloque curado en los hornos de vapor sellados a las presiones de vapor mayors a 120 psi y a temperatura de 350°F, es más duro y menos afectado por los cambios climáticos en comparación con el bloque curado en los hornos de vapor a las presiones atmosféricas y a las más bajas temperaturas.

**autogeneous healing** a unique characteristic of lime-based mortars that have the capability, by chemical reaction, to fill small voids that may occur in the joints.

**curación autogeneous** una característica especial de morteros basados en cal, que tienen la capacidad de llenar los pequeños vacíos que se produzcan entre las coyunturas, por la reacción química.

**backer wall** the wall to which the brick veneer is anchored.

**pared apoyante** la pared a la que la chapa del ladrillo está anclada.

**backfilling** replacing the earth against the outside of the foundation walls removed during construction.

**relleno de atrás** reemplazar la tierra en el exterior de las paredes de la fundación que se quita durante la construcción.

**bar reinforcement** steel rods placed in concrete for strengthening the concrete.

**reforzamiento de barras** barras de acero puestas en el hormigón para fortalecerlo.

**base** those components of a fireplace, including concrete footings, masonry foundation walls, ash pit, air inlet, and reinforced concrete hearth base.

**base** los componentes de un hogar, incluso los fundamentos de concreto, las paredes de fundación de albañilería, hoyo de ceniza, la entrada aérea, y la base reforzada del hogar de concreto.

**base flashing** flashing at the base of the wall just above the finished grade level.

**destello de la base** encender a la base de la pared anterior al nivel de calidad de acabado.

**basket weave pattern** a paving brick pattern where brick are paired perpendicular to adjoining pairs.

**diseño de tejido de cesto** un diseño del ladrillo pavimentado dónde el ladrillo se aparea perpendicular a los pares inmediatos.

**bat** another term for half of a full brick length.

**palo** otro término para la mitad de la longitud de un ladrillo.

**batter boards** pairs of horizontal boards nailed to wood stakes beyond each corner to which layout string lines are fastened.

**tablas de batido** los pares de tablas horizontales clavaron a las estacas de madera más allá de cada esquina a que se atan las líneas de cordón de diseño.

**bearing plates** steel plates with welded anchors on their bottom sides embedded in the tops of masonry load-bearing walls to which structural steel beams or joists are secured.

**placas de cojinetes** las placas de acero con las anclas soldadas en su fondo empotradas en las cabezas de las paredes de peso de albañilería que apoyan peso a las que las vigas de acero estructurales o viguetas están abrochadas.

**bench marks** points of known elevation used as a reference in determining other elevations.

**marcas del banco** los puntos de elevación conocida usados como una referencia para determinar otras elevaciones.

**bidder** one who submits a bid price proposal for work on a project.

**postor** uno que somete una propuesta de precio de oferta para el trabajo en un proyecto.

**bid price** the dollar amount of the proposal for supplying the materials and/or labor.

**precio de la oferta** la suma en dólares de la propuesta por proporcionar los materiales y/o mano de obra.

**bleeding** the result of excessive brick moisture content or mortar having too much water content, causing mortar stains on the face of walls as wet mortar gravitates from joints onto the face of walls.

**bleeding** el resultado de excesiva humedad en el ladrillo o mortero que contiene demasiado agua, que causa manchas de mortero sobre la fachada de paredes a medida que el mortero mojado gravita de las junturas hacia la fachada de las paredes.

**block jamb** a lead permitting block work along a wall in only one direction.

**jamba del bloque** una primacía que permite el trabajo del bloque a lo largo de una pared en sólo una dirección.

**block size** the nominal or approximate width of a block.

**tamaño del bloque** la anchura nominal o aproximada de un bloque.

**bond beam** a horizontal reinforced beam comprised of specially formed bond beam concrete masonry units, grout, and horizontally placed steel reinforcement bars designed to strengthen a wall, support loads above openings, or distribute imposed loads uniformly.

**viga del enlace** una viga reforzada horizontal que consta de unidades de la viga de enlace de albañería hechos especialmente del hormigón, la lechada, y horizontalmente puestas las barras de refuerzo de acero diseñadas para fortalecer una pared, para apoyar carga sobre las aperturas, o distribuir las cargas impuestas uniformemente.

**bond strength** a measure of the resistance to separation of mortar and masonry units.

**fuerza del enlace** una medida de la resistencia a la separación de mortero y unidades de la albañilería.

**bonded arch** an arch consisting of two or more rows of voussoirs forming the depth.

**arco unido** un arco que consta de dos o más filas de voussoirs que forman el fondo.

**border lines** solid, bold lines indicating the outer perimeters for any details intended for the drawing.

**líneas del marco** líneas sólidas, intrépidas que indican los perímetros exteriores para cualquier detalle dirigido para el dibujo.

**break lines** solid lines having diagonal lines extending beyond both sides of the lines, extending beyond a drawn object, indicating a break in the drawing with part of its scale size deleted.

**líneas de pausa** líneas sólidas que tienen líneas diagonales que extienden más allá de ambos lados de las líneas, extendiéndose más allá de un objeto dibujado, indicando una pausa en el dibujo con la parte de la escala de su tamaño eliminada.

**brick binding** forces created by differential movements, exerting forces on window frames and doorframes, often leaving them inoperable.

**encuadernación del ladrillo** fuerzas creadas por los movimientos del diferencial, que ejercen fuerzas por los marcos de ventanas y puertas, dejándolos a menudo inoperable.

**brick density** a measure of a brick's weight as compared to its size or volume.

**densidad del ladrillo** una medida del peso de un ladrillo en comparación con su tamaño o volumen.

**brick jamb** a lead permitting brickwork along a wall in only one direction.

**jamba del ladrillo** una primacía que permite el enladrillado a lo largo de una pared en sólo una dirección.

**brick mold** an exposed trim of door and window jambs adjacent to the ends of brick on a brick veneer wall.

**molde del ladrillo** una parte expuesta de las jambas de puerta y de la ventana adyacente a los fines de ladrillo sobre una pared de chapa de ladrillo.

**brick pavers** brick intended for either pedestrian or vehicular traffic, also called paving brick.

**pavimentadoras del ladrillo** el ladrillo para el tráfico pedestre o vehicular, también llamado el ladrillo pavimentador.

**brick shelf** a masonry ledge supporting the anchored brick veneer brick walls.

**estante del ladrillo** un anaquel de la albañilería que apoya las ladrillo chapa ladrillo paredes ancladas.

**building material symbols**
representations of common building
materials. There are two types of
building material symbols, elevations
and sections.

**símbolos de materiales de
construcción** las representaciones de
materiales comunes de construcción. Hay
dos tipos de símbolos de materiales de
construcción, elevaciones y secciones.

**camber** the rise above the spring line
for a jack arch.

**comba** el levantamiento sobre
la línea de muelle para un arco de
la sota.

**cap, also called capping,** an
architectural concrete or stone member
for the top of a pilaster or part of or an
entire wythe that is terminated below
the top of an adjoining wall.

**gorra, llamado también el
capping** un hormigón arquitectónico
o miembro de la piedra para la parte
arriba de una pilastra o parte de o un
wythe entero que se termina debajo
de la parte arriba de una pared
inmediata.

**capital** an architectural concrete or
stone top for piers, columns, and
pilasters that protects them from the
weather and enhances their appearance.

**capital** un hormigón arquitectónico o
cima de la piedra para los malecones,
columnas, y pilastras que los protegen
del clima y refuerzan su apariencia.

**cavity wall** a masonry wall consisting
of an inner and an outer wythe bonded
with corrosion-resistant metal ties and
separated by an air space not less than
2" nor more than 4½".

**pared de la cavidad** una pared de la
albañilería que consiste de un interno y
un wythe exterior unidos con los lazos
de metal resistentes a corrosión
separados por un espacio aéreo no
menos de 2" ni más de 4½."

**centerlines** lines appearing as a
repetitive pattern of a long broken
line followed by a shorter line or dot,
indicating the center of objects such as
doors, windows, and partition walls.

**líneas del centro** líneas que
aparecen como un diseño repetitivo de
una larga línea rota seguido por una
línea más corta o puntea, indicando el
centro de objetos como las puertas,
ventanas, y paredes de la partición.

**certification** assurance that the
contractor is in compliance with
applicable federal, state, and local
regulations pertaining to the specific
products or procedures.

**certificación** la convicción que el
contratista está en conformidad con las
reglas federales, del estado, y locales
aplicables que pertenecen a los
productos específicos o procedimientos.

**change order** a written change to the
original contract agreed upon by the
owner, architect, engineer, and
contractor. A change order identifies
the specific change to the contract,
changes in the monetary amount of the
contract, and changes in the time
allotted for completing the work in the
contract.

**orden de cambio** un cambio escrito
al contrato original confirmado por el
dueño, arquitecto, ingeniero, y
contratista. Un orden de cambio
identifica el cambio específico al
contrato, los cambios en la cantidad
monetaria del contrato, y cambios en
el tiempo repartido para completar
el trabajo en el contrato.

**chase** a recessed area in a wall
intended to provide space for electrical
panels and conduit, heating and cooling
ductwork, and plumbing.

**persecución** una área retirada en una
pared para mantener el espacio para los
tableros eléctricos y canalización,
calefacción y trabajo del conducto
refrescante, y aplomando.

**checking the range** confirming the
straight alignment of a wall.

**verificar el rango** confirmar la
alineación recta de una pared.

**chimney** that part of a fireplace
venting combustion byproducts to the
outside air while protecting the house
and its occupants from fire and the

life-threatening effects of carbon
monoxide poisoning.

**chimenea** la parte de un hogar
que da salida los subproductos de la
combustión al aire externo protegiendo
la casa y sus ocupantes del fuego y los
efectos vida-amenazantes de
envenenamiento del monóxido de
carbono.

**chimney base flashing** noncorrosive
metal flashing placed between rows of
shingles and turned upward against the
outside of the chimney walls.

**chimenea el destello bajo** flashing
de los metales non-corrosivos puesto
entre las filas de guijarros y tornado
arriba contra el exterior de las paredes
de la chimenea.

**chimney cap** the masonry element
atop a chimney designed to eliminate
water penetrating the top of the
chimney, made of reinforced concrete,
either prefabricated or cast-in-place.

**tapa de la chimenea** el elemento de
la albañilería encima de una chimenea
diseñado para eliminar agua que
penetra la cima de la chimenea, hecho
de hormigón reforzado, fabricado de
antemano o lanzamiento-en-pone.

**circular arch** also called a bull's eye
arch, a 360-degree round opening
framed with voussoirs.

**arco redondo** también llamado el
arco del ojo de un toro, un 360-grado
que la apertura redonda ideado con
el voussoirs.

**classification MX paving brick**
brick pavers used in climates where
freezing does not occur.

**clasificación MX de ladrillo
pavimentador** los ladrillos
pavimentadores usados en climas dónde
no nieve.

**classification NX paving brick** brick
pavers designed for interior use only.

**clasificación NX del ladrillo
pavimentador** los ladrillos
pavimentadores diseñados sólo para el
uso interior.

**classification SX paving brick** paving brick designed to withstand freezing temperatures while saturated.

**clasificación SX del ladrillo pavimentador** los ladrillos pavimentadores diseñados para resistir las temperaturas heladas al saturar.

**cleanout** a door-enclosed opening below the thimble permitting access within the flue lining for removing combustion products and other accumulated debris.

**limpieza** una apertura puerta-adjunta debajo del dedal que permite el acceso dentro del cañón alineado por quitar productos de la combustión y otras ruinas acumulado.

**climate control symbols** representations of heating, air conditioning, and ventilation system components.

**símbolos de mando de clima** las representaciones de calentar, el aire acondicionado, y componentes de sistema de ventilación.

**closure brick** the last brick laid in each course.

**ladrillo del cierre** el último ladrillo puso en cada serie.

**CMU** the abbreviation for concrete masonry unit, also called a block.

**CMU** la abreviación para la unidad de hormigón de la albañilería, también llamado un bloque.

**cold weather construction** construction executed when either air or material temperatures are below 40°F.

**construcción de tiempo fría** la construcción ejecutada cuando las temperaturas aéreas o materiales están debajo de 400°F.

**column** a masonry wall whose width does not exceed four times its thickness and whose height exceeds four times its least lateral dimension.

**columna** una pared de la albañilería cuya anchura no excede cuatro veces su espesor y de quien la altura excede cuatro veces su dimensión lateral.

**combustion chamber** also called the firebox, the enclosed space beyond the fireplace face where combustion occurs.

**cámara de la combustión** también llamado el fuego-caja, el más allá espacial adjunto la cara del hogar dónde la combustión ocurre.

**compass brick** also called radial brick, curved brick that are used for curved brickwork.

**ladrillo del compás** también llamado el ladrillo radial, ladrillo encorvado que se usa para el enladrillado encorvado.

**completed operations insurance** insurance providing coverage for damages that either products or services may cause after the work is completed.

**seguro de los funciones acabados** seguro con tal que el fondo para daño y perjuicios o productos o servicios que pueden ocurrir después de que el trabajo se complete.

**composite masonry wall** a masonry wall consisting of two wythes of different masonry units having different strength characteristics, connected with corrosion-resistant wire reinforcement or brick headers and acting as a single wall in resisting forces and loads.

**pared de la albañilería compuesta** una pared de la albañilería que consiste en dos wythes de unidades de la albañilería diferentes que tienen las características de fuerza diferentes conectados con refuerzo del alambre corrosión-resistente o títulos del ladrillo y actuando como una sola pared resistiéndose fuerzas y cargas.

**compression** the weight or forces above an arch which it must be designed to resist.

**condensación** el peso o fuerza sobre un arco para diseñarse para resistirse.

**compressive strength** a measure of the vertical weight or force that a wall can support.

**fuerza del compressive** una medida del peso vertical o fuerza que una pared puede apoyar.

**concentrated load** the weight supported by another structural component.

**carga concentrada** el peso apoyado por otro componente estructural.

**construction contract** a written agreement executed between the property owner of the proposed job or their representative and the contractor responsible for executing the job giving the terms, rates or prices, for doing the job.

**contrato de la construcción** un acuerdo escrito ejecutado entre el dueño de propiedad del trabajo propuesto o su representante y el contratista responsable para ejecutar el trabajo que da las condiciones, proporciones o precios, para trabajar.

**construction details** also called **details**, drawings at a larger scale and showing in greater detail than the working drawings significant parts that require more information than is provided on the working drawings.

**detalles de la construcción** también llamados los detalles, dibujos a un tamaño más grande mostrando en mayor detalle que los dibujos activos, las partes significantes que requieren más información de la que se proporciona en los dibujos activos.

**construction documents** comprised of the working drawings and the specifications, which are the conditions of a contract upon which the owner and the contractor agree.

**documentos de la construcción** comprendido de los dibujos activos y las especificaciones que son las condiciones de un contrato en que el dueño y el contratista están de acuerdo.

**construction manual log** a daily record of progress, including the number of workers on the job, the number of hours worked, the quantity of units laid, the wall or walls being constructed, and the daily weather.

**registro manual de la construcción** un registro diario de progreso, incluso el número de obreros en el trabajo, que el número de horas trabajó, la cantidad de unidades puso, la pared o paredes que se construyen, y el tiempo diario.

**contract** a written agreement by which a contractor is compensated at an agreed upon rate or price for completing specific work.

**contrato** un acuerdo escrito porque un contratista se compensa a un convenido en la proporción o precia por completar el trabajo específico.

**contract bond** an approved form of security guaranteeing complete execution of the contract and the payment of all debts pertaining to the construction of the project.

**acuerdo del contrato** un formulario aceptado de seguridad que garantiza ejecución completa del contrato y el pago de todas las deudas que pertenecen a la construcción del proyecto.

**contractor** the person or persons undertaking the terms of the contract.

**contratista** la persona o personas que emprenden las condiciones del contrato.

**control joint** a joint controlling the location of separation in a concrete masonry wall resulting in the dimensional changes of the building materials.

**coyuntura controladora** una coyuntura que controla la situación de separación en una pared de la albañilería concreta que produce los cambios dimensionales de los materiales del edificio.

**coping** the projecting top cap of a wall.

**cobertura** la tapa de la cima de una pared que se proyecta.

**coping brick** specially shaped brick intended for the outer edge of a brick paving project.

**ladrillo que cubre** el ladrillo especialmente formado para el borde exterior de un ladrillo que pavimenta el proyecto.

**corbelling** the process of setting the face of brick out beyond the face of the brick below it.

**corbelling** el proceso de poner la cara de ladrillo fuera el más allá la cara del ladrillo debajo de él.

**cored brick** a brick whose cross-section is partially hollow.

**ladrillo vacío** un ladrillo cuyo cruz-sección es parcialmente vacía.

**corner** two walls connected at ends, normally forming a right angle.

**esquina** dos paredes conectadas a los fines normalmente formando un ángulo recto.

**corner of the lead** that part of the corner where the two walls connect.

**esquina de la primacía** la parte de la esquina dónde las dos paredes se conectan.

**corrugated steel ties** sheet metal anchors for securing brick veneer to wood framing.

**lazos de acero contraídos** metal en plancha fija por afianzar la chapa del ladrillo al madera idear.

**cost estimate** a calculated cost of construction materials and/or labor, not a legally binding proposal.

**estimación del coste** un coste calculado de materiales de la construcción y/o mano de obra, no una propuesta legalmente obligatoria.

**counter-flashing** a noncorrosive metal having the shape of an inverted "L," placed in the chimney walls and turned down over base flashing.

**contador-brillante** un metal no corrosivo que tiene la forma de un "L" invertido, puesto en las paredes de la chimenea y vuelto abajo sobre del destello de la base.

**crawl space** the area between the ground surface and floor joists of houses without basements.

**espacio de arrastramiento** el área entre la superficie molida y vigas del suelo de casas sin los sótanos.

**creepers** the brick adjacent to the arch voussoirs.

**enredaderas** el ladrillo adyacente al voussoirs del arco.

**cricket** a sloped section of roof designed to divert water from behind the upper side of a chimney.

**grillo** una sección inclinada de tejado diseñó para desviar el agua del trasero el lado superior de una chimenea.

**cross-sectional area** the area as calculated from inside the walls of the flue lining.

**área cruz-particular** el área como calculado de dentro de las paredes del cañón linear.

**crowding the line** an expression describing brick that are too close to or possibly touching the line.

**apiñando la línea** una expresión que describe ladrillo cerca de que también es o posiblemente referente a la línea.

**damp proofing** the process of hindering the absorption and passage of water through the foundation walls below grade level.

**contra humedad** el proceso de impedir la absorción y pasaje de agua a través de las paredes de la fundación debajo del nivel de calidad.

**dead loads** permanent loads, such as structural framing, wallboard, floor and roof systems, and permanent attachments, such as heating, ventilation, and air conditioning components.

**cargas muertas** las cargas permanentes, como el idear estructural, pared-tabla, el suelo y sistema del tejado, y las ataduras permanentes, como calentar, la ventilación, y componentes del aire acondicionado.

**Dense Industrial 65** the industry standard recognition for sawn planking.

**Denso Industrial 65** la industria el reconocimiento normal por el sawn entablar.

**depth** the height of the brickwork forming the arch ring.

**profundidad** la altura del enladrillado que forma el anillo del arco.

**differential movements** unequal movements of building materials caused by changing temperatures or moisture contents of materials.

**movimientos del diferencial** los movimientos desiguales de construir los materiales causados cambiando temperaturas o volúmenes de humedad de materiales.

**dimension lines** solid lines having arrow heads, slashes, or dots at each end to indicate the terminations or end points between which the required measurement is given.

**líneas de la dimensión** líneas sólidas que tienen cabezas de la flecha, cuchilladas, o puntos a cada fin indicar las terminaciones o puntos del fin entre que la medida requerida se da.

**double-wythe wall** two adjacent masonry walls, joined together with brick or joint reinforcement.

**pared con doble-wythe** dos paredes de la albañilería adyacentes, unidas junto con ladrillo o refuerzo de la juntura.

**draft** a measure of the passage of air from the firebox to the smoke chamber.

**proyecto** una medida del pasaje de aire del firebox a la cámara de humo.

**drainage type wall system** a wall design that has a means of diverting water from the air space to the outside of the exterior, single-wythe masonry wall.

**desagüe tipo pared sistema** un plan de la pared que tiene un medios de desviar el agua del espacio aéreo al exterior del exterior, la pared con un wythe de la albañilería.

**drainage wall systems** masonry walls designed to divert water from

their air cavity to the exterior of the outer wythe.

**sistemas de pared de desagüe** las paredes de la albañilería diseñados para desviar el agua de su cavidad aérea al exterior del wythe exterior.

**drawing identification symbols** icons providing additional information for interpreting a set of drawings.

**símbolos de identificación de dibujos** iconos que mantienen la información adicional interpretando una serie de dibujos.

**drip** a cutout on the underside of the projection intended to prevent water from traveling beyond it and back to the face of the wall.

**goteo** un corte-exterior en la parte inferior de la proyección para impedir al agua viajar el más allá él y detrás de la cara de la pared.

**dry bonding** the process of establishing brick arrangement and the width of head joints.

**vinculación seca** el proceso de establecer el arreglo del ladrillo y la anchura de junturas de cabeza.

**Dutch corner** a styling for corners or ends of walls constructed in either the English bond pattern or Flemish bond pattern characterized by an approximately 6"-long brick at the end.

**esquina holandesa** un nombre para esquinas o fines de paredes o construidos en el modelo de la enlace inglés o diseño del enlace flamenco caracterizado al final por un ladrillo aproximadamente 6"-largo.

**eaves** those parts of the roof projecting beyond the exterior face of the wall.

**aleros** las partes del tejado que proyecta más allá de la cara exterior de la pared.

**efflorescence** a deposit of soluble salts appearing on masonry walls.

**efflorescence** un depósito de sales solubles que aparecen en las paredes de la albañilería.

**egress** a path of exit or rescue.

**salida** un camino de salida o rescate.

**elasticity** a property of mortar that enables it to retain its original size and shape after compaction.

**elasticidad** una propiedad de mortero que le permite retener su tamaño original y formar después de la consolidación.

**electrical symbols** representations of electric fixtures, switches, receptacles, panel boxes, and other wiring systems and accessories.

**símbolos eléctricos** las representaciones de adornos eléctricos, interruptores, receptáculos, cajas del tablero, y otros sistemas de la instalación eléctrica y accesorios.

**elevations** two-dimensional graphic scale representations of the front, sides, and back of a structure.

**elevaciones** las representaciones de la balanza gráficas bidimensionales del frente, lados, y parte de atrás de una estructura.

**employee** one whose personal services are rendered to comply with instructions about when, where, and how work is done that is assigned by the employer.

**empleado** uno cuyo se dan los servicios personales para obedecer las instrucciones sobre cuando, dónde, y cómo el trabajo se hace que se asigna por el patrón.

**employee benefits** such amenities as medical insurance, paid holidays, paid vacation time, paid sick leave, and retirement income received by the employee at the expense of the employer.

**beneficios del empleado** los tales conveniencias como seguro médico, fiestas pagadas, tiempo de la vacación pagado, licencia enferma pagada, y el ingreso jubilatorio recibidas por el empleado al gasto del patrón.

**employer** the one or more individuals licensed as a business responsible for the hiring, training, safety, supervising, and paying of wages for the employee.

**patrón** el uno o más individuos autorizaron como un negocio responsable para el contratar, entrenando, la seguridad, dirigiendo, y pagando de sueldos por el empleado.

engineered brick also called oversize brick, brick whose approximate dimensions are 7⅝' long, 3½" wide, and a 2⅝" face height.

**ladrillo diseñado** también llamado el ladrillo del sobretamaño, ladrillo cuyas dimensiones aproximadas son 7⅝' largo, 3½" ancho, y una 2⅝" altura de la cara.

English bond a pattern consisting of alternating courses of brick laid as all stretchers and all headers.

**enlace inglés** un diseño que consiste en cursos alternos de ladrillo puesto como todas horizontales y todas verticales.

English corner also called a queen closure, a styling for corners or ends of walls constructed in either the English bond pattern or Flemish bond pattern characterized by an approximately 2"-long brick piece adjacent to the end of the first brick of a course.

**esquina inglesa** también llamado un cierre de la reina, un llamando para esquinas o fines de paredes o construyeron en el modelo de la atadura inglés o el flamenco atadura modelo caracterizó por un pedazo del ladrillo aproximadamente 2"-largo adyacente al fin del primer ladrillo de un curso.

estimate a judgment of construction costs.

**estimación** un juicio de costos de la construcción.

estimator one who judges or calculates either or both the materials and labor required for completing a given job.

**estimador** uno que juzga o calcula cualquiera o los materiales y labor requirieron por completar un trabajo dado.

expansion joint a horizontal or vertical separation completely through a brick wythe that is filled with an elastic substance permitting the expansion of brick walls caused by thermal movements or increasing volume of brick.

**juntura de la expansión** una separación horizontal o vertical completamente a través del ladrillo wythe que están lleno con una substancia elástica que permite la expansión de paredes del ladrillo causados por movimientos termales o el volumen creciente de ladrillo.

experience modification rating also called EMR, the basis for adjusting worker's compensation premiums on the contractor's record of injury frequency and costs.

**rango para modificar experiencia** también llamado EMR, la base para otorgar los premios de la compensación obrera en el registro del contratista de frecuencia de la lesión y costos.

exposed aggregate CMUs also called aggregate faced units, block having the natural texture of crushed stone or small pebbles embedded in one or more faces.

**CMUs agregado expuesto** el agregado también llamado enfrentó las unidades, bloque que tiene la textura natural de piedra aplastada o los guijarros pequeños empotró en uno o más caras.

extension lines lines identifying the boundary of the dimension indicated by a dimension line.

**líneas de la extensión** líneas que identifican el límite de la dimensión indicadas por una línea de la dimensión.

extrados the imaginary curved line at the top edge of an arch's voussoirs.

**trasdós** la línea encorvada imaginaria al borde de la cima del voussoirs de un arco.

extruded brick brick that are shaped by forcing wet clay through a die, a form used to impress or shape an object.

**ladrillo empujado fuera** ladrillo que se forma forzando la arcilla mojada a través de un dado, un formulario impresionaba o formaba un objeto.

face brick brick intended to be used on the exposed surface of walls.

**ladrillo de la cara** el ladrillo pensó ser usado en la superficie expuesta de paredes.

face-shell spreading a term describing the bedding of mortar along the tops of both face sides of a block.

**cara-cáscara extendiendo** un término que describe la ropa de cama de mortero a lo largo de las cimas de los dos los lados de la cara de un bloque.

facing the block aligning the bottom edge of each block's face with the top edge of the block faces on the course below it.

**ante el bloque** encuadrando el borde del fondo de la cara de cada bloque con el borde de la cima del bloque enfrenta en el curso debajo de él.

facing the brick aligning the bottom edge of each brick's face with the top edge of the brick faces on the course below it.

**ante el ladrillo** encuadrando el borde del fondo de la cara de cada ladrillo con el borde de la cima del ladrillo enfrenta en el curso debajo de él.

fascia a board installed on the exposed ends of rafters or roof trusses.

**fascia** una tabla instaló en los fines expuestos de vigas o bragueros del tejado.

field pattern the brick pattern or configuration between the borders.

**diseño del campo** el modelo del ladrillo o configuración entre las fronteras.

film-formers water repellants adhering to the surface of masonry walls.

**agentes para creación de películas** riegue repellants que adhiere a la superficie de paredes de la albañilería.

**fireblocking** also called fire stopping, a noncombustible material preventing air drafts and the potential spread of smoke and fire between floors or ceilings.

**bloque de fuego** el fuego deteniendo también llamado, un material incombustible que previene los proyectos aéreos y el cobertor potencial de humo y dispara entre suelos o techos.

**firebox** also called the combustion chamber, the enclosed space beyond the fireplace face where combustion occurs.

**caja de fuego** también llamado la cámara de la combustión, el el más allá espacial adjunto la cara del hogar dónde la combustión ocurre.

**fireclay** clay capable of withstanding high temperatures used for chimney flue liners and chimney thimbles.

**arcilla de fuego** la arcilla capaz de resistir temperaturas altas usadas para los transatlántico de cañón de chimenea y dedales de la chimenea.

**fireplace brick** also called firebrick, refractory brick that has thermal resistance and thermal stability for lining the fireplace firebox.

**ladrillo de lugar de fuego** el firebrick también llamado, ladrillo terco que tiene resistencia termal y la estabilidad termal por linear el firebox del hogar.

**fireplace surround** the face of the fireplace immediately surrounding the opening.

**lugar de fuego rodea** la cara del hogar que rodea la apertura inmediatamente.

**fire stopping** also called fireblocking, a noncombustible material preventing air drafts and the potential spread of smoke and fire between floors or ceilings.

**fuego deteniendo** también llamado el fireblocking, un material incombustible que previene los proyectos aéreos y el cobertor potencial de humo y dispara entre suelos o techos.

**fire wall** a fire-resistant rated wall usually built within a structure from the foundation and extending above the roof to restrict the spread of fire from one part of a structure to another.

**dispare la pared** una pared tasada fuego-resistente normalmente construido dentro de una estructura de la fundación y extendido hasta el tejado anteriormente para restringir el cobertor de fuego de una parte de una estructura a otro.

**flashed brick** brick showing color variations as a result of firing the brick with alternately too much or too little air.

**ladrillo encendido** ladrillo que muestra las variaciones coloridas como resultado de disparar el ladrillo con alternadamente demasiado o demasiado poco el aire.

**flashing** a water impermeable material installed in the wall system, diverting collected water to the exterior of the wall.

**encendiendo** una agua que el material impermeable instaló en el sistema de la pared, mientras desviando el agua reunido al exterior de la pared.

**Flemish bond** a brick pattern consisting of alternating stretchers and headers for each course.

**enlace flamenca** un modelo del ladrillo que consiste en camillas alternas y títulos para cada curso.

**Flemish garden wall bond** a pattern of brickwork, typically having three headers between each stretcher, creating the appearances of diagonal lines and diamond patterns.

**anlace de pared de jardín flamenca** un modelo de enladrillado, que tiene tres títulos típicamente entre cada camilla, creando las apariencias de líneas diagonales y diseños del diamante.

**flexible base** a base below mortarless brick paving consisting of compacted sand above a compacted crushed stone base.

**base flexible** una base debajo de la enladrilla pavimentada sin mortar que consiste en arena apretada sobre una base de la piedra aplastada apretada.

**flexural strength** the measure of a material's capability to withstand bending.

**fuerza del flexural** la medida de la capacidad de un material para resistir el torcimiento.

**floor plans** scale drawings showing the overall length and width of the floor framing for a given floor level and the location of both exterior and interior walls and openings within the walls.

**planes del suelo** descascare dibujos que muestran la longitud global y anchura del suelo que idea para un nivel del suelo dado y la situación de paredes exteriores e interiores y aperturas dentro de las paredes.

**flue lining** a rectangular or round, hollow chimney lining intended to contain the combustion products and protect the chimney masonry walls from heat and corrosion.

**cañón lineando** un chimenea alineado en rectangulares o redondos, sin substancia para contener los productos de la combustión y proteger las paredes de albañilería de chimenea del calor y corrosión.

**fluted CMUs** also called ribbed or scored block, CMUs whose one or more faces have vertical, machine-molded grooves.

**CMUs estriado** también llamado rebordes o anotó bloque, CMUs cuyas uno o más fachadas tienen las ranuras verticales, máquina-amoldadas.

**footings** that part of the foundation supporting the foundation walls and transmitting the structure's weight to the soil.

**fundamentos** esa parte de la fundación apoyando las paredes de la fundación y transmitiendo el peso de la estructura a la tierra.

**formed footings** footings resulting from the placement of concrete within temporary or permanent forms.

**fundamentos formados** fundamentos que son el resultado de la colocación de hormigón dentro de los formularios temporales o permanentes.

**foundation** that part of a building below the first floor framing.

**fundación** la parte de un edificio debajo del primero suelo idear.

**foundation plans** scale drawings showing the foundation walls for the intended structure.

**planes de la fundación** descascare dibujos que muestran las paredes de la fundación para la estructura intencional.

**framing plans** scale drawings showing floor framing and roof framing, which are the frameworks that serve as the skeleton of a structure to which coverings can be applied.

**planes del marco** descascare dibujos que muestran suelo que idea y cubra ideando, qué es los armazones que sirven como el esqueleto de una estructura a que pueden aplicarse los techado.

**frieze board** a board extending from the soffit to a point below the top of the anchored brick veneer wall.

**tabla del friso** una tabla que se extiende del soffit a un punto debajo de la cima de la pared de chapa de ladrillo anclada.

**frog** a recessed area formed on the bottom side of wood-mold brick, created at the bottom of a brick mold when the mold is filled with wet clay.

**rana** una área retirada formó en el lado del fondo de ladrillo del madera-molde, creó al fondo de un molde del ladrillo cuando el molde está lleno con la arcilla mojada.

**frost line** the greatest depth to which ground may be expected to freeze.

**línea de escarcha** la más gran profundidad a que puede esperarse que la tierra hiele.

**gable** the triangular area below a roofline.

**aguilón** el área triangular debajo del marco del tejado.

**garden walls** brick walls or fencing typically built to add elegance to outdoor living areas.

**paredes del jardín** paredes del ladrillo o típicamente construidos para agregar la elegancia a las áreas vivientes al aire libre.

**gauged brick** a brick that has been tapered to be used as a voussoir.

**ladrillo calibrado** un ladrillo que se ha adelgazado para ser usado como un voussoir.

**general conditions** that part of the specifications defining the rights and the responsibilities of all parties involved in the work of the project.

**condiciones generales** esa parte de las especificaciones que definen los derechos y las responsabilidades de todas las fiestas involucró en el trabajo del proyecto.

**glazed CMUs** block with a glazing compound permanently molded to one or more faces, becoming an integral part of the unit and providing an impervious finish highly resistant to staining, impact, and abrasion.

**CMUs vidriado** bloque permanentemente con un compuesto de vidriado amoldado a uno o más caras, mientras volviéndose una parte íntegra de la unidad y proporcionando un acabado impenetrable muy resistente a manchar, impacte, y abrasión.

**Gothic arch** an arch having a rise equal to or greater than its span formed by two overlapping circles of equal length radii coming to a point at its center.

**arco gótico** un arco que tiene un igual del levantamiento a o mayor que su palmo formado por dos círculos solapando de radios de longitud iguales que vienen hasta su centro.

**granular fill insulations** lightweight, inorganic, perlite or vermiculite granules treated for water repellency and used as wall insulation.

**aislamientos de hartura granulares** el peso ligero, inorgánico, que perlite o gránulos del vermiculite trataron para repellency de agua y usado como el aislamiento de la pared.

**ground face CMUs** concrete masonry units with faces ground smooth and highly polished, exposing the natural colors of the aggregates and reflecting light.

**molido la cara CMUs** las unidades de la albañilería concretas con caras conectadas con tierra liso y favorablemente pulido, mientras exponiendo los colores naturales de los agregados y reflejando la luz.

**grout** a mixture of cement and aggregates to which sufficient water is added to produce a pouring consistency without separation of the ingredients, used to strengthen masonry walls.

**lechada** una mezcla de cemento y agregados a que se agrega el agua suficiente para producir una consistencia vertiendo sin la separación de los ingredientes, usados para fortalecer las paredes de la albañilería.

**grouting** the process of manually pouring or machine-pumping grout in walls such as the hollow cells of CMU walls.

**lechada** el proceso de verter por mano o máquina-bombear la lechada en las paredes como las células sin substancia de paredes de CMU.

**hanging the line** attaching a mason's line to the leads at opposite ends of a wall.

**colgar la línea** atando la línea de un albañil a las primacías a los fines opuestos de una pared.

**head** the top of an opening such as above doors and windows.

**cabeza** la cima de una apertura como sobre las puertas y ventanas.

**header** the position for a brick having its end oriented with the face of a wall and its bottom side bedded in mortar.

**título** la posición para un ladrillo que tiene su fin orientado con la cara de una pared y su lado del fondo plantada en un macizo en el mortero.

**head flashing** flashing above openings.

**destello de la cabeza** las aperturas anteriores encendiendo.

**hearth** the floor of the firebox.

**hogar** el suelo del firebox.

**hearth base** the structural support for the firebox floor.

**base del hogar** el apoyo estructural para el suelo del firebox.

**heavy traffic** brick paving designs intended for vehicular traffic such as roadways and loading docks.

**tráfico pesado** ladrillo que pavimenta los planes pensó para el tráfico vehicular como las carreteras y los andenes cargantes.

**heavyweight CMUs** block containing primarily Portland cement, crushed limestone, and sand.

**peso pesado CMUs** bloque que contiene cemento de Portland, caliza aplastada, y arena principalmente.

**heel** the end of a mason's trowel nearest the handle.

**talón** el fin de la paleta de un albañil más cercano el asa.

**herringbone pattern** a brick paving pattern where each brick is aligned perpendicular to that brick beside it.

**modelo del herringbone** un ladrillo que pavimenta modelo dónde cada ladrillo se alinea perpendicular a ese ladrillo al lado de él.

**hidden lines** lines used to indicate objects hidden from view.

**líneas ocultas** las líneas indicaban objetos escondidos de la vista.

**high-lift grouting** grouting done once a wall is built to its story height or final height, whichever is less, in a single pour.

**lechada del alto-alzamiento** lechada hecha una vez que una pared se construye a su altura de la historia o último altura, quienquiera es menos, en una sola lluvia.

**holding bond** maintaining a plumb aligned bond or brick pattern.

**atadura sosteniendo** manteniendo un plomo encuadrado atadura o modelo del ladrillo.

**hollow brick** brick whose combined surface area of the holes is more than 25% of the total bed surface area of the brick.

**ladrillo sin substancia** ladrillo cuyo combinó área de la superficie de los agujeros está más de 25% del área de superficie de cama total del ladrillo.

**hollow masonry pier** a pier constructed of masonry units supporting no weight other than its own.

**malecón de la albañilería sin substancia** un malecón construyó de unidades de la albañilería que no apoyan el peso de otra manera que su propio.

**hollow unit** a CMU in which the cross-sectional area of the cells is more than 25% of the overall cross-sectional area of the unit.

**unidad sin substancia** un CMU en que el área cruz-particular de las células está más de 25% del área cruz-particular global de la unidad.

**horseshoe arch** a circular arch greater than the 180-degree design of a semicircular arch but less than the 360-degree design forming a circular arch.

**arco herradura** un arco redondo mayor que el plan del 180-grado de un arco semicircular pero menos del plan del 360-grado que forma un arco redondo.

**hot weather construction** construction executed when either air or material temperatures are above 100°F.

**construcción de tiempo caliente** la construcción ejecutó cuando las temperaturas aéreas o materiales son anteriores 100°F.

**initial rate of absorption (IRA)** a test determining the moisture content of brick.

**proporción inicial de absorción (IRA)** una prueba que determina la humedad madura de ladrillo.

**inner hearth** the floor of the combustion chamber.

**hogar interno** el suelo de la cámara de la combustión.

**intrados** the imaginary curved line at the bottom edge of an arch, the edge between the vertical face and horizontal soffit.

**intrados** la línea encorvada imaginaria al borde del fondo de un arco, el borde entre la cara vertical y el soffit horizontal.

**jack arch** an arch design whose rise above the spring line is very little or nonexistent.

**arco de la sota** un plan astuto cuyo sube la línea de la primavera anteriormente es muy pequeño o inexistente.

**jamb** the exposed sides of a door or window frame.

**jamba** los lados expuestos de una puerta o marco de la ventana.

**job foreman** one responsible for supervising a group of workers.

**capataz del trabajo** uno responsable por dirigir un grupo de obreros.

**job superintendent** one responsible for the operations required for the completion of a job.

**superintendente del trabajo** uno responsable para los funcionamientos requeridos para la realización de un trabajo.

**journey-level worker** also called a journeyman, a competent, experienced worker possessing the skills required to perform without supervision the tasks for a particular trade.

**obrero jornada-nivelado** también llamado a un jornalero, un obrero competente, experimentado que posee las habilidades exigió realizar sin la vigilancia las tareas para un comercio particular.

**keystone** the voussoir at the center of an arch.

**clave** el voussoir al centro de un arco.

**labor constant** the amount of labor required to perform a specific amount of work.

**constante obrera** la cantidad de labor exigió realizar una cantidad específica de trabajo.

**ladder-type wire reinforcement** masonry wall reinforcement consisting of two or more longitudinal rods welded to perpendicular cross rods, forming a ladder design.

**refuerzo de alambre de escalera de mano-tipo** refuerzo de pared de albañilería que consiste en dos o las barras más longitudinales soldadas a las barras de la cruz perpendiculares, formando un plan de la escalera de mano.

**lateral strength** a measure of the horizontal force that a wall is capable of resisting.

**fuerza lateral** una medida de la fuerza horizontal que una pared es capaz de resistirse.

**layout** the brick pattern and joint spacing to satisfy the intended wall construction.

**diseño** el modelo del ladrillo y juntura que espacian para satisfacer la construcción de la pared intencional.

**lead** that part of the wall first constructed, becoming a guide for building the remaining wall.

**primacía** la parte de la pared construido primero, mientras volviéndose una guía para construir la pared restante.

**leader lines** solid lines used where space does not permit showing dimensions between extension lines, ending with an arrow head identifying the notation at its opposite end.

**líneas del líder** las líneas sólidas usados donde el espacio no permite las dimensiones de la exhibición entre las líneas de la extensión, mientras acabando con una cabeza de la flecha que identifica la anotación a su fin opuesto.

**level** perpendicular to the force of gravity.

**nivel** el perpendicular a la fuerza de gravedad.

**liability insurance** insurance providing protection for personal injuries and damage to the property of others arising from operations performed by the contractor.

**seguro de obligación** seguro con tal que protección para las lesiones personales y daña a la propiedad de otros que se levanta de los funcionamientos realizada por el contratista.

**licensed contractor** a contractor whom government authority has granted permission to engage in contracting.

**contratista autorizado** un contratista quien la autoridad gubernamental ha concedido al permiso para comprometer acortando.

**lift** the measured height from a mason's trowel blade to the center of the butt-end of the handle.

**alzamiento** la altura moderada de la hoja de la paleta de un albañil al centro de la culata del asa.

**light traffic** brick paving designs intended for residential pedestrian use only.

**tráfico ligero** ladrillo que pavimenta los planes sólo pensó para el uso del peatón residencial.

**lightweight CMUs** block containing expanded shale, clay, and slate or pumice to reduce the weight of the block and improve its performance.

**CMUs ligero** bloque que contiene esquisto extendido, arcilla, y pizarra o limpiar con piedra* pómez reducir el peso del bloque y mejorar su actuación.

**lime run** a white or gray, crusty calcium carbonate deposit on the faces of masonry walls as a result of any of several calcium compounds in solution with water, brought to the surface of the masonry through an opening, where the solution reacts with carbon dioxide in the air.

**carrera de la cal** un blanco o encanece, depósito de carbonato de calcio costroso en las caras de paredes de la albañilería como resultado de cualquiera de varios compuestos del calcio en la solución con agua, traída a la superficie de la albañilería a través de una apertura dónde la solución reacciona con el anhídrido carbónico en el aire.

**lipping** a condition where the bottom of a brick unintentionally extends beyond the face of the brick below it.

**lipping** una condición dónde el fondo de un ladrillo extiende involuntariamente más allá de la cara del ladrillo debajo de él.

**live loads** nonpermanent loads, such as occupants, furnishings, rain, and snow.

**cargas vivas** la carga temporal, como los ocupantes, muebles, lluvia, y nieve.

**load-bearing wall** a wall supporting its own weight and the structural loads, weights, and forces to which the structure is subjected.

**pared carga-productiva** una pared que apoya su propio peso y las cargas estructurales, pesos, y fuerzas a que la estructura se sujeta.

**loose fill** soil placed without the compaction necessary for structural support.

**hartura suelta** la tierra puso sin la consolidación necesario para el apoyo estructural.

**low-lift grouting** grouting done in multiple pours as the wall is built.

**lechada de lechada** de bajo-alzamiento hecha en las

lluvias múltiples como la pared se construye.

**major arch** an arch having a span greater than 6' or rise-to-span ratio greater than 0.15.

**arco mayor** un arco que tiene un palmo mayor que 6' o proporción del levantamiento-a-palmo mayor que 0.15.

**markup** the dollar amount over and beyond material and labor costs needed for business operations.

**encarecimiento** el dólar la cantidad encima de y más allá del material y el coste obrero necesitados para los funcionamientos comerciales.

**masonry cements** a mixture of Portland or blended hydraulic cement (a cement that hardens under water) and plasticizing materials (materials such as limestone, hydrated or hydraulic lime intended to enhance workability) whose air content varies widely between manufacturers and are recognized for their good workability.

**cementos de la albañilería** una mezcla de Portland o el cemento hidráulico mezclado (un cemento que endurece bajo el agua) y plastificando los materiales (los materiales como la caliza, hidrató o la cal hidráulica pensó reforzar la laborabilidad) de quien el volumen aéreo varía ampliamente entre los fabricantes y se reconoce para su laborabilidad buena.

**masonry column** a masonry wall whose width does not exceed four times its thickness and whose height exceeds four times its least lateral dimension.

**columna de la albañilería** una pared de la albañilería cuya anchura no excede cuatro veces su espesor y de quien la altura excede cuatro veces su dimensión lateral.

**masonry contractor** one or more individuals whom government authority has granted permission to engage in masonry construction.

**contratista de la albañilería** uno o más individuos quienes la autoridad gubernamental ha concedido al permiso para comprometer en la construcción de la albañilería.

**masonry instructor** employed by industry, organized labor, local governments, or career and technical training centers, one teaching others masonry trade-specific technical information and procedures.

**instructor de la albañilería** empleado por la industria, gobiernos obreros, locales organizados, o carrera y centros de entrenamiento técnicos, uno que enseña la albañilería a otros la información técnica comercio-específica y procedimientos.

**mason's scales** graduated markings on mason's rules and tapes permitting different course spacing of masonry units.

**balanzas de albañil** las señales graduadas en las reglas de albañil y cintas que permiten curso diferente que espacia de unidades de la albañilería.

**masonry sills** masonry installed below window and door openings, usually brick rowlocks or precast stone.

**umbrales de la albañilería** la albañilería instaló debajo de la ventana y aperturas de la puerta, normalmente sardineles del ladrillo o piedra del precast.

**Material Safety Data Sheets** also called MSDS, printed information available at the jobsite, including emergency and first aid procedures as well as storage and disposal information for every product used at the jobsite.

**Hojas de Datos de Seguridad Materiales** MSDS también llamado, la información impresa disponible al jobsite, incluso la emergencia y procedimientos del primeros auxilios así como el almacenamiento e información de la disposición para cada producto usado al jobsite.

**medium traffic** brick paving designs for public pedestrian traffic and limited vehicular traffic.

**tráfico elemento** ladrillo que pavimenta los planes para el tráfico del peatón público y el tráfico vehicular limitado.

**minor arch** an arch whose span does not exceed 6' with maximum rise-to-span ratios of 0.15.

**arco menor** un arco cuyo palmo no excede 6' con las proporciones de levantamiento-a-palmo de máximo de 0.15.

**modular masonry construction** a construction design permitting different types of masonry units being used to build walls to the same module or repetitive dimension.

**construcción de la albañilería modular** un plan de la construcción que permite tipos diferentes de unidades de la albañilería que se usan para construir las paredes al mismo módulo o la dimensión repetitiva.

**modular masonry unit** a masonry unit permitting modular masonry construction.

**unidad de la albañilería modular** una unidad de la albañilería que permite la construcción de la albañilería modular.

**modular paving brick** paving brick measuring 3⅝" wide by 7⅝" long.

**ladrillo pavimentador modular** ladrillo pavimentador que mide 3⅝" mucho tiempo ancho por 7⅝."

**mortar** a mixture of cementitious materials, sand, and water bonding individual brick or concrete masonry units to create masonry walls.

**mortero** una mezcla de materiales del cementitious, arena, y agua que unen ladrillo individual o las unidades de la albañilería concretas para crear las paredes de la albañilería.

**mortared brick paving** brick paving bedded in mortar on a rigid base and having mortared joints between the paving brick.

**pavimento de ladrillos con mortar** los ladrillos pavimentadores plantados en un macizo en el mortero en una base rígida y las junturas del mortared teniendo entre el ladrillo pavimentando.

**mortar bridging** the result of excessive mortar protruding from joints and contacting the backer wall, blocking the air space between the brick veneer wall and the backer wall.

**mortero ponteando** el resultado de mortero excesivo destacándose de las junturas y avisando el más atrás la pared, bloqueando el espacio aéreo entre la pared de chapa de ladrillo y el más atrás la pared.

**mortar cements** similar to propriety masonry cements, except they have lower air content and include a minimum flexural bond strength requirement.

**cementos del mortero** similar a los cementos de albañilería de conveniencia exceptúe ellos que tienen el más bajo volumen de aire e incluyen un flexural mínimo une el requisito de fuerza.

**mortar collection systems** also called mortar deflection systems or mortar breaks, products designed to prevent mortar from obstructing the performance of weep holes, weep vents, or wick ropes.

**sistemas de colección de mortero** sistemas de desviación de mortero también llamados o descansos del mortero, productos diseñados para impedir al mortero obstruir la actuación de lloran los agujeros, llore las aberturas, o sogas de la mecha.

**mortar protrusions** that mortar squeezed beyond the face and backside of the brick as the brick is bedded in mortar.

**protrusiones del mortero** ese mortero apretado más allá de la cara y espalda del ladrillo como el ladrillo se planta en un macizo en el mortero.

**mortarless brick paving** brick paving without mortar joints.

**pavimento de ladrillo sin mortar** ladrillo que pavimenta sin las junturas del mortero.

**multicentered arch** also called an elliptical arch, an arch having more than one radius.

**multicentered arquean** también llamado un arco elíptico, un teniendo astuto más de un radio.

**multi-face fireplace** a fireplace with openings on two or more sides.

**hogar del multi-cara** un hogar con las aperturas en dos o más lados.

**multiwythe grouted masonry wall** a multiwythe wall whose air cavity is filled with grout.

**multi-wythe la pared de albañilería de grouted** una pared del multiwythe cuya cavidad aérea está llena con la lechada.

**muriatic acid** also called hydrochloric acid, a water-based solution of hydrogen chloride gas recommended as a cleaning agent for removing mortar stains from only a few types of brick.

**ácido del muriatic** el ácido clorhídrico también llamado, una solución basada en agua de gas de cloruro de hidrógeno recomendada como un agente de limpieza por quitar el mortero mancha de sólo unos tipos de ladrillo.

**new building bloom** efflorescence appearing soon after construction because of the increased presence of water during construction.

**nueva flor del edificio** efflorescence que aparece la construcción poco después debido a la presencia aumentada de agua durante la construcción.

**nominal size** an expression of an object's approximate size rather than its actual size.

**tamaño nominal** una expresión del tamaño aproximado de un objeto en lugar de su tamaño real.

**non-load-bearing wall** a wall supporting its own weight and no other weights or forces.

**pared no productiva** una pared que apoya su propio peso y ningún otro peso o fuerzas.

**object lines** solid lines representing the outline shape of objects.

**líneas del objeto** líneas sólidas que representan la forma del contorno de objetos.

**outer hearth** also called hearth extension, that part of the fireplace extending beyond the fireplace face in front of the firebox floor.

**hogar exterior** la extensión del hogar también llamada que la parte del lugar de fuego que se extiende más allá de la cara de lugar de fuego delante del suelo de caja de fuego.

**oversize brick** brick whose approximate dimensions are 7⅝" long, 3½" wide, with a 2⅝" face height.

**ladrillo del sobretamaño** ladrillo cuyas dimensiones aproximadas son 7⅝" largo, 3½" ancho, con una 2⅝" altura de la cara.

**parapet wall** that part of a wall extending above the roofline.

**pared del parapeto** esa parte de una pared que extiende el tejado-línea anteriormente.

**parging** a cement-sand mixture coating applied to the exterior of foundation walls below finished grade.

**parging** un cemento-arena mezcla cubriendo aplicaron al exterior de paredes de la fundación debajo de la calidad acabada.

**pattern bond** the pattern created by the arrangement of masonry units on a masonry wall.

**enlace del diseño** el diseño creado por el arreglo de unidades de la albañilería en una pared de la albañilería.

**paving brick** abrasion-resistant brick intended for pedestrian and/or vehicular traffic.

**ladrillo pavimentador** el ladrillo abrasión-resistente pensó para el peatón y/o el tráfico vehicular.

**penetrants** water repellant coatings that penetrate the surface of masonry walls.

**penetrants** riegue capas del repellant que penetran la superficie de paredes de la albañilería.

**pier** a support below another structural component.

**malecón** un apoyo debajo de otro componente estructural.

**pigments** insoluble fine powders added to masonry cements to obtain a desired mortar color.

**pigmentos** los polvos finos insolubles agregaron a los cementos de la albañilería para obtener un color del mortero deseado.

**pilaster** a columnar projection from a masonry wall.

**pilastra** una proyección columnar de una pared de la albañilería.

**pilaster block** a concrete masonry unit designed for constructing concrete masonry pilasters.

**bloque de la pilastra** una unidad de la albañilería concreta diseñó por construir las pilastras de la albañilería concretas.

**plasticizers** admixtures that result in mortars that are easier to spread and that adhere to the trowel and masonry units better.

**plasticizers** mezclas que producen morteros que son más fáciles extender y ese adhiera bien a la paleta y unidades de la albañilería.

**plumb** parallel to the force of gravity.

**plomo** parangone a la fuerza de gravedad.

**plumbing symbols** representations of hot and cold water lines, waste lines, vent stacks, floor drains, gas lines, water meters, and a variety of plumbing fixtures and accessories.

**símbolos aplomando** las representaciones de líneas de agua calientes y frías, líneas desechadas, pilas de la abertura, desagües del suelo, líneas de gas, metros de agua, y una variedad de aplomar adornos y accesorios.

**point up** the process of filling mortar joints.

**apunte a** el proceso de llenar las junturas del mortero.

**porch** a sheltered area at the entrance of a building.

**porche** una área protegido a la entrada de un edificio.

**potential expansive soils** those soils that increase in volume significantly when wet.

**tierras expansivas potenciales** las tierras que aumentan significativamente en el volumen cuando mojado.

**presentation drawings** perspective drawings, drawings appearing in a three-dimensional form as they would appear to the eye.

**dibujos de la presentación** los dibujos en perspectiva, dibujos que aparecen en un formulario tridimensional cuando ellos aparecerían al ojo.

**profit** the amount of money remaining after all expenses have been met.

**ganancia** la cantidad de dinero que permanece después de que todos los gastos se han reunido.

**project manager** one representing the building's owner and supervising all or parts of the construction details.

**proyecte a gerente** uno representando al dueño del edificio y dirigiendo todos o partes de los detalles de la construcción.

**project representative** one assigned to a job site to oversee the construction contract.

**representante del proyecto** uno asignó a un sitio del trabajo para vigilar el contrato de la construcción.

**propriety compounds** chemicals containing organic and inorganic acids, wetting agents, and inhibitors recommended for removing a variety of stains from brick, especially those susceptible to metallic oxidation staining.

**compuestos de conveniencia** químicos que contienen los ácidos orgánicos e inorgánicos, mojando a agentes, y los inhibidores recomendaron por quitar una variedad de manchas del ladrillo, sobre todo esos susceptible al oxidación manchar metálico.

**quoined corners** patterns of brick projecting beyond the wall line creating block designs.

**quoined acorrala** los modelos de ladrillo que proyecta más allá de la línea de la pared que crea los planes del bloque.

**rack of the lead** the brick alignment at the tail end of the courses on the lead.

**percha de la primacía** la alineación del ladrillo al fin de la cola de los cursos en la primacía.

**racking** the process of intentionally setting back or stepping back the brick from the course below.

**atormentar** el proceso de poner atrás intencionalmente o andar el ladrillo atrás del curso debajo.

**radial pier** a round pier.

**malecón radial** un malecón redondo.

**rain screen wall** a masonry cavity wall containing protected openings permitting the passage of air but not water into the cavity, permitting equal air pressures of the outside air and that within the air cavity.

**pared como pantalla de lluvia** una pared de cavidad de albañilería que contiene aperturas protegido que permiten el pasaje de aire pero no el agua en la cavidad, permitiendo presiones atmosféricas iguales del aire externo y que dentro de la cavidad aérea.

**raising the line** repositioning the line to the next course of the lead upon completing each course of brick.

**levantar la línea** recalibrar la línea al próximo curso de la primacía al completar cada curso de ladrillo.

**rake board** trim enclosing the open space at the ends of the brick coursing and the sloping roof line in a gable wall.

**tabla del rastro** en buen estado que adjunta el espacio abierto a los fines del curso del ladrillo y la línea del tejado inclinándose en una pared del aguilón.

**reinforced masonry pier** a pier strengthened with concrete grout and steel reinforcing bars.

**malecón de la albañilería reforzado** un malecón fortalecido con la lechada concreta y acero que refuerzan las barras.

**repressed paving brick** brick pavers first extruded and then compressed to increase density.

**ladrillo Reprimido pavimentador** las pavimentadoras del ladrillo empujaron fuera primero y entonces comprimieron para aumentar la densidad.

**retarders** admixtures used to delay the mortar's rate of hydration.

**retarders** las mezclas tardaban la proporción del mortero de hydration.

**retempering** also called tempering, the process of adding water to the mortar to replace water lost by evaporation.

**re-templando** también llamado templando, el proceso de agregar el agua al mortero reemplazar el agua perdido por la evaporación.

**rigid base** a base below brick pavers consisting of a concrete slab, compacted crushed stone sub-base, and a compacted subgrade.

**base rígida** una base debajo de pavimentadoras del ladrillo que consisten en una tabla concreta, subalterno-base de la piedra aplastado apretado, y un subgrade apretado.

**rigid insulation** polystyrene materials produced from extruded foam or molded bead processes placed between the wythes of cavity walls as wall insulation.

**aislamiento rígido** los materiales del polystyrene producidos de la espuma empujada fuera o amoldados procesos de la cuenta puestos entre el wythes de paredes de la cavidad como el aislamiento de la pared.

**rise** the vertical height between two adjoining horizontal step treads.

**levantamiento** la altura vertical entre dos bandas de rodadura del paso horizontales uniendo.

**riser** the vertical component of a step.

**sublevación** el componente vertical de un paso.

**Rosin fireplace** a single-face fireplace with a curved, back-wall design.

**lugar de fuego de colofonia** un hogar de la cara única con un encorvado, plan de la parte atrás de la pared.

**Rumford fireplace** a single-face fireplace design with a tall and shallow firebox with significantly flared sides and a vertical back for radiating more heat than a conventional fireplace.

**hogar Rumford** un plan de hogar de cara única con un firebox altos y poco profundos con los lados significativamente señalados con luz y una parte de atrás vertical por radiar más calor que un hogar convencional.

**running bond** a brick pattern where every brick is laid in the stretcher position, and the brick of alternate courses forms a uniform overlap with the brick below.

**enlace corriente** un modelo del ladrillo dónde cada ladrillo se pone en la posición de la camilla, y el ladrillo de formularios de los cursos alternados un traslapo uniforme con el ladrillo debajo de.

**running bond pattern** brick rows having each brick offset half its length with the brick in adjacent rows.

**diseño del enlace corriente** filas del ladrillo que tienen cada ladrillo compensado medio su longitud con el ladrillo en las filas adyacentes.

**sawn planking** scaffold boards sawn from timber wood.

**sawn entablando** el andamio aborda el sawn de madera de madera.

**scaffold buck** another term for a scaffold end frame.

**ciervo del andamio** otro término para un marco de fin de andamio.

**scale drawing** a drawing showing all of the construction details smaller than actual size so that the project can be represented on paper.

**dibujo a la escala** una exhibición del dibujo que toda la construcción detalla más pequeño que el tamaño real para que el proyecto pueda representarse en el papel.

**screen wall** a brick wall constructed in the Flemish bond pattern with the headers omitted to create openings through the wall.

**pared de pantalla** una pared del ladrillo construida en el modelo de la atadura flamenco con los títulos omitió para crear las aperturas a través de la pared.

**section** a drawing showing the vertical composition of a specific part of a working drawing.

**sección** una exhibición del dibujo la composición vertical de una parte específica de un dibujo activo.

**section lines** solid, parallel, slanting lines indicating a break in a vertical, cut-through part of a drawing.

**líneas de la sección** el sólido, parangone, líneas sesgado que indican un descanso en un vertical, corte-a través de la parte de un dibujo.

**section reference line** also called a cutting-plane line, a line drawn for reference to a cross-section, a section perpendicular to the drawing shown elsewhere.

**línea de referencia de sección** también llamado una línea del cortante-avión, una línea atraída para la referencia a un cruz-sección, un perpendicular de la sección al dibujo mostrado en otra parte.

**segmental arch** an arch formed by one radius line less than a semicircle.

**arco segmentario** un arco formado por una línea del radio menos de un semicírculo.

**seismic loads** loads or forces acting on buildings due to the action of earthquakes.

**cargas sísmicas** cargas o fuerzas que actúan en los edificios debido a la acción de terremotos.

**self-employed** one who works for another on a contract basis rather than as an employee.

**auto empleado** uno que trabaja para otro en una base del contrato en lugar de como un empleado.

**semicircular arch** an arch appearing as half of a circle, formed by one radius line and 180 degrees of rotation.

**arco Semi-redondo** un apareciendo astuto como la la mitad de un círculo, formó por una línea del radio y 180 grados de rotación.

**semi-rigid base** a base below brick pavers consisting of asphalt and compacted crushed stone.

**base semi-rígida** una base debajo de pavimentadoras del ladrillo que consisten en asfalto y apretó la piedra aplastada.

**set-back** a condition exhibited by a brick when its bottom edge is unintentionally back of the face of the brick below.

**choque** una condición exhibida por un ladrillo cuando su borde del fondo regresa involuntariamente de la cara del ladrillo debajo.

**shelf angles** steel structural members on which brick are bedded for wall support.

**ángulos del estante** acero miembros estructurales en que el ladrillo se planta en un macizo para el apoyo de la pared.

**shiner** the position for a brick having its face side or back side bedded in mortar with its top side or bottom side oriented with the face of the wall.

**ojo a la funerala** la posición para un ladrillo que tiene su lado de la cara o atrás el lado plantó en un macizo en el mortero con su lado de la cima o el

lado del fondo orientó con la cara de la pared.

**sill flashing** flashing below masonry sills at windows and doorframes.

**umbral encendiendo** encendiendo debajo de los umbrales de la albañilería a las ventanas y doorframes.

**single-face fireplace** a fireplace with an opening on only one side.

**lugar de fuego de cara única** un lugar de fuego con una apertura en sólo un lado.

**single-wythe brick wall** a brick wall with a bed depth equivalent to the width of one brick.

**pared del ladrillo wythe único** una pared del ladrillo con una profundidad de la cama equivalente a la anchura de un ladrillo.

**skewback** that part of an abutment supporting the voussoir at the end of an arch.

**skewback** esa parte de un estribo que apoya el voussoir al final de un arco.

**skid resistance** a measure of vehicular traction on a wet surface.

**resistencia del rodillo** una medida de tracción vehicular en una superficie mojada.

**slab footing** a concrete footing spanning the entire area below the structure it supports, continuing beyond its exterior walls a minimum of 6" on each side.

**fundamento de la tabla** un fundamento concreto que mide por palmos el área entera debajo de la estructura él apoya, mientras continuando el más allá sus paredes exteriores un mínimo de 6" en cada lado.

**slip resistance** a measure of pedestrian traction on a wet surface.

**resistencia del resbalón** una medida de tracción pedestre en una superficie mojada.

**slump** a measure of the collapse of fresh concrete below the 12" formed by the slump cone.

**depresión** una medida del derrumbamiento de hormigón fresco debajo de los 12" formados por el cono de la depresión.

**slump cone** a 12"-tall formed cone with a base diameter of 12" and a top diameter of 4" used to form a sample of fresh concrete for conducting a slump test.

**cono de la depresión** un cono formado 12"-alto con un diámetro bajo de 12" y un diámetro de la cima de 4" formaba una muestra de hormigón fresco por dirigir una prueba de la depresión.

**slump test** an assessment of the consistency of freshly placed concrete.

**prueba de la depresión** una valoración de la consistencia de hormigón frescamente puesto.

**smoke chamber** an inverted funnel-shaped area extending from the level of the throat damper to the beginning of the chimney flue lining.

**cámara del humo** una área embudo-formada invertido que se extiende del nivel del apagador de la garganta al principio del chimenea cañón linear.

**smoke shelf** the area at the base of the smoke chamber, directly behind the throat damper opening.

**estante del humo** el área a la base de la cámara de humo, directamente detrás de la garganta la apertura más húmeda.

**snap header** half of a brick length laid in the header position for 4"-wide Flemish bond walls.

**título instantáneo** la mitad de una longitud del ladrillo puso en la posición del título para las paredes de la atadura flamencas 4"-anchas.

**soffit** the horizontal underside of an arch, perpendicular to the face of a wall, below an arched opening.

**soffit** la parte inferior horizontal de un arco, el perpendicular a la cara de una pared, debajo de una apertura arqueada.

**soffit board** a board attached to the underside of the rafters or roof trusses extending beyond the face of the exterior wall.

**placa soffit** una tabla ató a la parte inferior de las vigas o bragueros del tejado que se extienden más allá de la cara de la pared exterior.

**soft joints** horizontal expansion joints minimizing wall cracks by permitting both the expansion of brick masonry walls and the deflection of the shelf angles.

**junturas suaves** junturas de la expansión horizontales que minimizan la pared crujen permitiendo ambos la expansión de paredes de albañilería de ladrillo y la desviación de los ángulos del estante.

**soldier** the position for a brick having its end bedded in mortar and its face side oriented with the face of the wall.

**soldado** la posición para un ladrillo que tiene su fin plantado en un macizo en el mortero y su lado de la cara orientados con la cara de la pared.

**solid masonry unit** a masonry unit whose net cross-sectional area parallel to its bedding area is 75% or more of its gross cross-sectional area measured in the same plane.

**unidad de la albañilería sólida** una unidad de la albañilería cuya teje una malla el área cruz-particular paralelo a su área de la ropa de cama es 75% o más de su área cruz-particular gruesa midió en el mismo avión.

**solid unit** a CMU either having no cells or the cross-sectional area of the cells are 25% or less than the overall cross-sectional area of the unit.

**unidad sólida** un CMU cualquiera teniendo ninguna célula o el área cruz-particular de las células son 25% o menos del área cruz-particular global de la unidad.

**sound-absorbing CMUs** also called acoustical masonry units, structural concrete masonry units designed to reduce sound transmission.

**CMUs legítimo-absorbente** también llamadas las unidades de la albañilería acústicas, las unidades de la albañilería concretas estructurales diseñaron para reducir la transmisión legítima.

**spalling** the cracking and flaking of the bricks' surface due to freezing water.

**spalling** el agrietamiento y dividiendo en hojuelas de la superficie de los ladrillos debido al agua helada.

**span** the horizontal distance between the supporting abutments at each end of an arch opening.

**palmo** la distancia horizontal entre los estribos de apoyo a cada fin de una apertura astuta.

**special conditions** conditions of the contract other than general conditions or supplementary conditions setting forth requirements unique to the project.

**condiciones especiales** las condiciones del contrato de otra manera que condiciones generales o la escena de las condiciones suplementaria adelante los requisitos único al proyecto.

**specifications** a written description of the conditions of the contract.

**especificaciones** una descripción escrito de las condiciones del contrato.

**split-face CMUs** block produced by mechanically splitting multiple-molded units, having one or more rough, exposed aggregate, three-dimensional faces.

**raja-cara CMUs** bloque producido hendiéndose las unidades múltiple-amoldadas mecánicamente, mientras teniendo uno o las caras agregado, tridimensionales más ásperas, expuestas.

**spring line** the imaginary line at which each end of an arch begins.

**línea del muelle** la línea imaginaria a que cada fin de un arco empieza.

**stack bond** also called the stacked bond, a pattern consisting of brick aligned vertically in each consecutive course, in straight columns with no overlap.

**enlace de la pila** también llamado el enlace apilado, un diseño que consiste en ladrillo alineados verticalmente en cada curso consecutivo, en las columnas rectas sin el traslapo.

**stacked bond pattern** a brick pattern where all joints between the brick form continuous straight lines perpendicular to each other.

**diseño del enlace apilado** un modelo del ladrillo dónde todas las junturas entre el formulario del ladrillo el perpendicular de las líneas recto continuo a nosotros.

**standard size brick** brick whose approximate dimensions are 7⅝" long, 3½" wide, with a 2¼" face height.

**ladrillo del tamaño normal** ladrillo cuyas dimensiones aproximadas son 7⅝" largo, 3½" ancho, con una 2¼" altura de la cara.

**steel angle lintel** a steel angle iron supporting the brickwork above an opening.

**dintel de ángulo de acero** un hierro de ángulo de acero que apoya el enladrillado sobre una apertura.

**stone-face CMUs** block having the natural look of hand-chiseled stone.

**piedra-cara CMUs** bloque que tiene la mirada natural de piedra mano-cincelada.

**stoop** an unsheltered platform at the entrance of a building.

**inclinación** una plataforma desabrigado a la entrada de un edificio.

**stretcher** the position for a brick observed in a wall having its bottom side bedded in mortar and its face side oriented with the face of the wall.

**camilla** la posición para un ladrillo observado en una pared que tiene su lado del fondo plantada en un macizo en el mortero y su lado de la cara orientados con la cara de la pared.

**structural engineer** one meeting requirements for licensing to design a structure or parts of it to withstand all weights and forces that it may have to support.

**ingeniero estructural** un requisitos encontrándose por autorizar para diseñar una estructura o partes de él para resistir todo los pesos y fuerzas que puede tener que apoyar.

**structural load** the weight or pressure exerted on a structure, including the weight of the building materials, occupants, furnishings, wind, snow, and rain.

**carga estructural** el peso o presión ejercieron en una estructura, incluso el peso de los materiales del edificio, ocupantes, muebles, viento, nieve, y lluvia.

**structural pier** a pier designed to support a concentrated load.

**malecón estructural** un malecón diseñó para apoyar una carga concentrada.

**suction rate** the rate at which a brick absorbs water.

**proporción de la succión** la proporción a que un ladrillo absorbe el agua.

**supplementary conditions** that part of the specifications that supplements or modifies provisions of the general conditions.

**condiciones suplementarias** la parte de las especificaciones que complementan o modifican comestibleses de las condiciones generales.

**supported scaffolds** platforms supported by legs, beams, or other approved rigid supports.

**andamios apoyados** plataformas apoyadas por las piernas, vigas, u otros apoyos rígidos aceptado.

**suspension scaffolds** platforms suspended by ropes or other nonrigid means from overhead.

**andamios de la suspensión** plataformas suspendidas por las sogas u otro nonrigid significa sobre la cabeza de.

**symbols** marks, letters, characters, figures, or combinations of these representing specific objects.

**símbolos** las marcas, cartas, carácteres, figuras, o combinaciones de éstos que representan los objetos específicos.

**tail of the lead** the stepped-back end of corners and jambs.

**cola de la primacía** el fin del caminar-parte de atrás de esquinas y jambas.

**tensile strength** the measure of a material's capability to withstand stretching.

**fuerza tensor** la medida de la capacidad de un material para resistir estirando.

**thimble** a round, fired clay chimney component used as a chimney inlet and attached horizontally from the flue lining to the face of the wall into which the appliance connector is inserted.

**dedal** una ronda, la arcilla disparada el componente de la chimenea usada como una entrada de la chimenea y ató horizontalmente del cañón que linea a la cara de la pared en que el conector del aparato se inserta.

**throat** the narrow passage between the combustion chamber and the smoke chamber.

**garganta** el pasaje estrecho entre la cámara de la combustión y la cámara de humo.

**throat damper** a cast iron or steel frame and valve plate control, controlling the burning rate by managing the passage of air from the firebox to the smoke chamber.

**apagador de la garganta** un hierro colado o el marco de acero y mando de plato de válvula, controlando la

proporción ardiente manejando el pasaje de aire del firebox a la cámara de humo.

**through-wall flashing** flashing extending completely across the air cavity separating masonry wythes and through the outer wall beyond the exterior face of the wall.

**destello del a través de-pared** encendiendo extendiéndose completamente por la cavidad aérea el wythes de la albañilería de separación y a través de la pared exterior más allá de la cara exterior de la pared.

**toe** the point of a mason's trowel at the opposite end of the handle.

**dedo del pie** el punto de la paleta de un albañil al fin opuesto del asa.

**toothing** continuing with courses on a lead where the tail of the lead normally limits additional courses by temporarily laying a half-brick or bat on the stretcher below.

**toothing** continuando con los cursos en una primacía dónde la cola de la primacía normalmente limita los cursos adicionales poniendo un medio-ladrillo temporalmente o mueve en la camilla debajo.

**topographical features** physical features of the land such as ground cover, wooded land, lakes, streams, and land contour.

**rasgos topográficos** los rasgos físicos de la tierra como la tapa de tierra, tierra arbolada, lagos, arroyos, y contorno de la tierra.

**topographical symbols** representations of ground vegetation, surface water, roadways, utilities, wells, septic systems, fencing, property lines and corners, surface contours, and benchmarks.

**símbolos topográficos** las representaciones de vegetación molida, agua de la superficie, las carreteras, las utilidades, los pozos, los sistemas sépticos, cercando, la propiedad linea y esquinas, contornos de la superficie, y referencias.

**tread** the horizontal surface of a step supporting pedestrian traffic.

**banda de rodadura** la superficie horizontal de un paso el tráfico del peatón de apoyo.

**tread depth** a measurement taken from the tread's front edge to its back edge adjoining the riser.

**profundidad de la banda de rodadura** una medida tomada del borde del frente de la banda de rodadura a su borde atrasado que une la sublevación.

**trench footings** footings resulting from the placement of concrete directly into excavated trenches.

**fundamentos de la trinchera** fundamentos que son el resultado directamente de la colocación de hormigón en las trincheras excavadas.

**triangular arch** an arched opening formed by two straight inclined sides.

**arco triangular** una apertura arqueada formada por dos lados inclinados rectos.

**trisodium phosphate** a strong cleaner once found in household laundry and dishwashing detergents that is used to clean brick.

**fosfato del trisodium** un limpiador fuerte encontrado una vez en el lavado de la casa y detergentes del dishwashing que se usan para limpiar el ladrillo.

**trowel blade** that part of a mason's trowel holding mortar.

**hoja de la paleta** esa parte de la paleta de un albañil que sostiene el mortero.

**trowel shank** that part of a mason's trowel extending from the trowel heel and into the handle.

**zanca de la paleta** esa parte de la paleta de un albañil que se extiende del talón de paleta y en el asa.

**truss-type wire reinforcement** masonry wall reinforcement consisting of two or more longitudinal rods welded to diagonally oriented cross rods, forming a truss design.

**refuerzo de alambre de braguero-tipo** refuerzo de pared de albañilería que consiste en dos o las varas más longitudinales soldó a las varas cruzadas diagonalmente orientadas, mientras formando un plan del braguero.

**Tudor arch** an arch having a rise less than its span formed by two overlapping circles of equal length radii coming to a point at its center.

**Tudor arquean** un teniendo astuto un levantamiento menos de su palmo formado por dos círculos solapando de radios de longitud iguales que vienen a un punto a su centro.

**twigging the line** clipping a twig, also known as a trig, onto a mason's line and positioning the twigged line above a brick bedded in mortar that is aligned level and plumb with the leads.

**twigging la línea** sujetando una ramita, también conocido como un arreglado, hacia la línea de un albañil y posicionando el twigged linean sobre un ladrillo plantado en un macizo en mortero que se alinea nivelado y aploma con las primacías.

**unbonded arch** an arch consisting of a single row of voussoirs.

**unbonded arquean** un consistiendo astuto de una sola fila de voussoirs.

**veneer wire anchoring system** a system for securing brick veneer to both wood-framed and metal-stud framed structures.

**alambre de la chapa que fija el sistema** un sistema por afianzar la chapa del ladrillo a los dos madera-idearon y el metal-montante ideó las estructuras.

**Venetian arch** a semicircular arch flanked by a narrower horizontal line of brickwork on both sides.

**arco veneciano** un arco semicircular flanqueado por un narrower la línea horizontal de enladrillado en ambos lados.

**vials** glass or acrylic liquid-filled tubes in mason's spirit levels permitting level and plumb alignments.

**redomas** vaso o acrílico los tubos líquido-llenos en el espíritu de albañil nivelan permitiendo nivel y alineaciones de plomo.

**voussoir** the name given each masonry unit forming an arch ring.

**voussoir** el nombre dado cada unidad de la albañilería que forma un anillo astuto.

**waterproofing** an approved membrane extending from the top of the footing to ground level, preventing water penetration of the foundation wall; required for foundation walls below finished grade wherever a high water table or other severe soil-water conditions exist.

**agua-corregiendo** una membrana aceptado que se extiende de la cima del fundamento a nivel conectada con tierra que evita la penetración de agua de la pared de la fundación requerida para las paredes de la fundación debajo acabado gradúa una mesa de agua alta dondequiera que u otras condiciones de tierra-agua severas existen.

**water repellants** liquids applied to masonry walls for the purpose of reducing water absorption.

**repelentes contra agua** los líquidos aplicados a las paredes de la albañilería con el propósito de reducir la absorción de agua.

**water retention** the ability of mortar to hold water when placed in contact with absorbent masonry units.

**repelentes contra retención** la habilidad de mortero de sostener el agua cuando puso en el contacto con las unidades de la albañilería absorbentes.

**weep holes** openings left in the head joints of brickwork to permit water drainage and allow air circulation in the air space.

**agujeros del desagüe** las aperturas que aparecen en las junturas de cabeza de enladrillado que permiten el desagüe de agua y permitir la circulación aérea en el espacio aéreo.

weep vents vented covers, corrugated materials, or open mesh polyester installed in open head joints to permit water drainage and to allow air circulation.

**aberturas del desagüe** las tapas dadas salida, materiales arrugados, o el poliéster de la malla abierto instalados en las junturas de cabeza abiertas permitir el desagüe de agua y permitir la circulación aérea.

white scum white or gray stains on the face of brick masonry, typically related to the cleaning of brickwork with unbuffered hydrochloric (muriatic) acid solutions or inadequate prewetting or rinsing of the brickwork during cleaning.

**escoria blanca** las manchas blancas o grises en la cara de albañilería del ladrillo, típicamente relacionado a la limpieza de enladrillado con el unbuffered clorhídrico (el muriatic) soluciones ácidas o prewetting inadecuados o enjuagando del enladrillado durante limpiar.

wick ropes pieces of ¼" to ⅜" diameter cotton rope routed from the air cavity through head joints beyond the face of the wall to divert water from the air cavity above the flashing.

**sogas de la mecha** los pedazos de ¼" a ⅜" soga de algodón de diámetro derrotados por la cavidad aérea a través de las junturas de cabeza más allá de la cara de la pared desviar el agua de la cavidad aérea sobre el destello.

wind loads calculated as wind pressures, increasing with wind velocity.

**cargas del viento** calculado como las presiones del viento aumentando con la velocidad del viento.

wood centering a temporary arch form.

**diseño central de madera** un formulario del arco temporal.

wood-mold brick brick whose shapes are created by placing wet clay in wooden forms or molds.

**ladrillo del molde de madera-** ladrillo cuyas formas son creadas poniendo la arcilla mojada en formularios de madera o moldes.

workability a measure of the ease with which the mortar can be mixed, placed, and finished.

**laborabilidad** una medida de la facilidad con que el mortero puede mezclarse, ser colocado, y acabado.

workers' compensation insurance covering employee accident, injury, or death.

**compensación de obreros** el accidente de empleado de techado de seguro, lesión, o muerte.

working drawing a scale drawing showing specific design details, including engineered specifications and required materials.

**dibujo activo** un dibujo a la escala que muestra los detalles del plan específicos, incluyendo las especificaciones diseñadas y requirió los materiales.

wythe a term used to express the bed depth of a masonry wall or thickness of a masonry wall in masonry units; each unit of bed depth is considered a wythe.

**wythe** un término expresaba la profundidad de la cama de una pared de la albañilería o espesor de una pared de la albañilería en las unidades de la albañilería; cada unidad de profundidad de la cama es considerada un wythe.

# Index